BUILDERS DETAIL SHEETS

Second Edition

Sam Smith

Edited by Phil Stronach

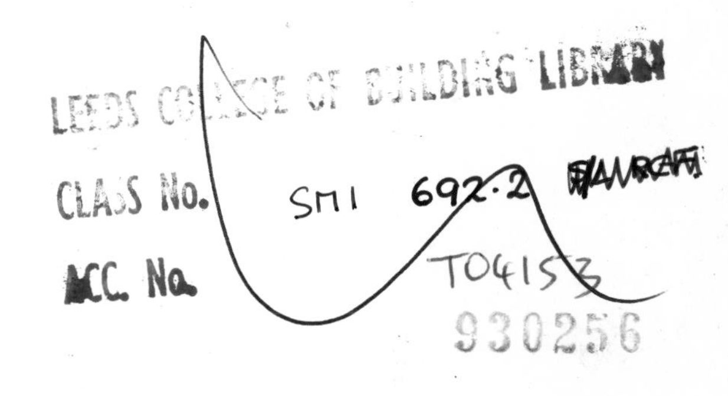

E & FN SPON
An Imprint of Chapman & Hall
London · New York · Tokyo · Melbourne · Madras

Published by E & FN Spon, an imprint of Chapman & Hall,
2 – 6 Boundary Row, London SE1 8HN

Chapman & Hall, 2 – 6 Boundary Row, London SE1 8HN, UK

Chapman & Hall Japan, Thomson Publishing Japan, Hirakawacho Nemoto Building, 7F, 1-7-11 Hirakawa-cho, Chiyoda-ku, Tokyo 102, Japan

Chapman & Hall Australia, Thomas Nelson Australia, 102 Dodds Street, South Melbourne, Victoria 3205, Australia

Chapman & Hall India, R. Seshadri, 32 Second Main Road, CIT East, Madras 600 035, India

Second edition 1986
Reprinted 1987, 1989, 1991

Printed in Great Britain by Ipswich Book Co, Ipswich

ISBN 0 419 15730 1

CONTENTS

PREFACE

These detail sheets were originally published in the 'Building Trades Journal' over a period of two-and-a-half years. During this time the change to S.I. continued apace and new editions of the Building Regulations have appeared. Changes also occurred in various British Standards, Codes of Practice etc.

The original sheets have been revised and brought up-to-date—S.I. units are used throughout, but here and there a few imperial units have been left as well, where their inclusion has not basically affected the details.

The original intention was to present a series of information sheets for the Builder and students of building, to show good practice and how the requirements of the Building Regulations may be complied with.

A sound knowledge of materials, their properties, limitations and correct use is essential for sound construction and these have been dealt with in some detail. British Standards, Codes of Practice and Building Research Establishment Digests contain a wealth of useful information, as does the Advisory Leaflets of the 'Department of the Environment' (formerly published by the Ministry of Public Building and Works). Much useful information is to be obtained from the publications of the various trade associations e.g. Cement and Concrete Association, TRADA, Lead Development Association etc. Where relevant, references have been given of these particular publications, so that readers who wish to pursue the subject in greater detail may do so.

References are as up-to-date, as far as possible at the time of writing, but readers are advised to check that they have the latest edition of any publication when pursuing any section, as additions and amendment are constantly appearing.

Since the publication of the original 'Builders' Detail Sheets' there has been the publication of the Building Regulations 1985. The detail sheets have now been amended to conform to the requirements of the current Regulations.

The Regulations Part L, deal with the thermal insulation of dwellings, and I would recommend readers who wish to study these requirements in detail to consult 'The Building Regulations 1985' by John Stephenson, published by Building Trades Journal Books 1986.

SAM SMITH

BUILDING PRELIMINARIES

GENERAL PLANNING
It is advisable for the site manager and other principals to visit the site and familiarise themselves with the layout and geography of the area. In particular position of roads, means of access, restrictions in force etc. A layout plan can be prepared, showing position of the building and runs of various services. Scale cut outs can then be used to site temporary hutting to the best advantage. Before the building commences certain formalities must be observed:-

1. Notify district surveyor or building inspector giving all necessary details.
2. Notify local authority and obtain licence if a hoarding is necessary. (Hoarding encroaching on footpath must have warning lights.)
3. Obtain permission from local authority for temporary access if required. Police may require details of vehicular access.
4. Enquire from L.A. times when excessive noise is restricted. e.g. compressors etc.

SITE PRELIMINARIES
Water:- Best way is to have the permanent supply brought to the site boundary. A stand pipe can then be set up for mixing water, and supplies taken to workmens accommodation and to latrines.
Latrines:- Ascertain if a temporary drain can be connected to existing sewer. If not make early arrangements for the permanent connection to the sewer for the site drainage, and run a temporary pipe to a junction connection for temporary use. In some cases a mobile amenity unit may provide the best solution. Chemical closets are useful on occasions, but as soon as possible a properly flushed system should be provided.
Electricity:- Make early arrangements for the company to run permanent supply to site. If considerable use of electrical power is envisaged, then a proper switchboard, fuse system and transformers may be required, which must be housed in a secure building.

Make sure that temporary wiring is carried out by a competent person and that no unauthorised person interferes with the system.
Site offices:- These should be sited so as to give as far as possible a good view of the site and the main entrance. No huts should be placed where they become obstructions and need moving e.g. over drainage runs etc. Preferably sectional huts used. Sometimes expedient to place huts on staging or gantry, on restricted sites. Telephone to be provided as early as possible.
Canteens:- Staff and operatives canteens can be served from one kitchen. On large contracts it is often best to let welfare services. Drying rooms should be close to canteen.
Welfare requirements:-

1. provision of shelter for inclement weather.
2. accommodation for clothing and facilities for drying wet clothes.
3. accommodation and provision for meals — sufficient tables and chairs or benches — facilities for boiling water and warming meals.
4. adequate supply of drinking water.
5. washing facilities.
6. tool stores or secure boxes.

First aid provisions:- (i) over 5 employees, suitable first aid boxes, in the charge of a responsible person. If over 50 employees boxes to be in the charge of a person trained in first-aid. (ii) over 25 employees, (a) notify local health authority of location of site, nature of operations and probable completion date. (b) provide a suitable stretcher, (c) an ambulance to be readily available. (iii) over 250 employees, a suitable first-aid room in the charge of a person trained in first-aid.
Storage:- Areas for storage of materials should be marked out and deliveries directed and controlled. All accommodation should be provided with locks.
Temporary roads:- Provided as required to give access to storage compounds and stacking areas. Sleepers may be used, hardcore laid, especially in areas to be paved or concreted later, or in some cases a portable track may be suitable.

CONCRETE MIXING
The cement store, stock piles of aggregate, stand pipe and plant for concreting should be situated in close proximity to one another so as to simplify operations.

A diagrammatic layout of plant and materials is shown. It is wise to arrange for an immediate change over of mixer should a failure occur, thus ensuring continuity of work.

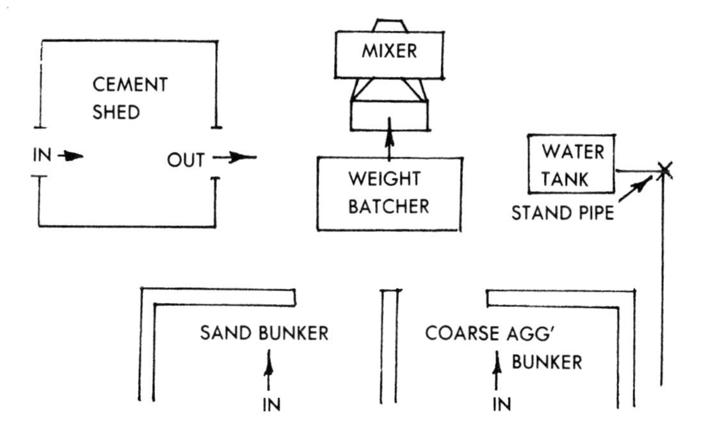

SETTING OUT I

PROCEDURE
I, ascertain position of the frontage line from the layout plan or block plan. This will be given relative to the road, adjacent buildings etc.
2, establish a base line (the frontage line) by driving stout pegs at each end 'A' & 'B', clear of the building position (fig I), marking the exact position of the line by nails in the tops of the pegs, and straining a line between (fig 2).
3, drive pegs 'C' & 'D' marking the position of the front quoins. A steel tape should be used for measuring.

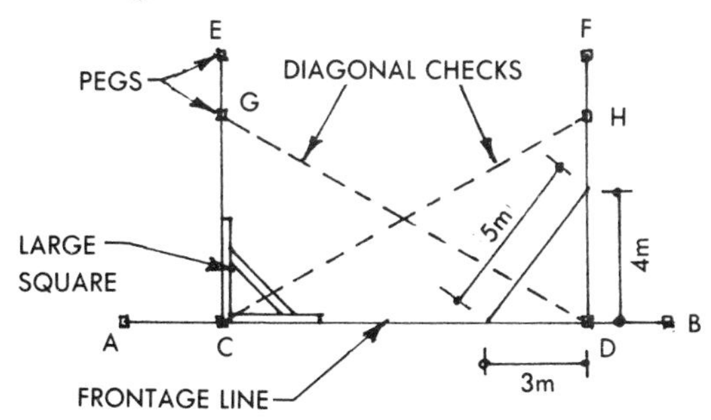

FIG 1

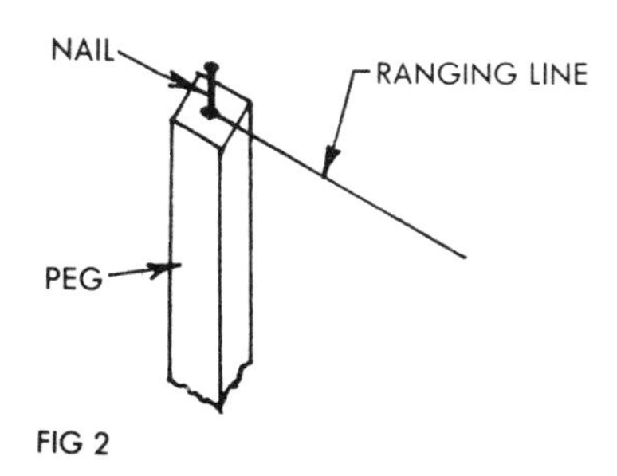

FIG 2

4, drive pegs 'E' & 'F' so that E-C and F-D are perpendicular to C-D. Right angles may be set out by (i) using a large builders square. (left hand corner fig I). (ii) by measurement, using the 3:4:5 method. This is an application of the theorem of pythagoras. A triangle having sides in the proportion 3:4:5 is measured out, this will be a right angled triange. One metre is a convenient unit to adopt for this purpose. (Right hand corner fig I). (iii) using a site square.
5, drive pegs 'G' & 'H' marking the position of the rear quoins.
6, the main rectangle should be checked for square by measuring opposite sides and the diagonals, i.e. C-D = G-H and C-H = D-G.
7, the positions of breaks, offsets and crosswalls can now be pegged out. N.B. When setting out a building where lengths exceed I5 metres it is advisable to use a theodolite.

PROFILES
When the pegging out has been checked, profiles are set up clear of the trench runs and the positions of the ranging lines transferred to them. Fig 3 profiles consist of horizontal boards fixed to stout posts, the position of the wall and trench being marked on the boards by sawcuts. Profiles are placed at all corners and at the ends of the cross walls Fig 4. The positions of trenches and walls can be obtained from lines strained between them Fig 5.

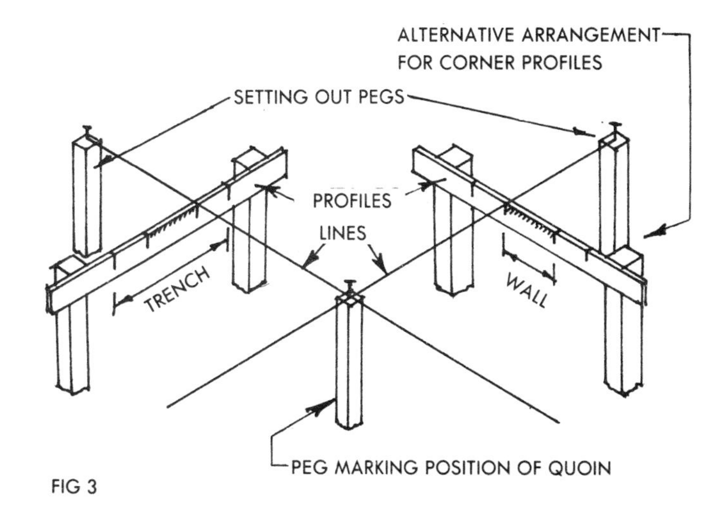

FIG 3

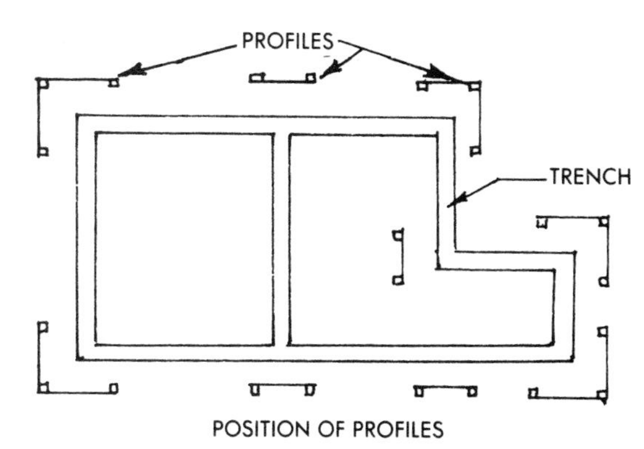

POSITION OF PROFILES

N.B. BEFORE SETTING OUT IT IS ADVISABLE TO CHECK THE DRAWINGS, ADDING THE OPENINGS AND PIERS ALONG EACH WALL AND COMPARING THE TOTAL WITH THE OVERALL LENGTH.

FIG 4

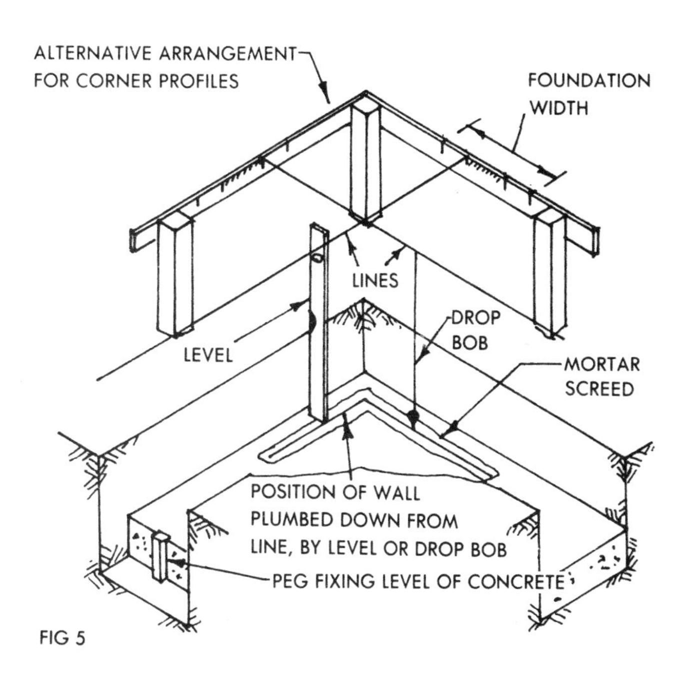

FIG 5

SETTING OUT 2

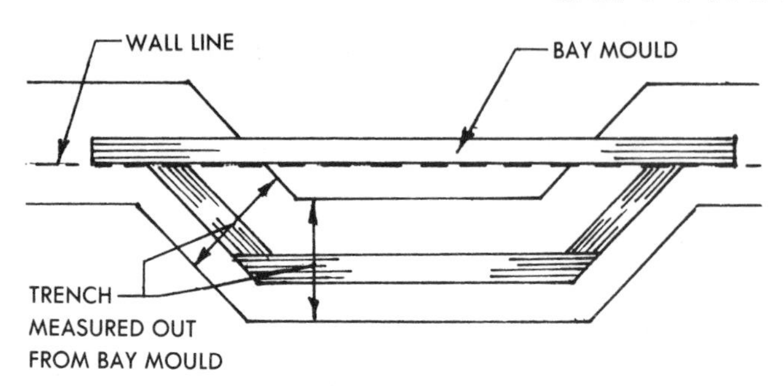

WHEN THE WALL REACHES GROUND LEVEL THE BAY MOULD IS PLACED ON TOP AS A FINAL CHECK.

FIG 1

SQUINT BAY

When setting-out squint bays, (sometimes known as 'cant' bays) it is common practice to make a light timber bay mould, the size and shape of the wall. The bay mould is then lined up with the main wall line and the position of the trench measured out from it. Fig I.

When the trench has been excavated and the foundation concrete placed, the bay mould is again placed in position (supported on two boards across the trench) and the line of the wall plumbed down.

When the wall reaches ground level the bay mould is placed on top as a final check.

CURVED WALLS

A curved bay may also be set out by using a bay mould. Fig 2. An alternative method is to use a radius rod as in Fig 3. A curved templet as shown will assist the bricklayer.

For curves having a long radius or where the striking point is at the ends of the curve and at the mid point. A light timber frame is then made as shown. If the peg at 'A' is removed and the frame moved across keeping it pressed against the pegs, then point 'A' will describe the required arc.

Large curves may be set out in a number of ways. One method is to erect calculated ordinates from a chord line, outlining the curve with pegs. Fig 5. A templet may be used between the pegs.

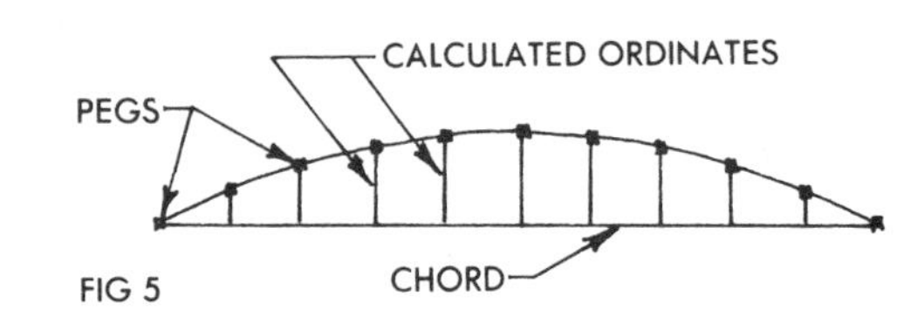

FIG 5

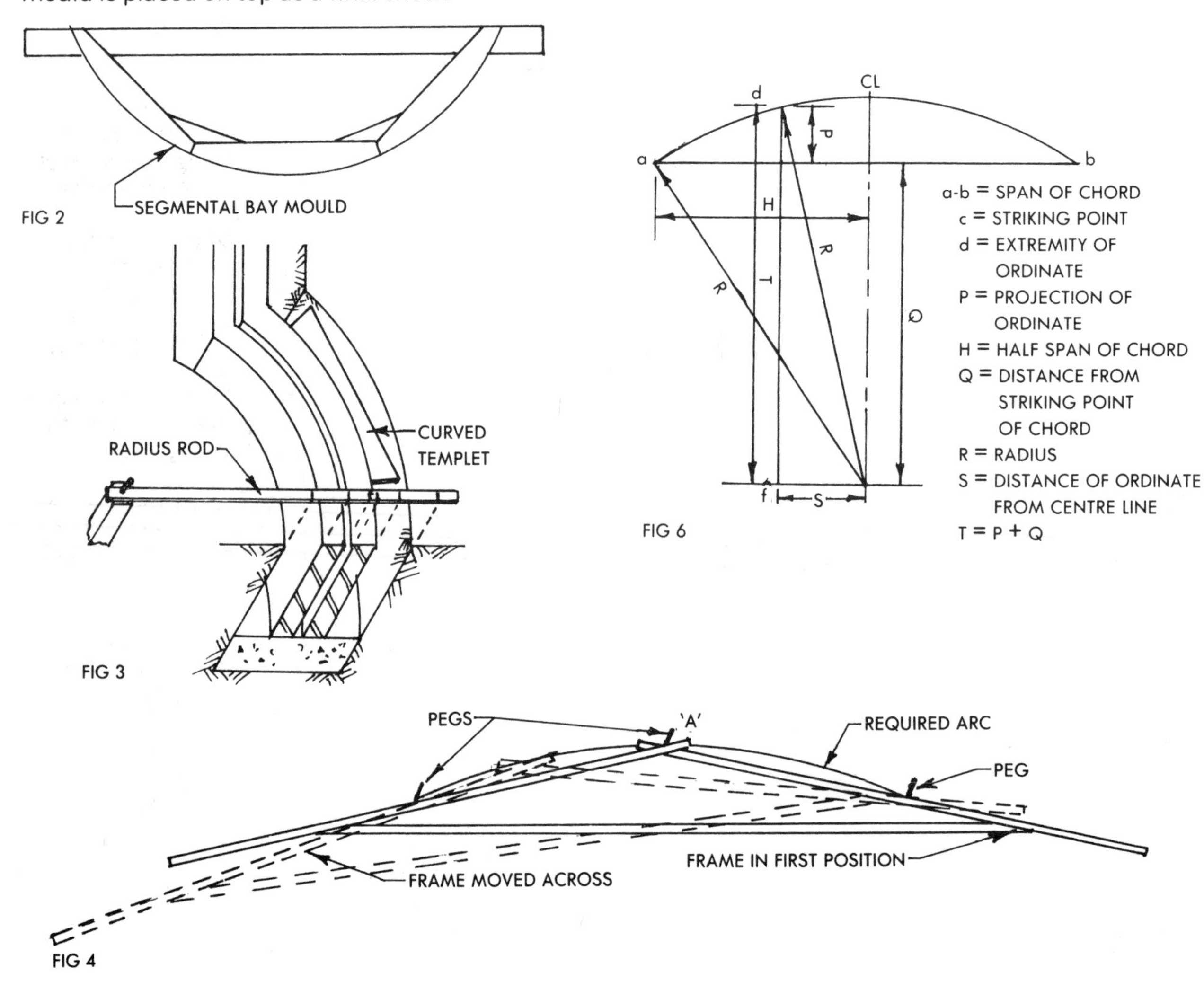

FIG 2

FIG 3

FIG 6

FIG 4

CALCULATING ORDINATES
A curve having chord a-b and radius 'R' is shown in Fig 6.
Required to calculate ordinate 'P'.

From pythagoras Q =
$R^2 - H^2$ in triangle cdf, cd = R, cf = s, df = T

$$T^2 = R^2 - S^2$$

$$T = R^2 - S^2$$

and $P = R^2 - S^2 - Q$

Example:- Radius 60m, chord 60m ordinate 6m from c.. i.e. S = 6m

$$Q = R^2 - H^2 = 60^2 - 30^2 = 51.962$$

$$P = R^2 - S^2 - Q$$

$$= 60^2 - 6^2 - 51.962$$

$$= 59.699 - 51.962$$

P = 7.737 metres

SETTING OUT 3

THE SITESQUARE
A useful instrument for setting out right angles, quick and easy to use.

The instrument (Fig I) has two fixed focus telescopes mounted at right angles to each other, each telescope able to swivel vertically through a wide arc. The instrument has a range of from 2m (6ft) to 91.5m (300ft).

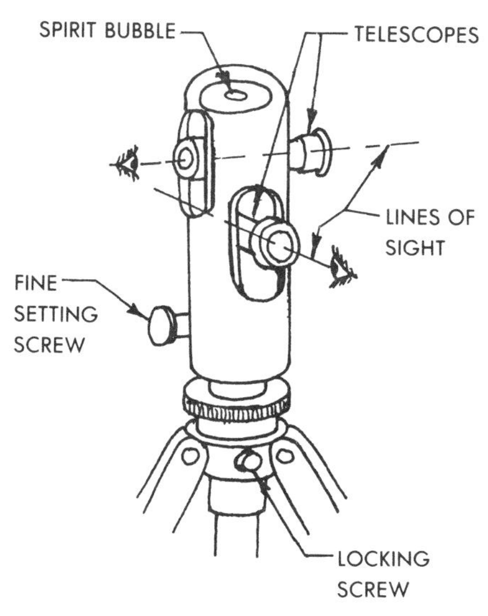

FIG 1 INSTRUMENT

SETTING UP THE INSTRUMENT
I. Before erecting the tripod check that the spike is correctly positioned. It is pointed at one end for use on a mark etc, the opposite end being hollow for use on the nail of a peg (Fig 3).
2. Set up the tripod, ensuring that the bolts joining the head and legs are tight. This avoids any 'shake' when using the instrument.
3. Place the tripod in position so that the datum rod is over the peg or mark which represents the corner point. (Fig 2)
4. Release the spike screw and extend the spike so that it is directly over the nail or mark. Tighten spike screw.
5. Screw the sightsquare on to the tripod and release the locking screw. By rotating the sightsquare, point the lower telescope along the frontage line. Tighten locking screw.
6. Release tripod leg screws and adjust the instrument so that the spirit bubble is central. Tighten tripod leg screw. Check bubble. Instrument is now ready for use.

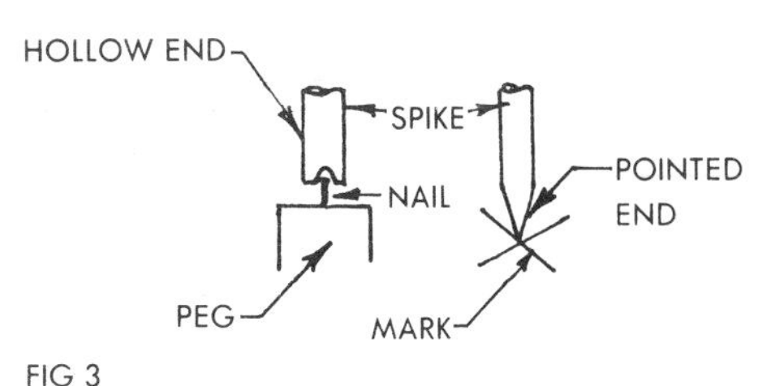

FIG 3

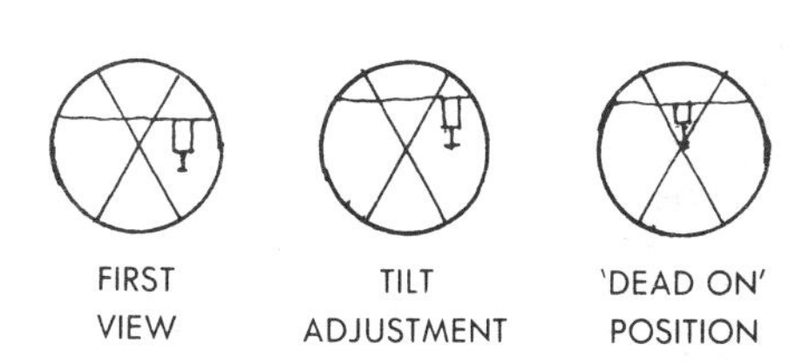

FIG 4 PEG & NAIL AS VIEWED THROUGH TELESCOPE

SETTING OUT
I. Sighting on to peg (I) through the lower telescope (Fig 5), obtain the 'dead on' position (Fig 4), by means of the fine setting screw which moves the telescope to the right or left and by tilting the telescope up or down.
2. When this position is obtained measure the distance required to peg (2), and by sighting through the top telescope, taking care not to rotate the instrument, direct assistant to move peg sideways until it is 'dead on'. Peg (2) is now positioned at an angle of 90° Fig 5.
3. The remaining corner peg of the rectangle can now be positioned, by moving the sitesquare to peg (2), 'lining up' on the corner peg and repeating the procedure.
Profiles :- Can be quickly positioned and marked by sighting the telescope on to the peg and then tilting the telescope upwards to transfer the point onto the profile board.

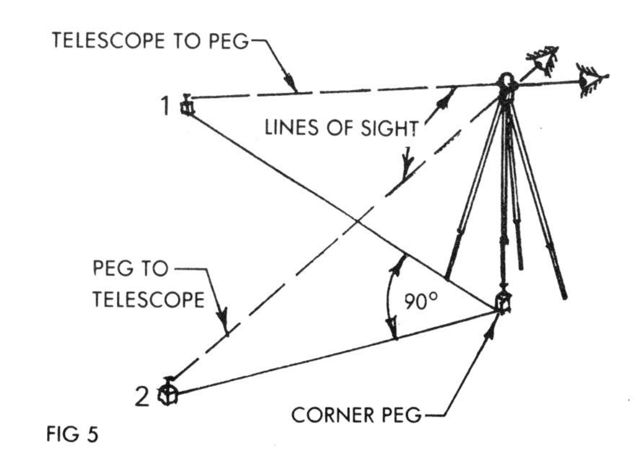

FIG 5

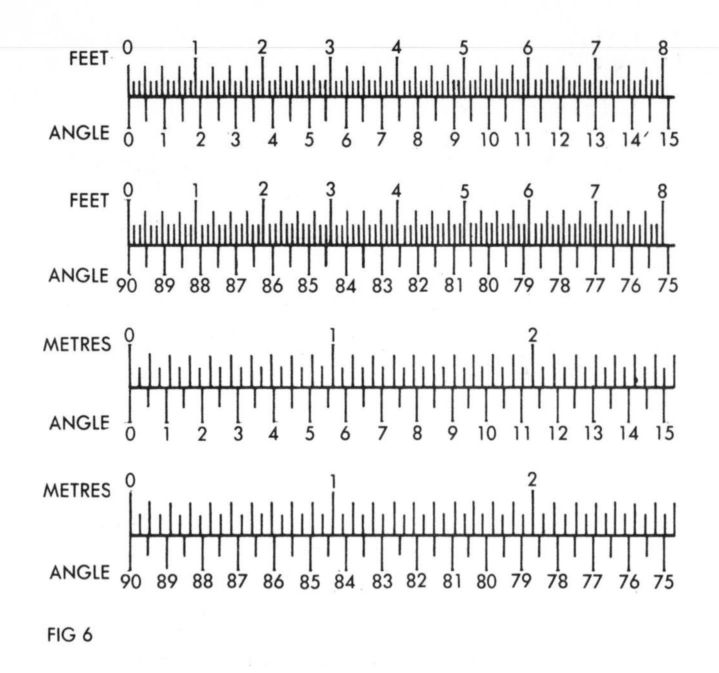

FIG 6

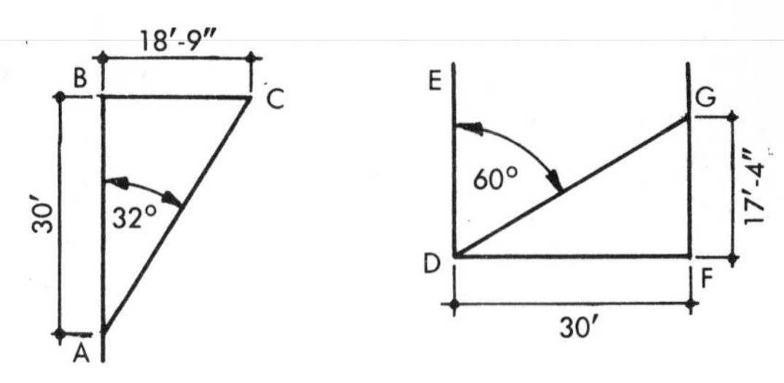

ANGLES OTHER THAN RIGHT ANGLES

A simple set of tables (in both metres and ft & ins) is supplied for setting out angles. Fig 6. *To set out an angle from 0° to 45°*:- Example:- to set off an angle of 32° from 'A' on line A-B. l. measure base line 30ft along A-B from 'A'. 2. set off B-C at right angles using sitesquare. 3. read offset distance from table, opposite 32° ie 18′-9″, and set off along B-C as shown. N.B. base line is constant at 30ft.

To set out an angle from 45° to 90°:- Example:- to set off an angle of 60° from 'D' on line D-E. l. set off line D-F 30ft long at 90° to D-E. 2. set off F-G at 90° to D-F. 3. read offset distance for 60° from tables ie 17′-4″. Set off along F-G as shown. N.B. for metric units use a 10m base line.

LEVELLING I

SITE DATUM

It is necessary to have a fixed point on site to which all levels can be related. This is the site datum and is fixed at a convenient height, usually ground floor level. The datum itself must be related to some fixed point, preferably an ordnance bench mark (O.B.M), or some other clearly defined point, e.g. a manhole cover. The level of an O.B.M. some distance away may be transferred to the site by means of a series of levels (known as flying levels) using a levelling instrument and staff. The site datum is marked by a peg or steel angle, concreted in, fenced off to protect it, and conveniently situated, preferably near the site office. Fig. l.

Ordnance bench mark, Fig 2, incised into walls, marks fixed height above ordnance datum. Ordnance datum is mean sea level at Newlyn, Cornwall.

LEVELLING BOARD

The simplest method of levelling is to use a straight edge about 3m long and 152mm x 25mm planed perfectly straight. This board is used in conjunction with a spirit level to level in a series of pegs as required for trenches etc. The level should be checked for accuracy by reversing it on the levelling board. If the level is accurate the bubble will come to rest in the same position. When levelling, the board and level should be reversed at each move; as indicated by positions of ends 'A' & 'B' in Fig 3, to minimise any error.

LINE LEVEL

A small level which can be hung on a line as shown. Figs 4 & 5. It can be used with reasonable accuracy provided that the level is kept in the centre of the line and the line is pulled tight.

WATER LEVEL

A length of hose with a transparent tube set in each end. The hose is filled with water, a cap or cork being provided to prevent spilling. Fig 5. Care must be taken to see that no air is trapped when filling. The water level works on the principle that water finds its own level. It can be used over distances from 30m or so, and is useful for marking a number of points quickly, especially around corners and obstructions Fig 6.

The trench bottom and pegs for concrete may be levelled by suing sight rails which have been carefully levelled and a boning rod. Fig 7.

For long lengths of kerb, drainage runs etc, a set of three rods may be used. Fig. 8.

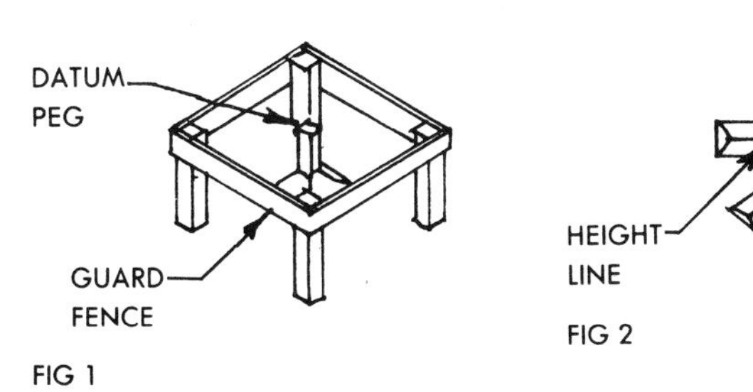

FIG 1

FIG 2

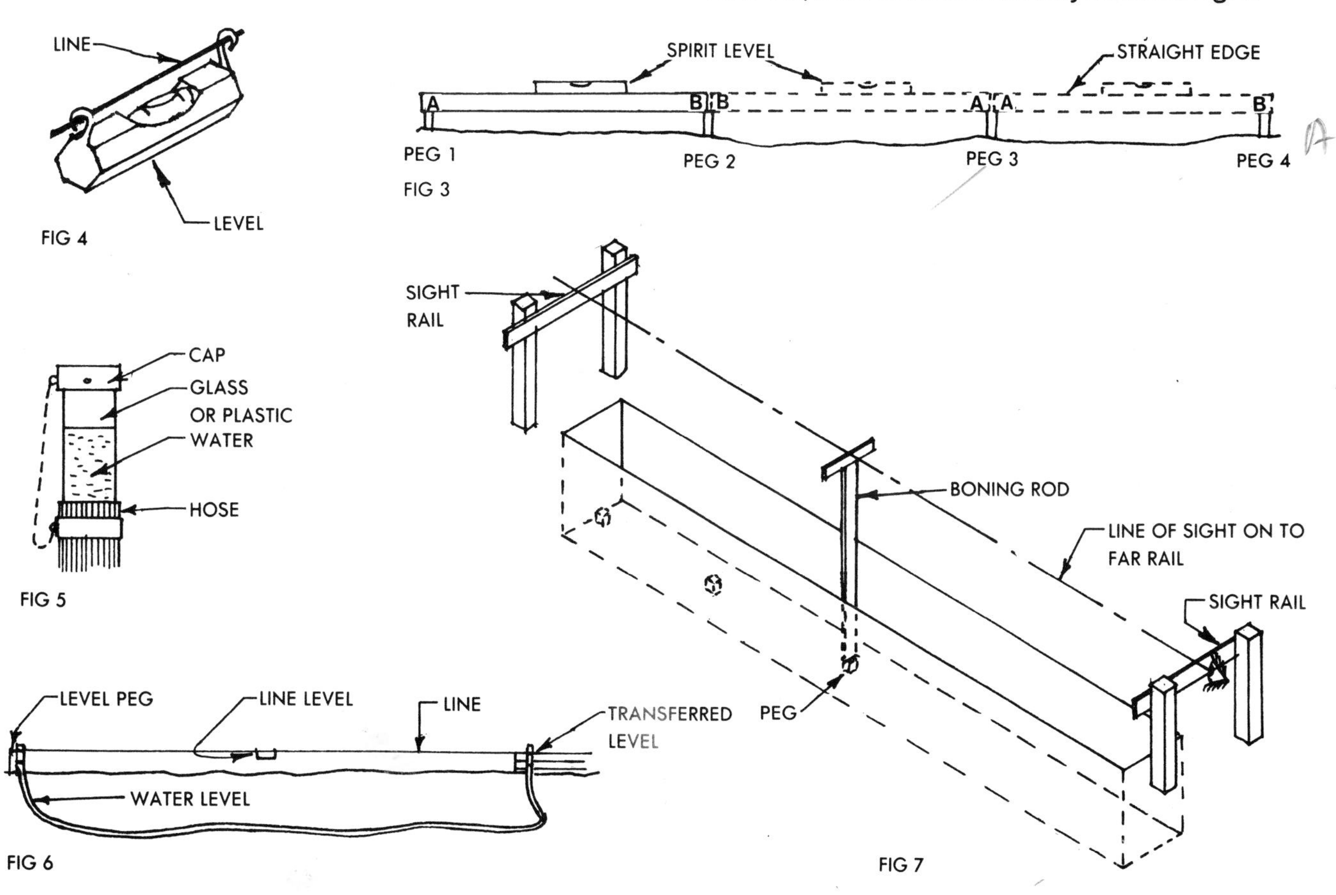

FIG 4

FIG 3

FIG 5

FIG 6

FIG 7

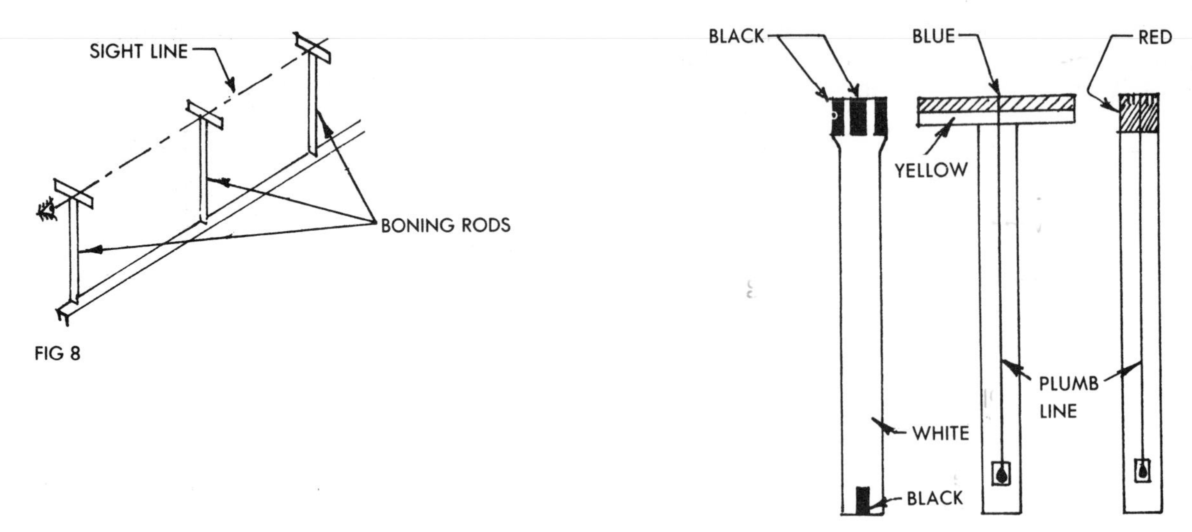

FIG 8

FIG 9 BONING RODS TO THE ROAD RESEARCH LABORATORY DESIGNS

LEVELLING 2

DUMPY LEVEL

A line diagram of a dumpy level is shown in Fig I. The instrument is set up on the tripod approximately level. It is then set accurately by means of the levelling screws as follows:-

Turn the telescope parallel to any two screws, e.g. 'A' & 'B' Fig 2. and adjust for level by turning screws 'A' & 'B'. Turn telescope through 90° and adjust for level by turning screw 'C' Fig 3. Turn back to original position and check.

N.B. when adjusting over two screws, the screws should be turned together, either both inwards or both outwards.

Focussing:- I. Put the telescope out of focus.
2. Bring the cross hairs into sharp focus by moving the eyepiece. (A piece of white card held at an angle to the object glass or a white handkerchief draped over the glass will help).
3. Bring the staff into sharp focus with the telescope focussing screw.

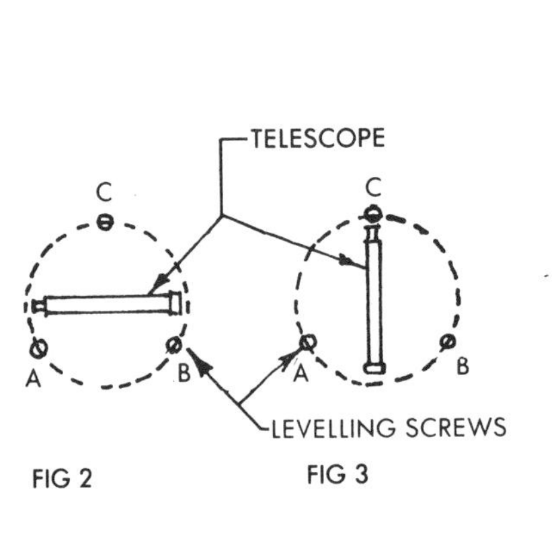

FIG 2 FIG 3

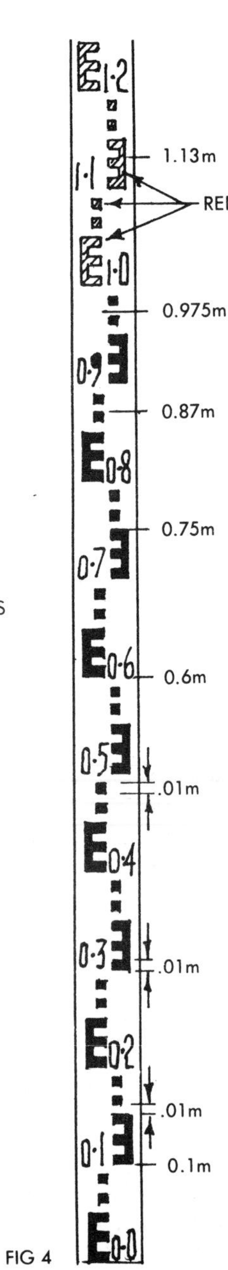

FIG 4

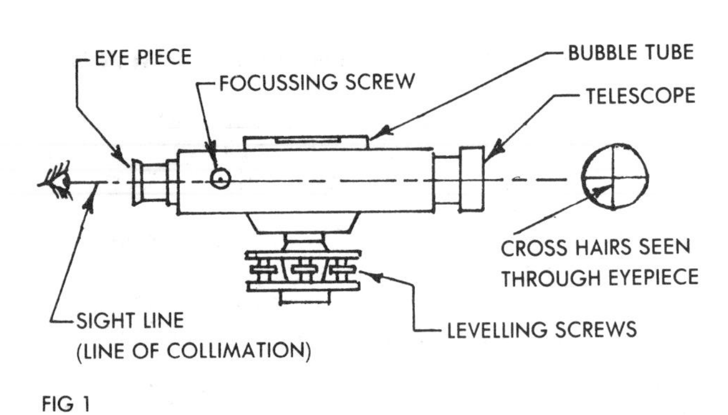

FIG 1

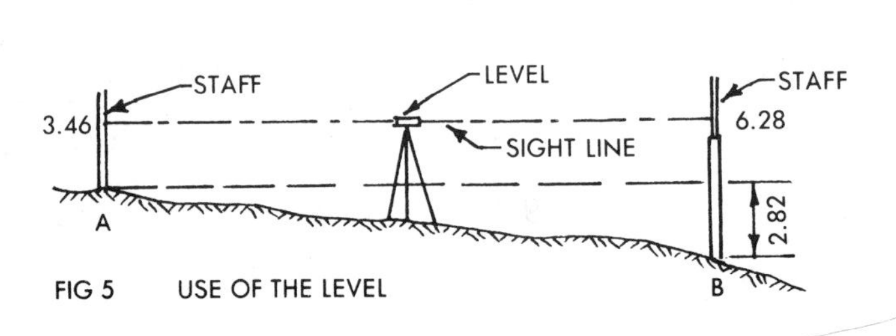

FIG 5 USE OF THE LEVEL

Metric staff:-A telescopic staff graduated in metric units. The spaces, dots and the bars of the 'Es' are all 10mm. All the figures are in black, but the markings for alternate metres are black and red. i.e. 0 to 1 and 2 to 3 are black, and 1 to 2 and 3 to 4 are red. A portion of the staff as viewed direct is shown in Fig 4. It would of course appear inverted when viewed through the levelling instrument.

Metric fascias are available for converting existing staffs.

Use of the level:- Required to find difference in level between points 'A' & 'B' Fig 5. The instrument is set up approximately midway between 'A' & 'B' Staff readings are then taken at 'A' & 'B' as shown. Difference in levels = 6.28 – 3.46 = 2.82. Note:-Increased staff reading shows a fall, decreased reading shows a rise.

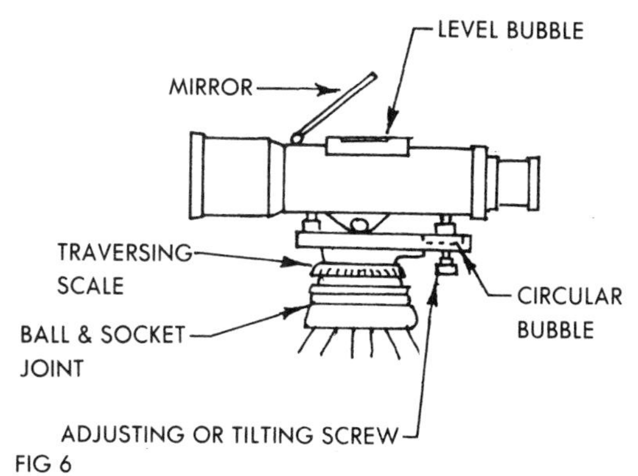

FIG 6

QUICKSET LEVEL

Is a development of the Dumpy level and as its name implies can be quickly and easily adjusted. Fig 6.

Operation:-Instrument is placed on its mounting adjusted so that the circular bubble is central and secured by the locking screw. Telescope is sighted on to staff and set horizontal by adjusting the tilting screw so that the level bubble viewed in the mirror is central. Telescope is focussed and the reading taken.

LEVELLING 3

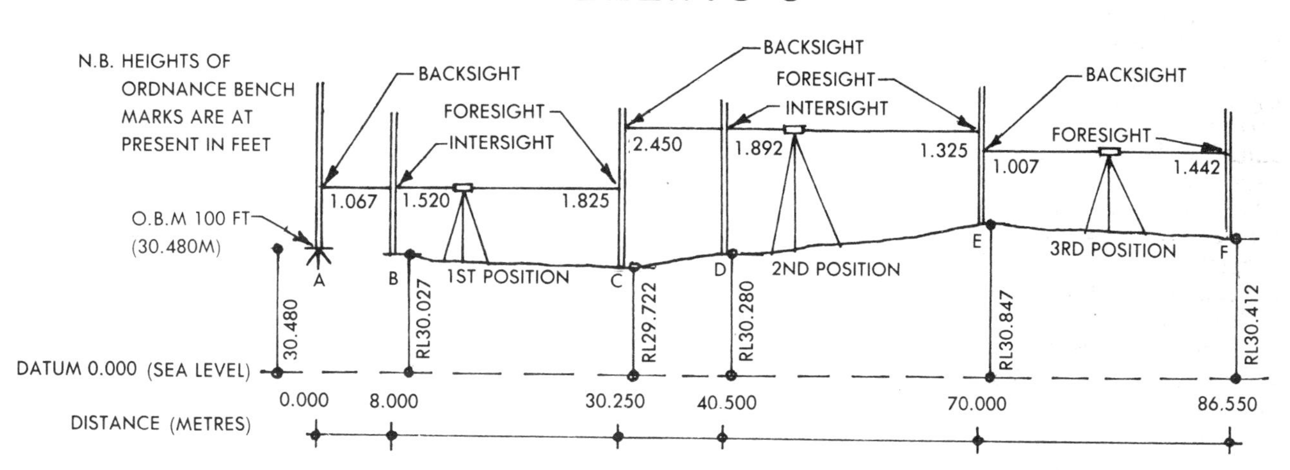

TERMS

Line of collimation:- The line of sight of the instrument.

Height of instrument:- Height of line of collimation above datum.

Back sight:- The first sight taken after the level has been set up in position.

Fore sight:- The last sight taken from a position before moving the level.

Intermediate sight:- Any other reading taken between backsight and foresight.

Change point:- A point where the staff is kept in position when the level is moved to the next position. The point at which both a foresight and then a backsight are taken.

Reduced levels:-Heights of points above datum. (A) to point (F).

USING THE LEVEL

An example using the metric staff is given. The method is the same if a sopwith staff is used.

Example:- Required to record levels from point (A) to point (F).

1. The levelling instrument is set up in the 1st position.
2. Staff is held at station (A) (in this case an O.B.M.) and the reading taken (backsight).
3. Staff is moved to station (B) and the reading taken. (Intersight)
4. Staff is moved to station (C) (a change point) and the reading taken (foresight).
5. The level is moved to the 2nd position, and the reading taken on the staff at station (C). Backsight.
6. Staff is moved to station (D) and the reading taken (intersight)
7. Staff is moved to station (E) (change point) and the reading taken (foresight).
8. The level is moved to the 3rd position, and the reading taken on the staff at station (E) (backsight).
9. Staff is moved to station (F) and the reading taken (foresight).

Back sight	Inter sight	Fore sight	Rise	Fall	Reduced level	Dist' (metres)	Remarks
1.067					30.480	0.000	A, O.B.M
	1.520			0.453	30.027	8.000	B
2.450		1.825		0.305	29.722	30.250	C.Ch'nge Pt
	1.892		0.558		30.280	40.500	D
1.007		1.325	0.567		30.847	70.00	E.Ch'nge Pt
		1.442	0.435	30.412	86.550	F	
4524 Check		4.592 4.524 0.068	1.125	1.193 1.125 0.068	First Last	R.L = R.L. =	30.480 30.412 0.068

Rise and fall method

BOOKING READINGS

When booking readings two methods are used.

I. Rise and fall method :-The readings are entered in columns as shown. The difference in adjacent staff readings will indicate the rise or fall from one station to the next. eg. 1.825 - 1.520 = 0.305, the fall from (B) to (C). A rise is added to the reduced level of the preceding station to give the reduced level of the station, a fall is substracted.

2. Collimation method :- Readings are entered in columns as shown. The height of collimination for each position of the instrument is obtained by adding the backsight to the reduced level of that particular station. Reduced levels are obtained by subtracting readings from the height of collimation.

CHECKING

The difference of the sum of the backsights and the sum of the foresights, should equal the difference of the sum of the rises and the sum of the falls, and also the difference between the first and last reduced levels.

The rise and fall method gives a check on the intersights, since they affect the rise and fall columns, whereas in the collimation method only the backsights and foresights are checked.

Back sight	Inter sight	Fore sight	Ht of col'mn	Reduced level	Dist (metres)	Remarks
1.067			31.547	30.480	0.000	A. O.B.M
	1.520			30.027	8.000	B
2.450		1.825	32.172	29.722	30.250	C.Ch'nge Pt
	1.892			30.280	40.500	D
1.007		1.325	31.854	30.847	70.000	E.Ch'nge Pt
		1.442		30.412	86.550	F
4.524 Check		4.592 4.524 0.068		First Last	R.L = R.L. =	30.480 30.412 0.068

Collimation method

LEVELLING 4

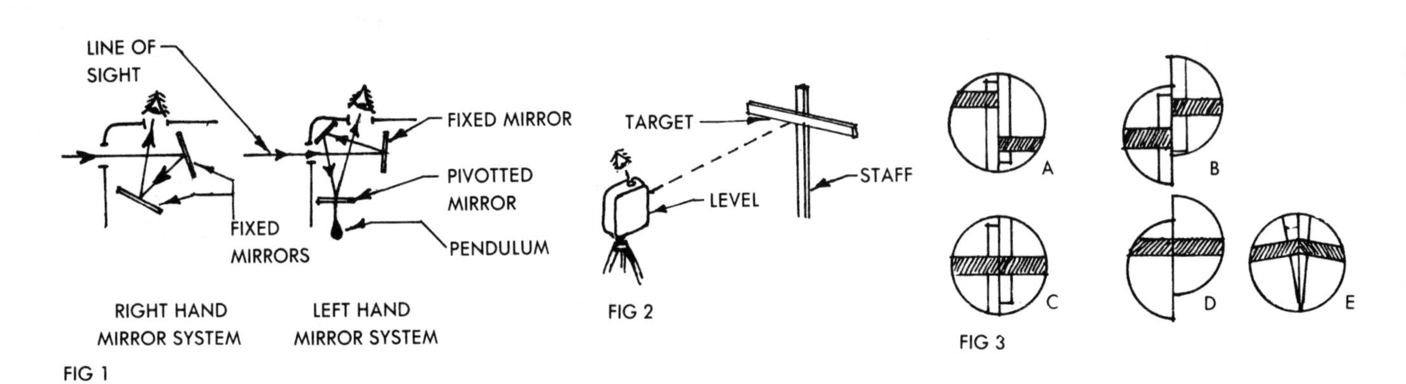

COWLEY LEVEL

The instrument consists of a compact metal case housing a dual system of mirrors. Fig 1. When the instrument is mounted on the spike at the top of its lightweight tripod a clamp is released and the level is ready for use. The level is used in conjunction with a staff having a sliding crosspiece or target. Fig 2.

OPERATION

1. Place level on tripod.
2. Sight instrument on to target.
3. If target is not level you will see either 'A' or 'B' Fig 3.
4. Move target up or down the staff until you see 'C' or 'D' Fig 3.
5. Staff holder can now take reading from staff.

N.B. The views at 'B' & 'D' indicate a forward tilt. If seen as view 'E' a sideways tilt in either case the level line is correct and the tilt can be ignored.

Staves :- Three staves are available as single 5ft staff, a short sliding staff (Figs 4 & 5) extending to 6ft 6in and along staff extending to 9ft 8in. The staves are marked in metres on one side (Fig 4) and in ft & ins on the other side. (Fig 5).

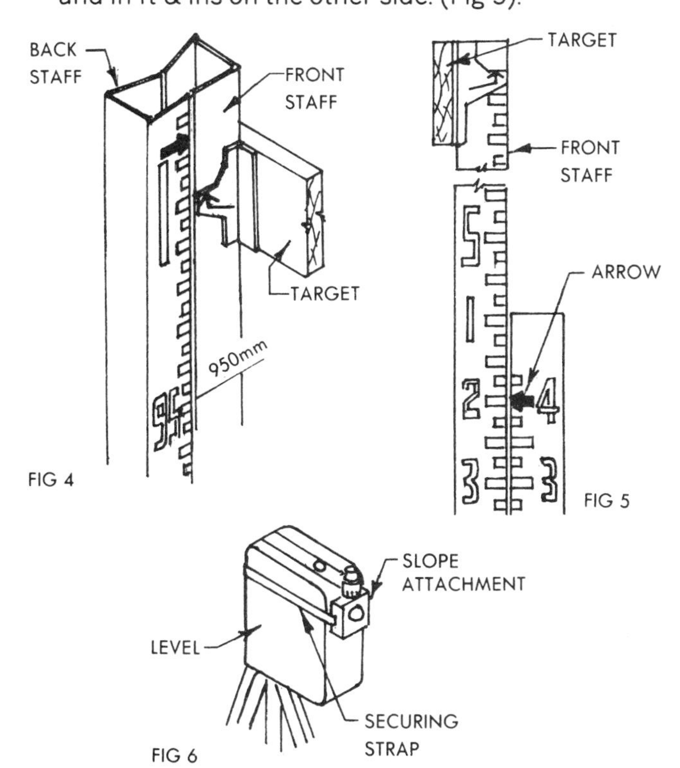

FIG 4

FIG 5

FIG 6

Using the staff illustrated in the range 0-1 metre, the target is attached to the front staff, with both sections clamped together in the closed position, readings are taken against the top edge of the black pointer, from the back. Fig 4 shows the staff reading 1 metre. (The figures on the front staff are omitted for clarity).

When reading from 1 metre upwards the target is clamped flush with the top of the front staff. The front section, (carrying the target) is raised to the required height and a direct reading taken off the front staff, opposite the arrow on the back staff. Fig 5 shows the staff viewed from the opposite side to Fig 4. The staff is extended and shows a reading of 5ft 2in.

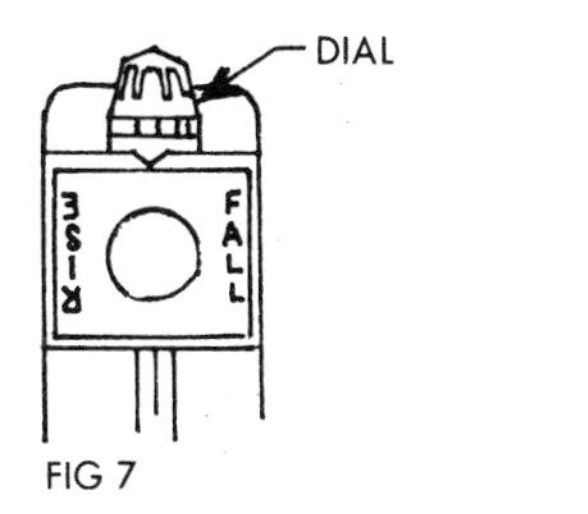

FIG 7

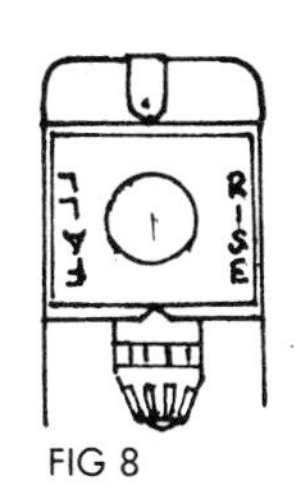

FIG 8

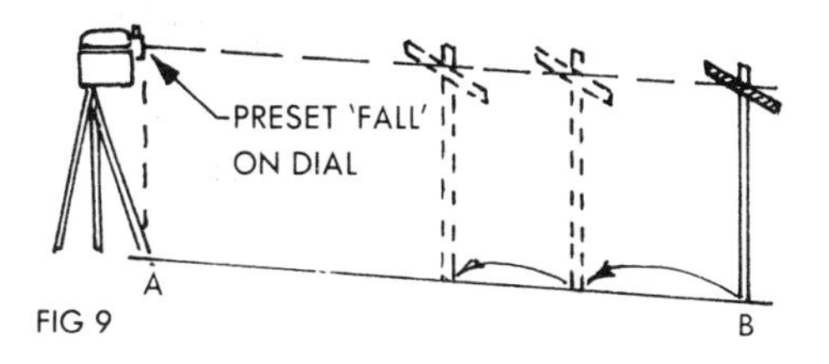

FIG 9

SLOPE ATTACHMENT

A small optical attachment that enables the level to be used to set out gradients. Two models are available with ranges of 1 in 10 to 1 in 50 and 1 in 60 to 1 in 250 respectively. Fig 6. The line of sight may be set to 'fall' or 'rise', by the position in which the attachment is fitted to the levels. Figs 7 & 8.

To set out a required gradient :-

1. Erect Cowley so as to sight along trench.
2. Set dial to gradient required.
3. Fit slope attachment in 'fall' or 'rise' position on the level.
4. Sighting through the level, the target is directed into coincidence and clamped.
5. Transfer staff to intermediate pegs in turn and adjust height of pegs so that target image is coincident with line of sight. Fig 9.

To set out a slope between two fixed points :-

1. Erect tripod over point 'A'.
2. Place Cowley 'spacer' on tripod pin and set target as shown. Fig 10.
3. Mount level on tripod. Fit slope attachment to level, dial reading zero.
4. Transfer staff to point 'B', sighting through the level rotate 'slope' to bring target into coincidence.
5. Transfer staff to intermediate pegs and adjust height of pegs so that target image is coincident with line of sight.

Bricklayer's stand :- Fig 11. A small stand to enable the level to be used on a wall.

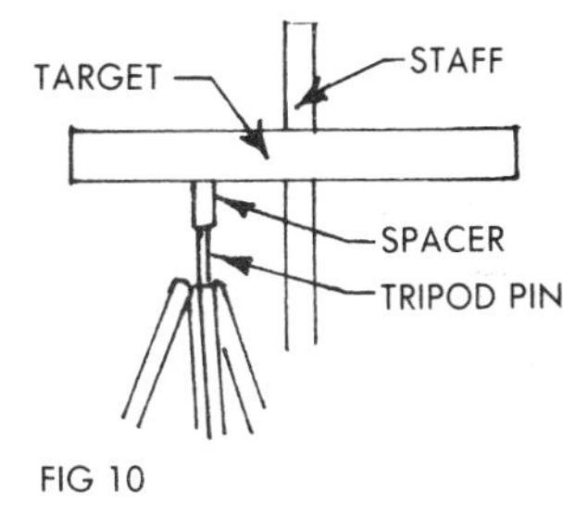

FIG 10

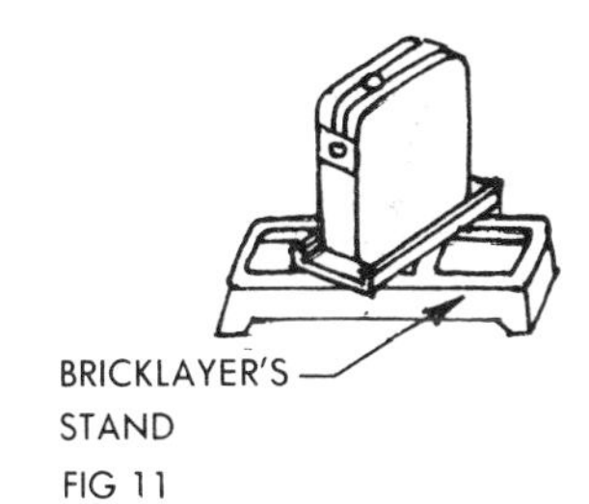

FIG 11

EXCAVATING

The method used will depend on a number of factors. For small quantities hand excavationi is usually cheaper, and it may be necessary on restricted sites. The final selection of plant will depend on:-

1. The nature of the excavation, e.g. deep or shallow trenches, basement etc.
2. The type of soil.
3. Volume of soil to be dug and transported.
4. Length of haul to spoil heaps and type of terrain.
5. Cost of transporting plant.
6. Possibility of keeping plant fully occupied.
7. Plant available.

EXCAVATING EQUIPMENT

General purpose excavators are most useful for excavating and loading vehicles. These machines have various attachments and may be rigged to operate as a backacter, face shovel, skimmer, dragline, crane and grab etc.

Face shovel:- Will excavate most materials and can handle rock and stone which has been previously loosened. Quick acting. Can excavate about 1m below ground level, to a height depending on size of machine. Produces a reasonably clean bottom. Vehicles for transporting are at same level as machine.

Backacter:- Used mainly for trenches and basement work. Machine works on surface digging below itself.

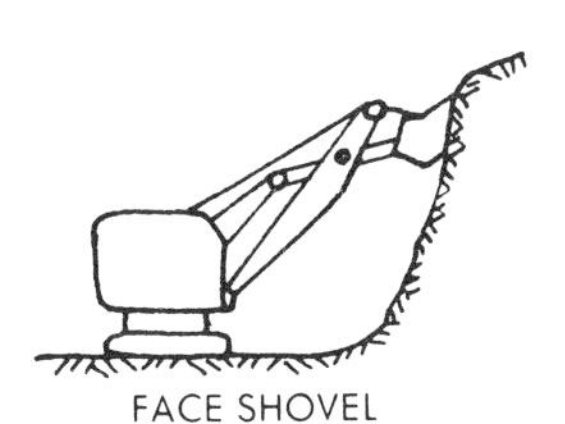

FACE SHOVEL

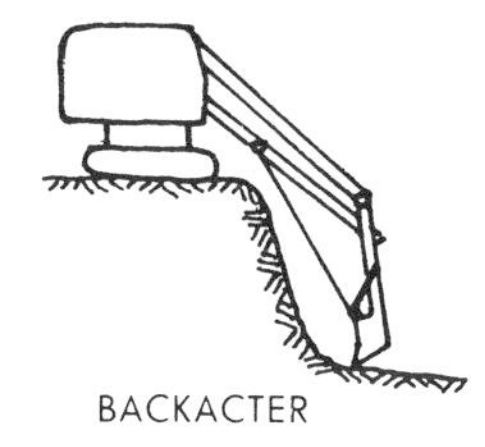

BACKACTER

Dragline:-Suitable for a range of material. Particularly useful in soft or waterlogged ground and when vehicles cannot operate from the bottom of the excavation. Filled by drawing bucket along ground. Operates from surface, digging below itself. Fitted with a long or boom giving it a long reach.

Skimmer:- For surface excavation and levelling up to about 300mm depth. Leaves a clean bottom. Can load.

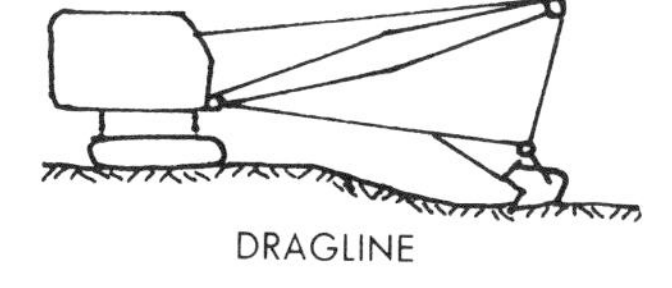

DRAGLINE

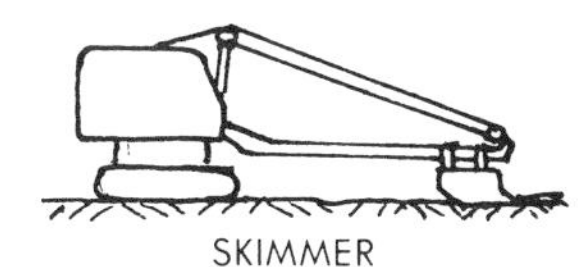

SKIMMER

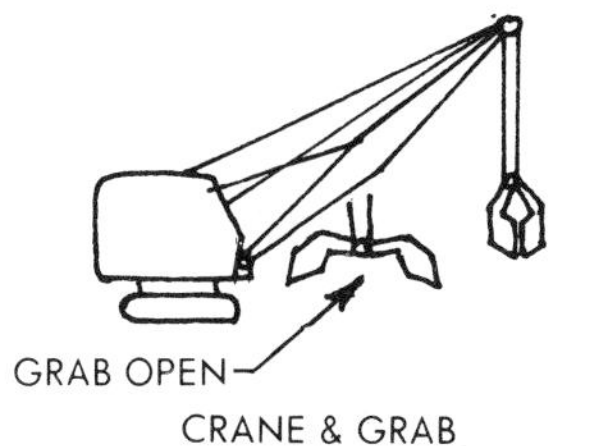

CRANE & GRAB

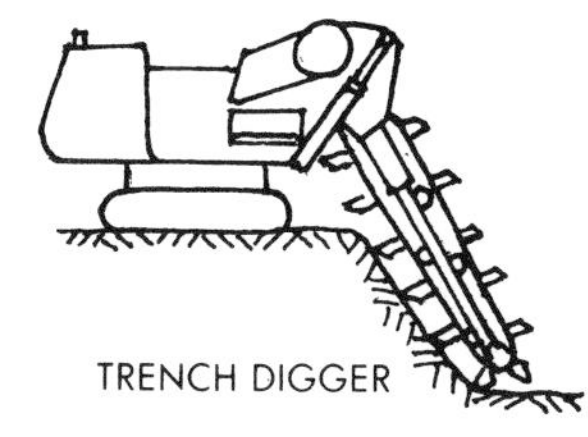

TRENCH DIGGER

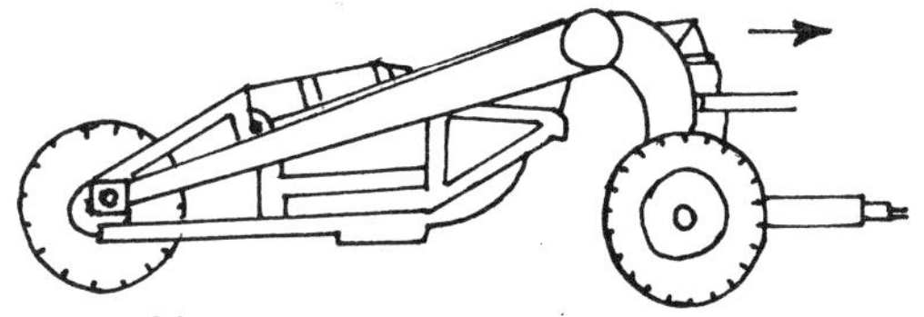

SCRAPER (TOWED BY TRACTOR)

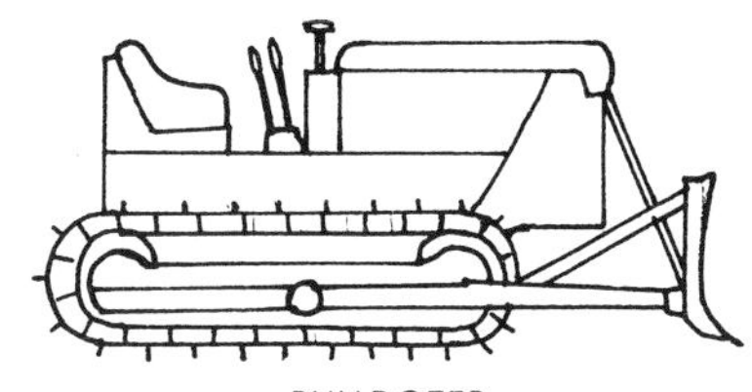

BULLDOZER

Crane & grab:- Suitable for soft materials, handling loose or broken material, or for underwater work. May be used to keep batching plant supplied from stock piles.

Combined digger & loading shovel:- For trench excavating, loading and clearing. Controls operated hydraulically. Oversize pneumatic tyres.

N.B. The advantage of crawler tracks is that although travel may be slower than with pneumatic tyres, the load is spread and the machine can work in very bad ground conditions.

Trench digger:- Has excavating buckets mounted on a chain or a wheel. Boom can be adjusted to required level. High output in soils free from large stones. Buckets discharge on to a conveyor which deposits spoil on either side of trench.

Scraper:- (Towed by tractor). Machine has a large open-fronted bucket which can be raised or lowered by cable control from the tractor, and is lowered to the ground and dragged forward for filling. When full the bucket is raised clear of the ground and towed to dumping area. For surface excavation over large areas.

Bulldozer:- Used for spreading and levelling, also for backfilling. Shallow cuttings may be excavated. The blade can be set to give required depth of cut.

Loading shovel:- Used for loading loose or broken materials. Most types can be used to strip top soil.

Grader:- Used for finishing the grade for roads and for spreading various sub-base and surfacing materials.

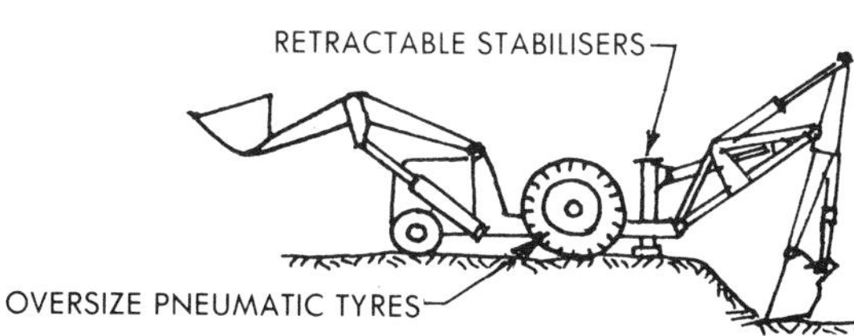

COMBINED DIGGER & LOADING SHOVEL

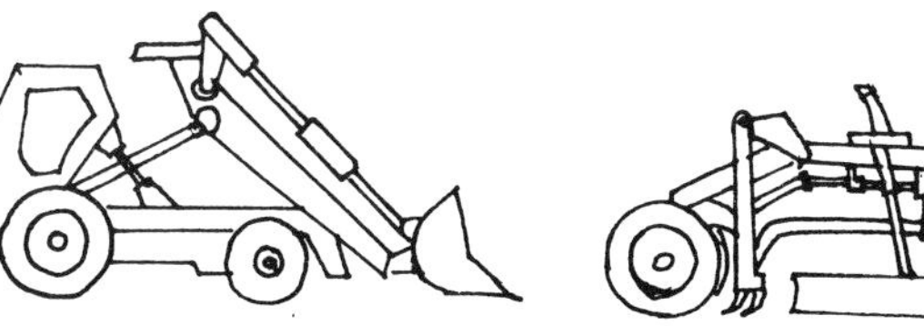

LOADING SHOVEL

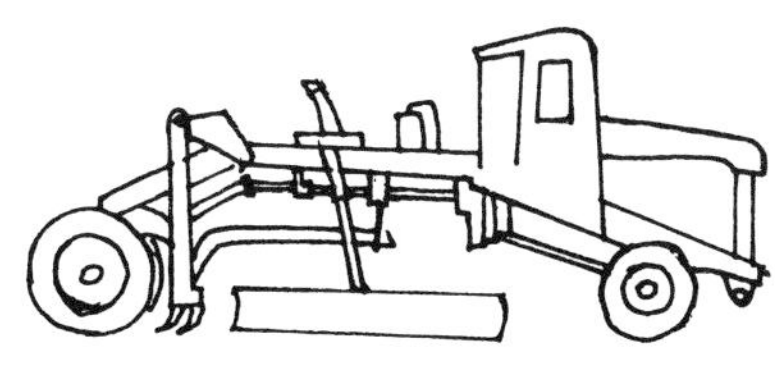

GRADER

TIMBERING TRENCHES I

FACTORS DETERMINING METHOD OF TIMBERING

1. Nature of the ground
2. Depth of trench.
3. Length of time trench is likely to remain open.
4. Presence of ground water.
5. Method of excavation, hand or machine.
6. Any nearby loads, brick stacks etc.
7. Proximity of existing buildings.
8. Vibration from traffic, railways etc.
9. Work to be carried out, e.g. asphalting.
10. Must timbering be placed as work proceeds or can it be placed after excavation.
11. Possibility of removing timbers when back filling.
12. Weather condition.

The following table from BS 6031 (1981) "Code of Practice for Earthworks" indicates support required for excavations with vertical sides in uniform ground.

*Open or close sheeting or sheet piling may be required if site conditions are unfavourable.

This table does not apply to complex ground conditions.

With varying ground strata, soil conditions towards the bottom of the trench must be considered. The ground here being most heavily stressed, slips into the bottom may occur and loss of stability due to water may develop.

In running sand additional measures may be necessary, de-watering, or consolidation by freezing, chemical means or cementation.

TYPE OF SOIL	DEPTH OF EXCAVATION		
	Up to (1.524m) SHALLOW	(1.524m to 4.572m) MEDIUM	Over (4.572m) DEEP
Soft peat	C	C	C
Firm peat	A	C	C
Soft clays & silts	C	C	C
Firm & stiff clays	A*	A*	C
Loose gravels & sands	C	C	C
Slightly cemented gravels & sands	A	B	C
Compact gravels & sands with or without claybinder	A	B	C
All gravels & sand below water table	C	C	C
Fissured or heavily jointed rocks (shales, etc)	A*	A*	B
Sound rock	A	A	A

A. No support required
B. Open sheeting should be used
C. Close sheeting or sheet piling should be used.

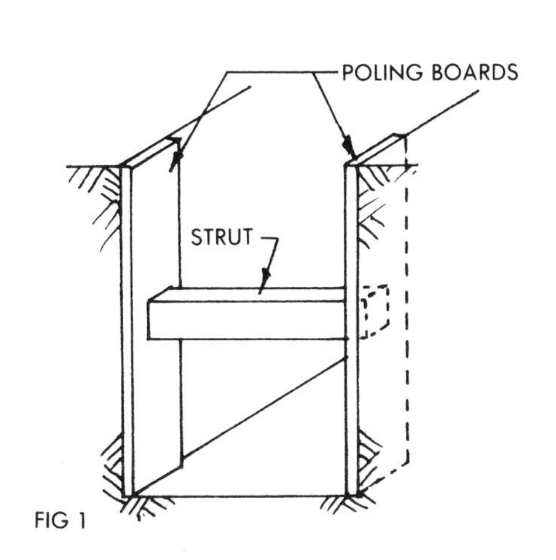

FIG 1

Pinchers:- Two poling boards strutted apart. Fig I. In good ground spaced approx 1.5m to 2m centres

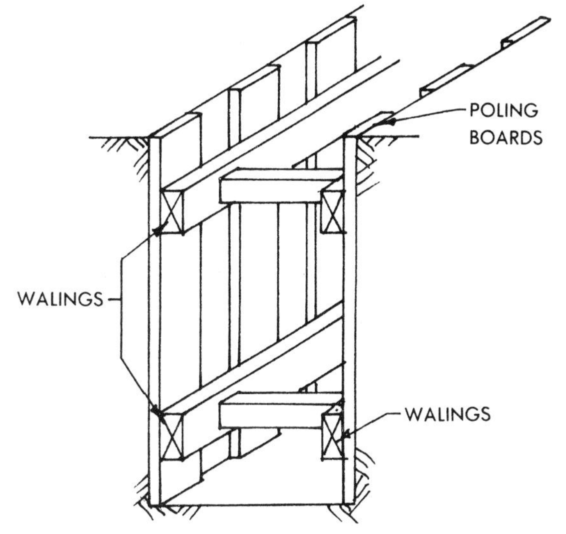

FIG 2 OPEN POLING IN MODERATELY FIRM GROUND

Poling boards:- Vertical boards 1m to 1.5m long, 150mm to 225mm wide, 31mm to 50mm thick supporting sides of excavation. Figs 1 to 3.

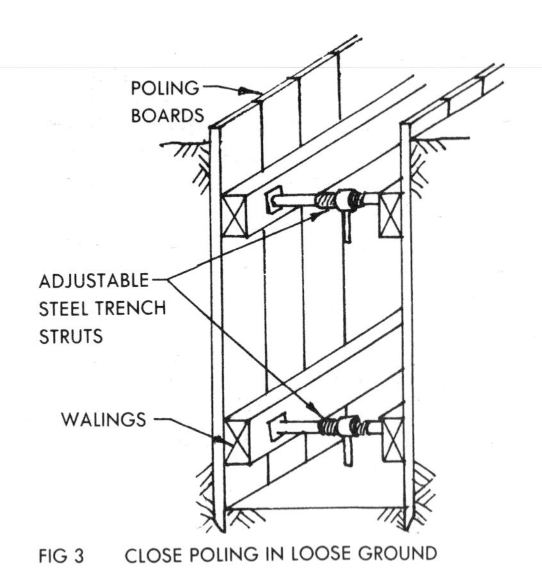

FIG 3 CLOSE POLING IN LOOSE GROUND

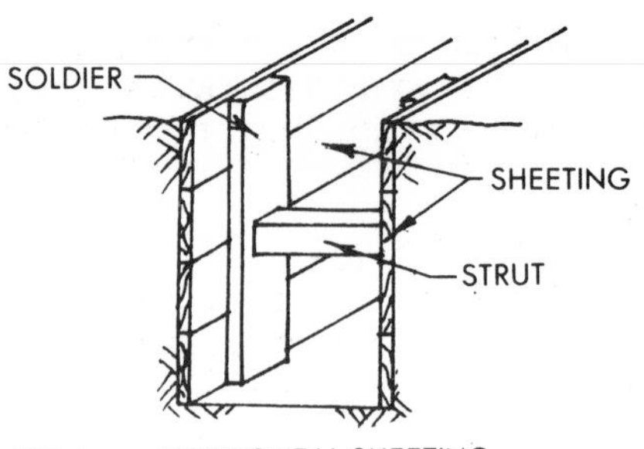

FIG 4 HORIZONTAL SHEETING IN LOOSE GROUND

Horizontal sheeting:- 225mm x 38mm or thicker and up to 4.267m long, used to support ground which will stand up to a face of 300mm to 600mm while timbers are being placed.
Struts:- Horizontal members resisting thrust from sides of excavation. Not less than 75mm x 75mm Fig I, 2 & 4. Steel struts are an alternative. Fig 3.

TIMBERING TRENCHES 2

Timbering should be inspected by a competent person at least once in every two days and after adverse weather conditions. If trench is over 2m deep, barriers should be erected as close as possible and at least 600mm high. Fig I. Spoil heaps should be kept clear of trenches, a minimum distance of l.2m is recommended.

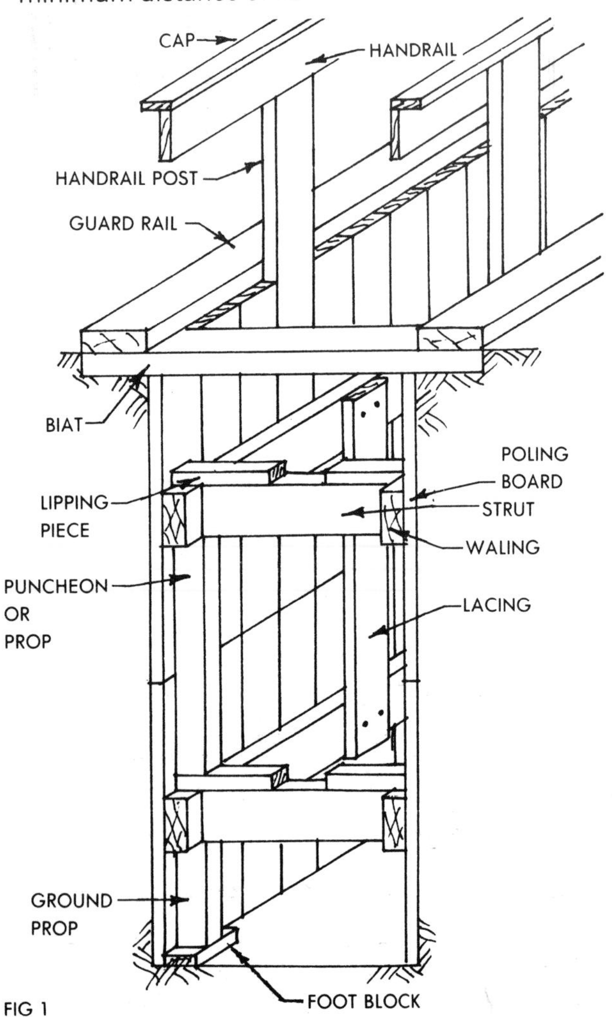

FIG 1

Biat:- Timber bearer giving support to guard rails, decking, walkways etc. Fig I.
Lip (Lipping piece):- Short length of timber fixed to upper edge of strut and projecting over the waling. Supports strut while wedges are driven.
Lacing:- A vertical timber spiked to walings or struts, tying them together to carry weight of the lower frames as excavation proceeds.
Puncheon:- Vertical prop to support a higher waling or strut from the one below.
Runners:- Long vertical timbers at least 50mm thick, with their lower end chisel shaped; used in unstable ground instead of poling boards, and driven down in advance of the excavation. Figs 3 & 4.

The lower ends of runners are usually sharpened to a chisel edge which may be splayed, as this tends to keep the joints tight when the runners are driven. Fig 4. If hard driving is needed the lower ends may be iron-shod as shown, and the heads bound with hoop iron.

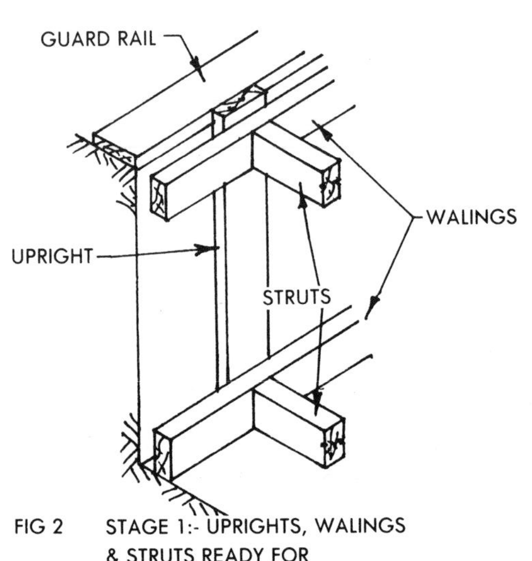

FIG 2 STAGE 1:- UPRIGHTS, WALINGS & STRUTS READY FOR PITCHING RUNNERS

HAND EXCAVATION

For hand excavation in deep trenches stages must be provided and the spoil shovelled from stage to stage. Fig 5.

STEEL TRENCH SHEETING

Interlocking steel trench sheeting is particularly useful in loose and waterlogged ground. There are a number of types available, the Acrow trench sheeting is shown in Fig 6. Advantages are, strength, long life and ease of driving. A steel driving cap is supplied and a hole near the top of the sheeting allows for easy withdrawal using a shackle and hoisting by crane or winch.

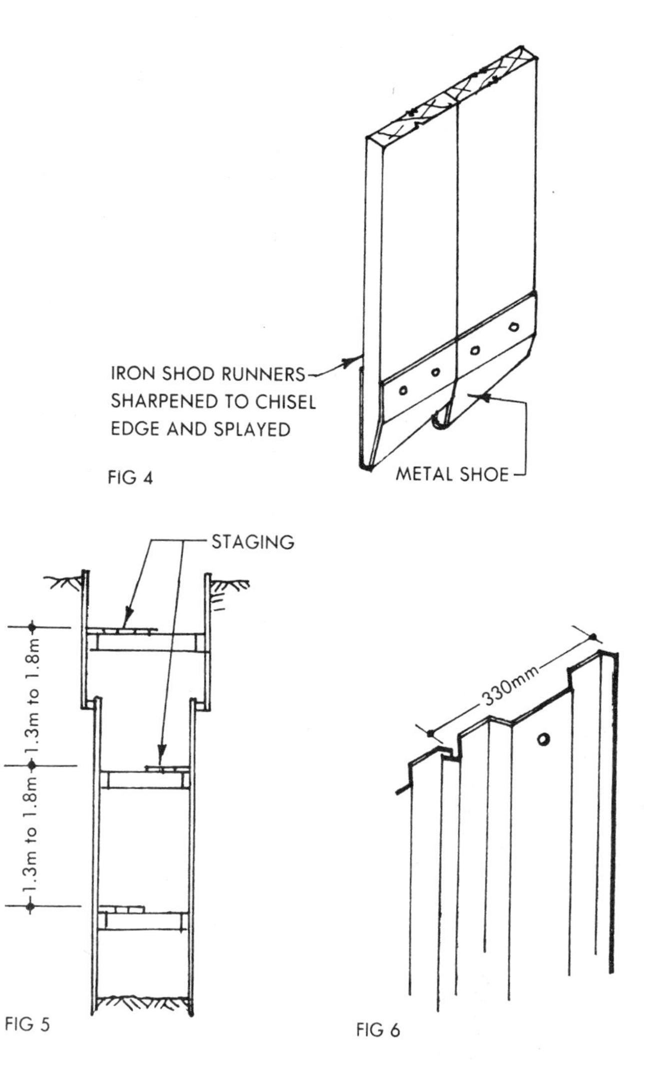

FIG 4

FIG 5

FIG 6

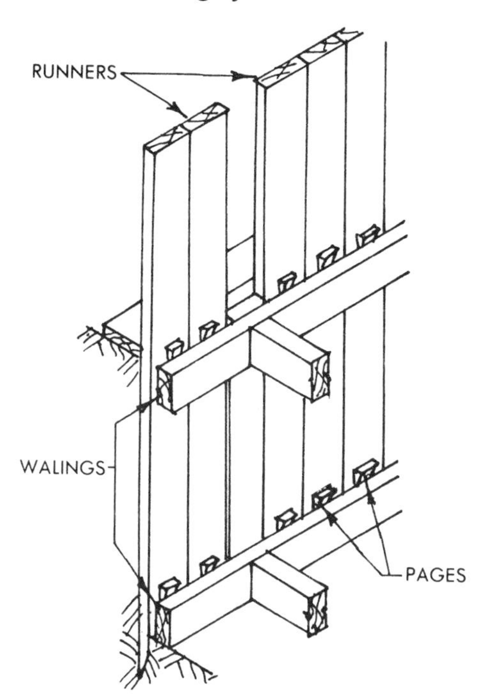

FIG 3 STAGE 2:- RUNNERS PITCHED READY FOR EXCAVATION

FOUNDATIONS 1

VEGETABLE SOIL
The surface layer of soft vegetable soil which covers most areas, varies in thickness, but the average depth is about 150mm. This soft soil must be stripped off as it is compressible and the organic content may have deleterious effects. This soil is usually kept and utilised for the garden, landscaping etc.

Building Regulation Approved Document C Section 1 states:-The site of any building, other than an excepted building, shall be effectively cleared of turf and other vegetable matter.

SUBSOILS
Rock:- Sound, solid rock usually only requires to be levelled and any cavities or fissures filled with concrete. Where rock outcrops on the slope or is exposed by excavation, careful examination must be made to determine if it will be necessary to excavate below possible slip planes or provide retaining walls.-
Chalk:- May deteriorate under the action of water or frost and protective layer of concrete should be laid as soon as the sub-foundation is 'bottomed up'.

Swallow holes are liable to develop in chalk or limestone. Cavities in the rock are formed by underground water dissolving the rock away and the overburden collapses into the cavity, forming a swallow hole. Since water movement causes these holes, soakaways should be kept at a safe distance away from buildings in such areas.

Where stanchions are supported on rock a calculated concrete base should be used to receive and distribute the load from the stanchions, to contain the anchor bolts and facilitate levelling of the stanchions.
Gravel:- Has high compressive resisting qualities. There may be a loss of fine particles in water bearing ground, and for this reason foundations on gravel or sand should be kept above the water table if possible. If not, water should be drained away from the excavation rather than towards it, to reduce risk of erosion at its face.
Sands:- Dense beds of sand, provided they are confined make a good sub-foundation, having high shear strength and being only slightly compressible. Care must be taken to avoid water from adjacent higher ground washing out fine particles. This scouring action can considerably weaken the sand and make it less stable. In some cases de-watering may be needed.
Frost heave:- In some soils e.g. good gravels water drains away quickly, some other soils e.g. fine sand and silt tend to retain water for sometime, and in freezing conditions ice lenses may form. The resulting pressure may lift the ground surface, and this is known as frost heave. The ground below a building is sheltered by it and this results in the setting up of unequal pressure arising from frost heave which can cause cracking and failure of foundations. In soils of this type foundations should be taken down to a depth of not less than 610mm below ground level. Such material should not be used as a filling underground floor slabs as these are particularly vulnerable especially in unfinished buildings.
Clays:- The strength and stability of clays are affected by water content. Clays shrink on drying and swell again when wetted. The drying out of clays may be accelerated by tree roots and extend deeper, and it is recommended that a building on shallow foundations should not be closer to a single tree than 1½ times the mature height of the tree, or twice the height for groups of trees. When a site has been cleared of trees there is a risk of the clay swelling and lifting the building, and time should be allowed for the clay to regain water. This may take several years. Alternatives are (1) to anchor the building by means of reinforced piles, sleeved from the ground over their top 3m and providing suspended floors. Beams spanning between the piles should be well clear of the ground surface. (2) Use flexible framed construction without brickwork or plastering. (3) Making the building rigid by either constructing a basement or reinforcing the foundation and brickwork.

Heating appliances and boilers have been known to cause excessive drying out of clays resulting in movement and cracking of the foundation slab. Adequate insulation is important in these circumstances. Clay soils on sloping sites are likely to move down hill albeit slowly if the angle of repose exceeds 10°. The minimum depth of strip foundations in clay is recommended to be 1m, although it may be necessary to go deeper in the vicinity of trees.
Made-up-ground:- Movement and settlement is likely and in general it is advisable to use deep foundations passing through the fill, or possibly a raft foundation. Similar precautions are required in peat.

FOUNDATIONS 2

PURPOSE
The purpose of the foundation is to spread the load from the structure over a safe bearing area of the subsoil, and to provide a stable, level base on which to build. The Building Regulations , require that the foundations shall:-

1. Safely sustain and transmit to the ground the combined dead load, imposed load and wind load in such a manner as not to cause any settlement or other movement which would impair the stability of, or cause damage to, the whole or any part of the building or any adjoining building or works; and
2. Be taken down to such a depth, or be so constructed, as to safeguard the building against damage by swelling, shrinking or freezing of the subsoil; and
3. Be capable of adequately resisting any attack by sulphates or any other deleterious matter present in the subsoil.

TYPES OF FOUNDATIONS

The design of the foundations for a particular building will depend upon the nature of the structure, the loads carried, the type of soil and its properties, any special conditions existing on the site and comparative costs. Depending on circumstances foundations may be:-

1. Concrete strip
2. Reinforced concrete
3. Pad foundations
4. Piles
5. Rafts.

STRIP FOUNDATIONS

This is the most common type of foundation and consists of a concrete strip beneath the walls, not less than 150mm thick and of sufficient width to adequately spread the load. For houses and similar structures the width of the foundation can be ascertained from the table given in Approved Document A, table E1 of the Building Regulations 1985 (see Sheet No. 14). For larger and heavier structures the foundations will need to be carefully designed and calculated.

THICKNESS OF CONCRETE FOUNDATIONS

Where the concrete foundation is wider than the wall, the thickness must be at least equal to the projection from the wall face. The angle of dispersion of the load in the conrete is such, that if the concrete should fail under the compression induced by the load, it will crack at an angle of 45° from the horizontal (Fig 1) thus, making the base thick enough to accommodate an angle of 45° from the base of the wall as shown (Fig 1), will ensure that the area bearing on the soil is retained.

Approved Document A, Diagram E1 of the Building-Regulations 1985 — Foundation Dimensions states the thickness 'T' of a concrete foundation shall be 150mm or 'P' is derived using Table E1.

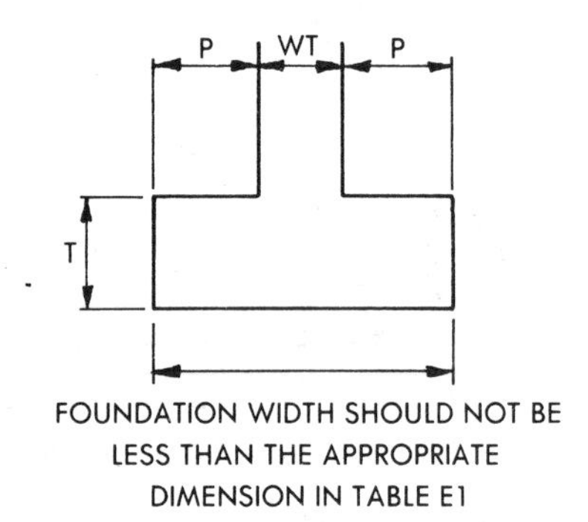

FOUNDATION WIDTH SHOULD NOT BE LESS THAN THE APPROPRIATE DIMENSION IN TABLE E1

FIG 1

FOOTINGS

A method used to widen the base of a wall and spread the load, by means of ¼ brick offsets or footings courses. Fig 2. The bottom course of footings is twice the width of the wall and the concrete foundation projects 150mm as shown. Heading bond is used for footings, but they are seldom used today.

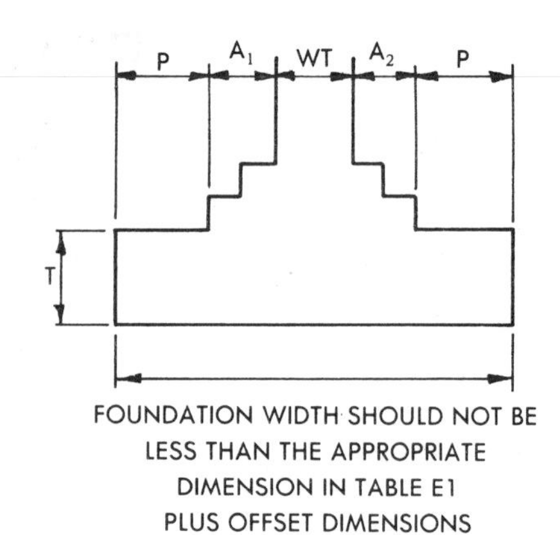

FOUNDATION WIDTH SHOULD NOT BE LESS THAN THE APPROPRIATE DIMENSION IN TABLE E1 PLUS OFFSET DIMENSIONS A_2 and A_2

FIG 2

STRENGTH OF CONCRETE

Concrete for foundations should be composed of cement and fine and coarse aggregate conforming to BS 882: 1983 in the proportion of 50kg of cement to not more than 0.1m^3 of fine aggregate and 0.2m^3 of coarse aggregate. N.B. A stronger mix may be used, but not a weaker mix.

STEPPED FOUNDATIONS

On sloping sites strip foundations should be on a horizontal bearing and stepped. At each step the higher foundation must extend over and unite with the lower foundation for a distance of not less than the thickness of the foundation, and in no case less than 300mm. Figs 3 and 4.

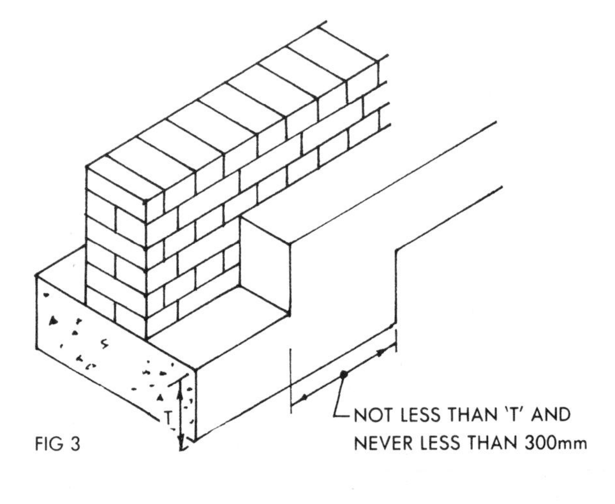

FIG 3

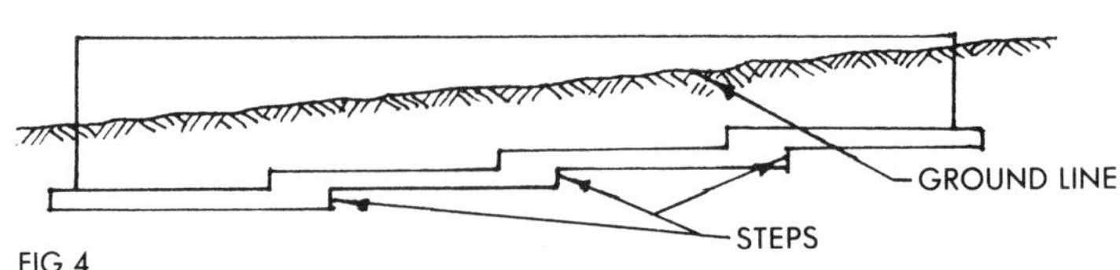

FIG 4

FOUNDATIONS 3

The width of concrete strip foundations must be not less than indicated in the table given. Approved Document A, Table EI, The Building Regulations 1985.								
Type of Subsoil	Condition of Subsoil	Field test applicable	Minimum width in millimetres for total load in kilonewtons per lineal metre of loadbearing walling of not more than.					
			20kN/m	30kN/m	40kN/m	50kN/m	60kN0m	70kN/m
Rock 1	Not inferior to standstone, firm chalk or limestone	Requires at least a pneumatic or other mechanically operated pick for excavation	In each case equal to the width of the wall					
Gravel 2 Sand	Compact Compact	Requires pick for excavation. Wooden peg 50mm square in cross section hard to drive beyond 150mm	250	300	400	500	600	650
Clay 3 Sandy clay	Stiff Stiff	Cannot be moulded with the fingers and requires a pick or pneumatic or other mechanically operated spade for its removal	250	300	400	500	600	650
Clay 4 Sandy clay	Firm Firm	Can be moulded by substantial pressure with the fingers and can be excavated with graft or spade	300	350	450	600	750	850
Sand 5 Silty sand Clayey sand	Loose Loose Loose	Can be excavated with a spade. Wooden peg 50mm square in cross section can be easily driven	400	600	Note:- In relation to types 5, 6 and 7, foundations do not fall within Do not fall within the provisions of Approved Document A. If the total load exceeds 30kN/m			
Silt 6 Clay Sandy clay Sily clay	Soft Soft Soft Soft	Fairly easily moulded in the fingers and readily excavated	450	650				
Silt 7 Clay Sandy clay Silty clay	Very soft Very soft Very soft Very soft	Natural sample in winter conditions exudes between fingers when squeezed in fist	600	850				

WIDE STRIP FOUNDATIONS
Where the bearing capacity of the ground is such that extra spread of the load is required, wide strip foundations may be used. Suitable transverse and longitudinal reinforcement should be provided to withstand the tensions induced, and an example is shown in Fig I.
Minimum cover to reinforcement:- Steel reinforcement in concrete below ground should have a minimum cover of 50mm. When cast against rough timbering or in direct contact with the soil an increase in the nominal cover is necessary to ensure the minimum is maintained and 75mm is common. Fig I.

PROJECTIONS
Diagram E3 of Approved Document A of the Building Regulations 1985 states that the foundation of piers, buttresses and chimneys should project as indicated in diagram E3 and the project 'X' should never be less than 'P'.

CONCRETE MIX
Part E2 (c) design provisions require that concrete for foundationis shall not be weaker than a mix having proportions of 50kg of cement to not more than 0.1m³ of fine aggregate, and 0.2m³ of coarse aggregate.

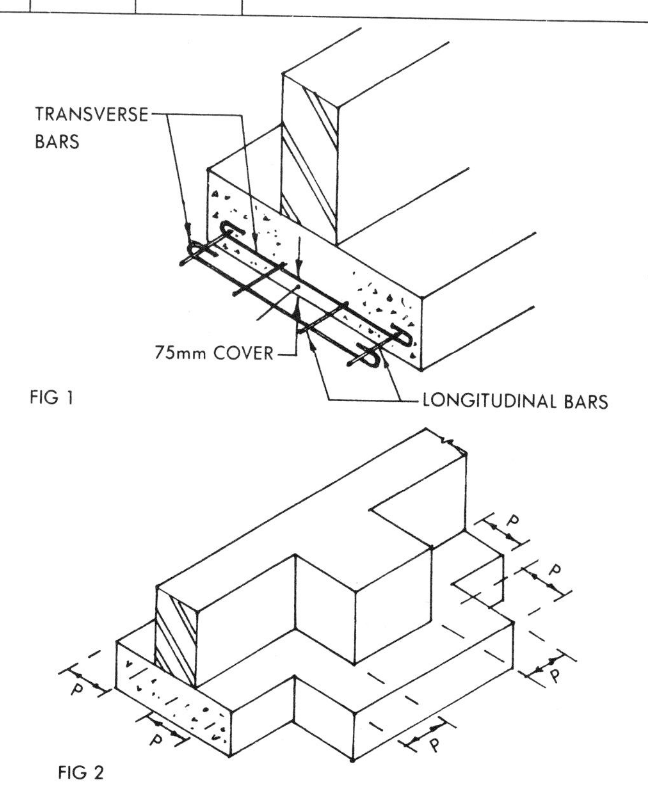

FIG 1
FIG 2

FOUNDATIONS 4

FOUNDATIONS ON CLAY
The movement of clay soils due to moisture changes, was mentioned in Sheet No. 12. The effect is most marked towards the outer periphery of the foundation and the movement of the clay may cause the foundation slab to tilt and the wall to crack. Fig 1. The volume changes become less with depth and below 0.9m the movement is so slight that it will not seriously effect foundations. Thus the minimum depth of normal strip foundations is 0.9m.
Note:- This may not be effective against severe drying shrinkage caused by trees, shrubs etc, and in this case it will be necessary to go deeper.

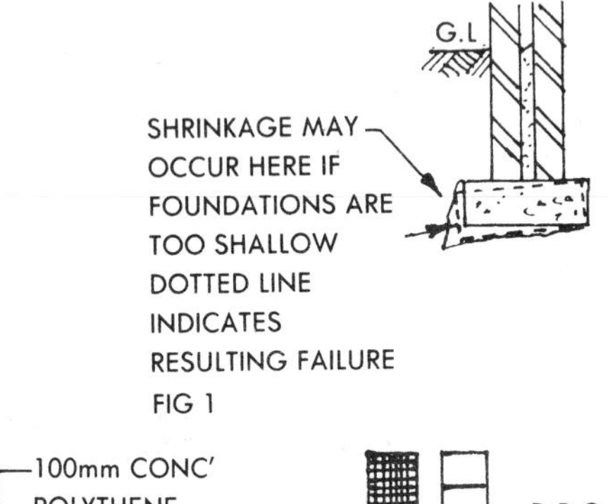

FIG 1

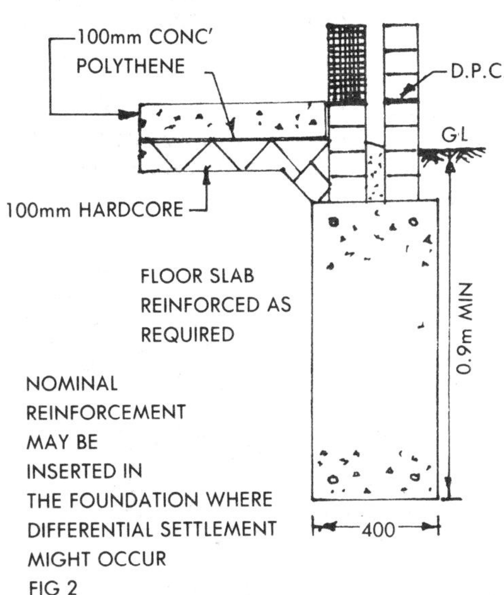

FIG 2

NARROW STRIP FOUNDATIONS
Fig 2. This type of construction is suitable for low-rise buildings on shrinkable clays. The trench need be only 400mm wide and is excavated by machine because of the narrow width. The trench is filled with concrete to just below ground level, and this overcomes the difficulty of bricklayers working in a narrow trench.

A narrow strip foundation as shown of 1:9 concrete should be cheaper than a traditional strip foundation. There are labour savings in trimming trench sides and bottom. To achieve maximum efficiency and economy this type of foundation needs greater accuracy in line and levels to ensure the wall is level and does not overhang the foundation. Care must be exercised in positioning services which pass through the foundation as their position cannot easily be adjusted.

It is simple to insert reinforcement in this type of foundation, and this is particularly important where local soil variations might tend to cause unequal settlement.

SHORT BORED PILE FOUNDATIONS
To avoid excessive excavation in clay soils a system of short bored piles, coupled with RC ground beams has been developed. Figs 3 & 4. Where trees have been removed the upper part of the pile shaft should be sleeved in polythene sheets or cardboard tubes to reduce uplift forces from the swelling clay subsequently Fig 4. A layer of compressible material, e.g. polystyrene can be placed beneath the ground beams. Fig 3.

There is risk of damage where the height of a tree exceeds its distance from a building.

Pile lengths of 4.5m or so are generally satisfactory.

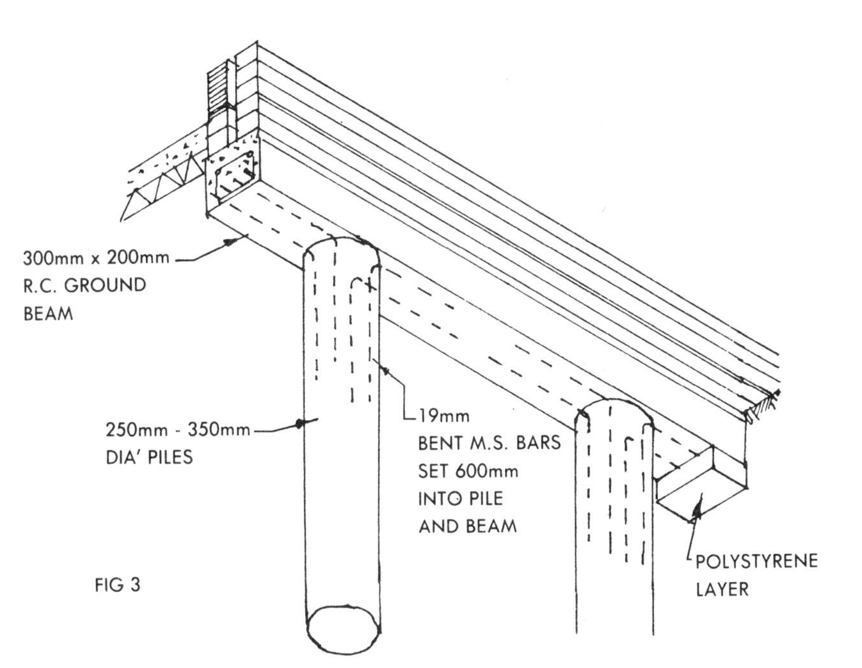

FIG 3

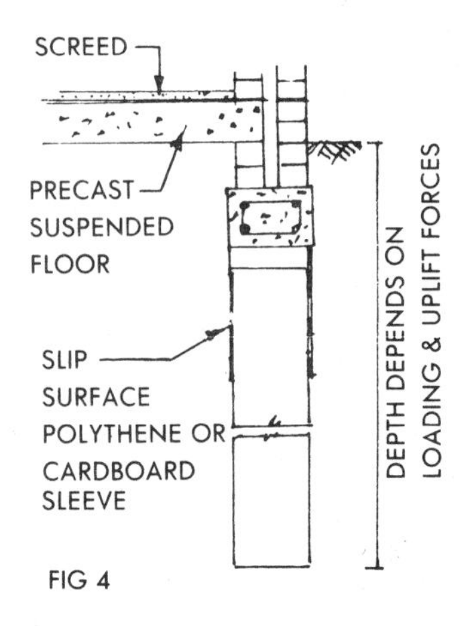

FIG 4

FOUNDATIONS 5

PAD FOUNDATIONS
These are square or rectangular concrete slabs used where loads are carried by concrete columns, stanchions or isolated brick piers. The size of the pad depends upon the load on the column base and the bearing capacity of the soil. On soils of good bearing capacity a mass concrete base may be economical. Fig I shows an example of a mass concrete pad foundation to a stanchion. In this case the thickness of the pad 'T' must be at least equal to the projection 'P' from the edge of the steel base plate.

Where heavy loads are carried and the spread is such that mass concrete is uneconomical, a reinforced pad foundation is used. Fig 2 shows a reinforced concrete pad foundation to a reinforced concrete column.

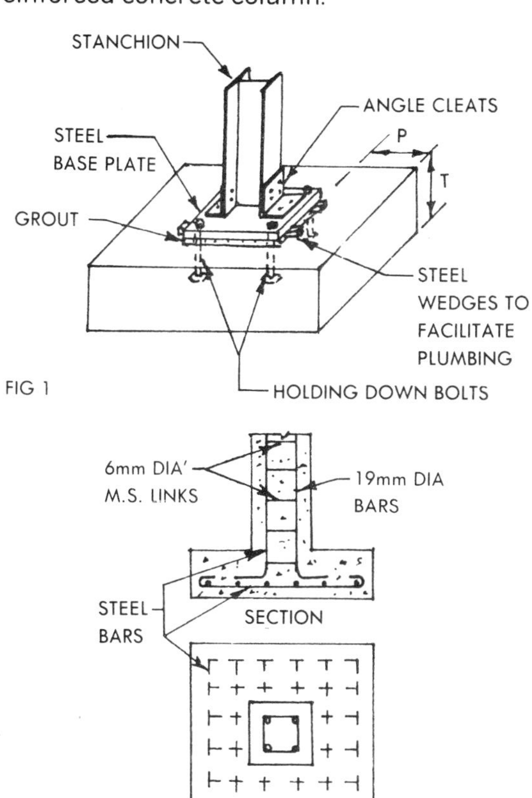

FIG 1

FIG 2

RAFT FOUNDATIONS
A reinforced concrete foundation slab, often thickened under the walls, which covers the area beneath the building and may extend beyond the outer walls. The reinforcement may be mild steel bars laid in both directions, or in some cases welded steel fabric.

Raft foundations may be used where the subsoil is of low bearing capacity or where relative settlement may occur.

A raft should be shaped so that the centre of gravity of the imposed load is over the centre of area but the design of rafts is complex and expert advice should be sought.

Ground movement:- The level of raft foundations is usually near the surface and precautions must be taken against swelling and shrinking of the ground under the raft. It may be necessary to extend the raft so that a protective apron extends beyond the effective ground bearing.

Services:- The positioning of service pipes, ducts and drains should be planned so that the strength of the raft is not unduly reduced by holes and penetrations. Provision for maintenance and access to services must also be considered.

An example of a light raft suitable for a smaller building is shown in Fig 3.

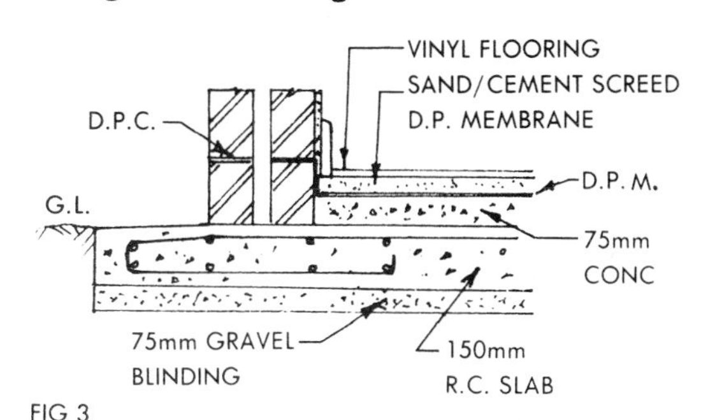

FIG 3

Heavy duty rafts:- Raft foundations are sometimes used for heavy buildings and in this case a beam and slab raft may be used. Fig 4.

Cellular raft foundations
For basement construction in very poor soil cellular rafts may be used. These have floors and walls in reinforced concrete forming a cellular box construction. Fig 5.

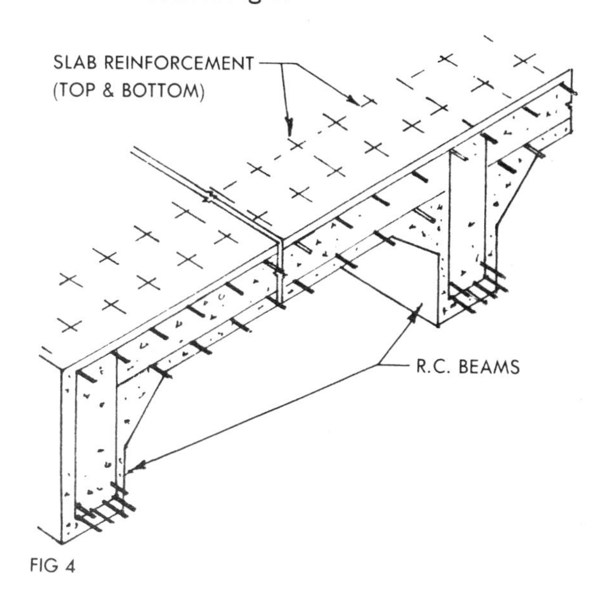

FIG 4

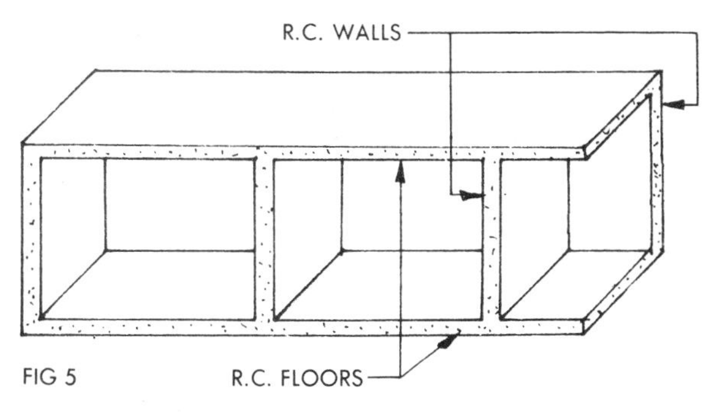

FIG 5

BASEMENTS I

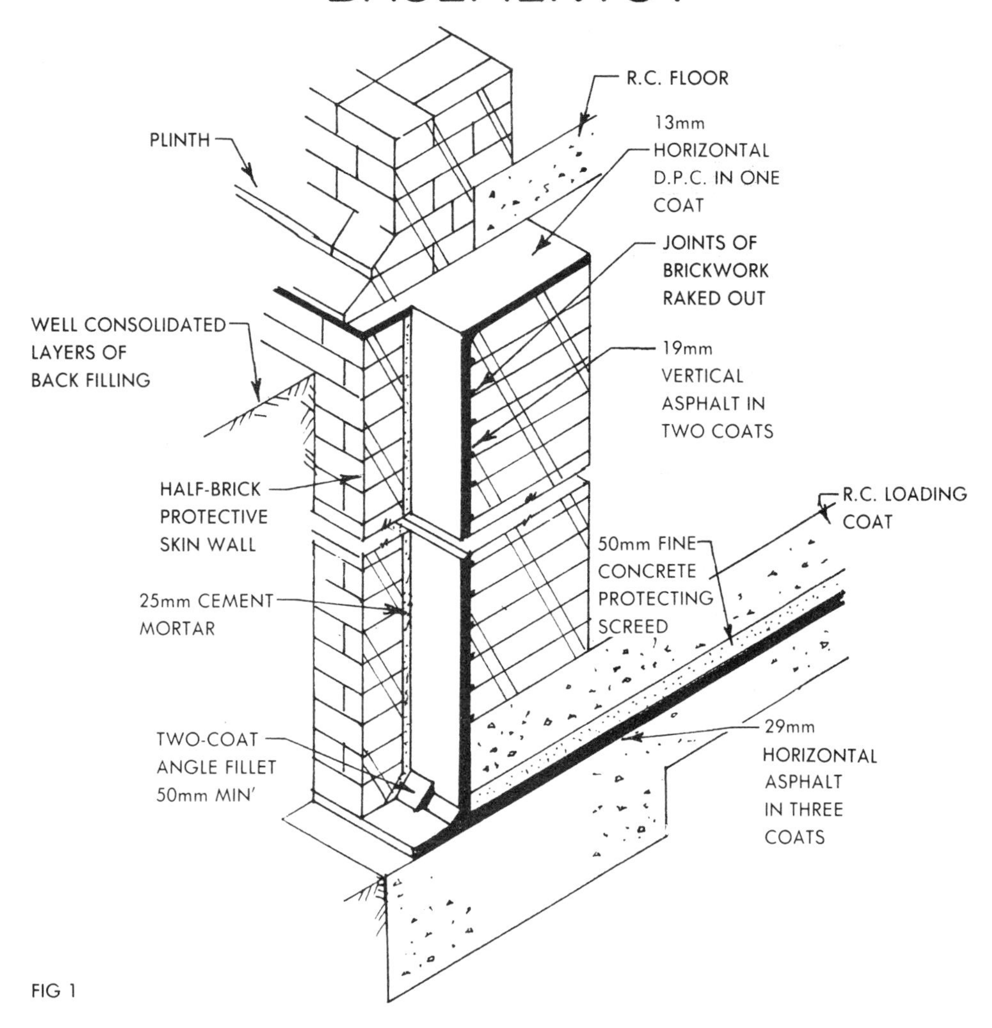

FIG 1

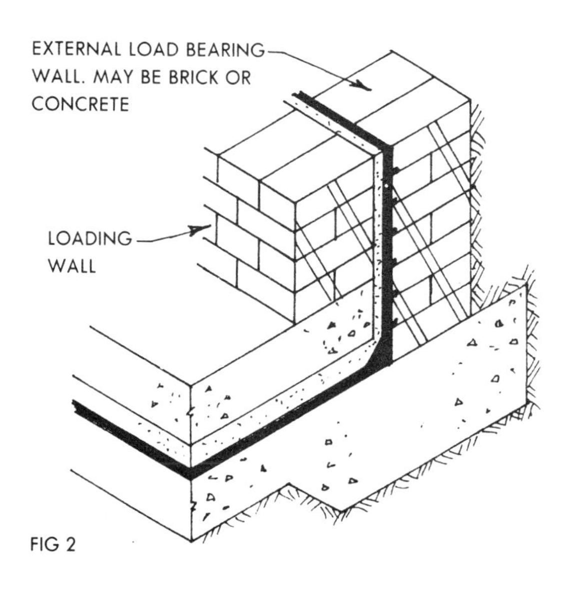

FIG 2

ASPHALT TANKING

Where there is sufficient room to allow working space for the asphalters the vertical mastic asphalt may be applied externally. Fig I.

On restricted sites where the 600mm min working space cannot be provided, or in existing buildings, the vertical asphalt may be applied internally. Fig 2. In this case it is necessary to build a loading wall against the asphalt to resist pressure from ground water.

The horizontal asphalt is in three coats finishing 48mm thick, and the vertical asphalt in three coats finishing I9mm thick.

PRECAUTIONS

1. Vertical surfaces must provide a good key for the asphalt. Joints of brickwork should be raked out and well brushed down. Vertical concrete surfaces must be keyed by hacking, grooving etc.
2. Immediately upon completion the horizontal asphalt should be protected by covering with a fine concrete screed, not less than 50mm thick.

3. Vertical asphalt must be protected as quickly as possible by the erection of skin walls or main structural walls as the case may be. The wall should be kept at least 25mm clear of the asphalt and the space well grouted in as the work proceeds.
4. Two-coat fillets should be provided at all internal angles.
5. It is essential that pumping operations be maintained on wet sites until protective loading coats and walls are completed and fully set.
6. Care must be taken to ensure that the asphalt membrane is not damaged in any way, e.g. by being punctured by nails, reinforcing rods etc, and that strutting is not placed directly on the membrane.

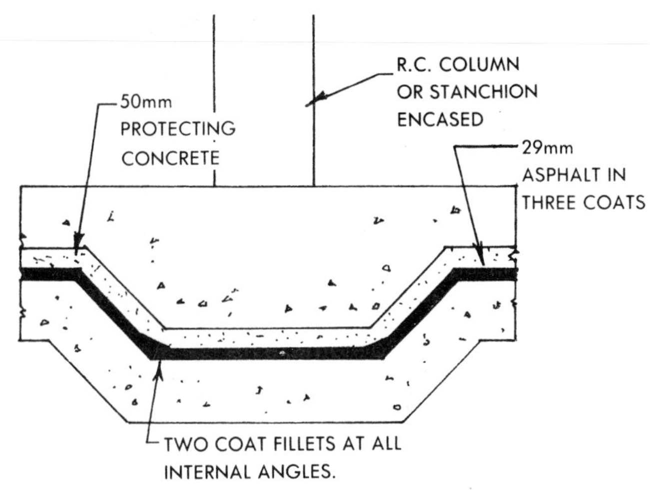

FIG 3 ISOLATED COLUMN OR STANCHION DETAIL

BASEMENTS 2

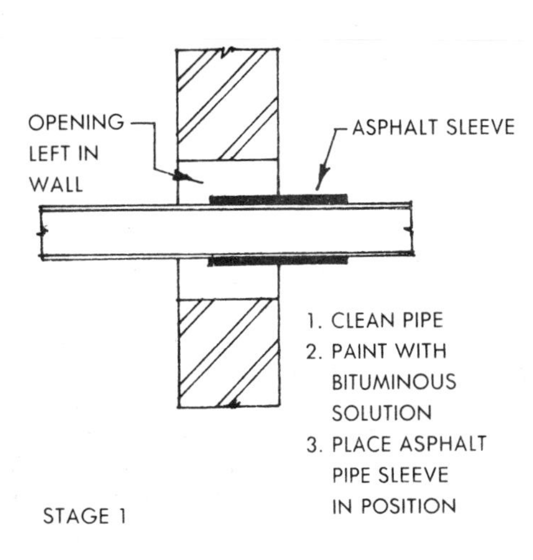

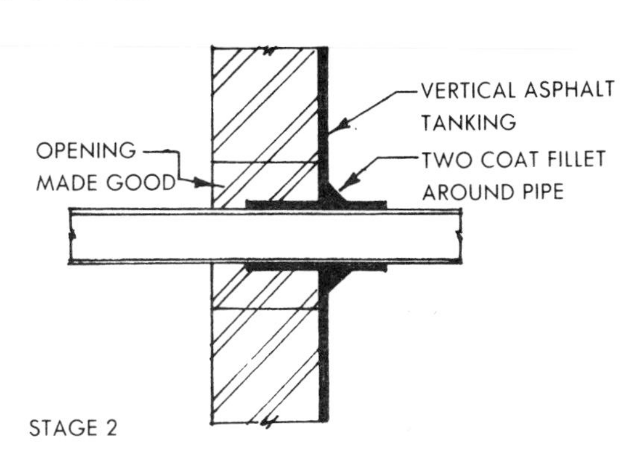

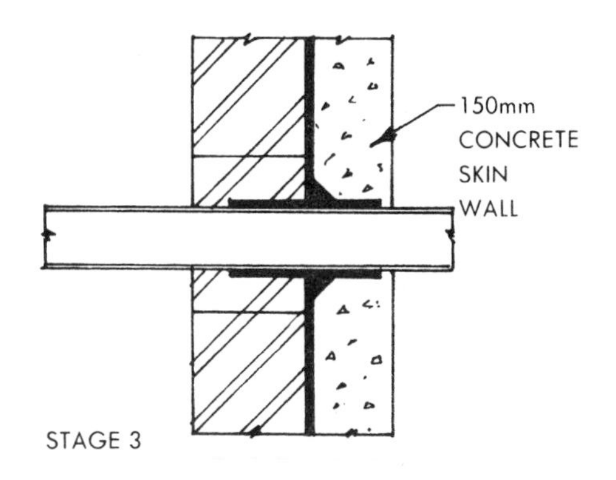

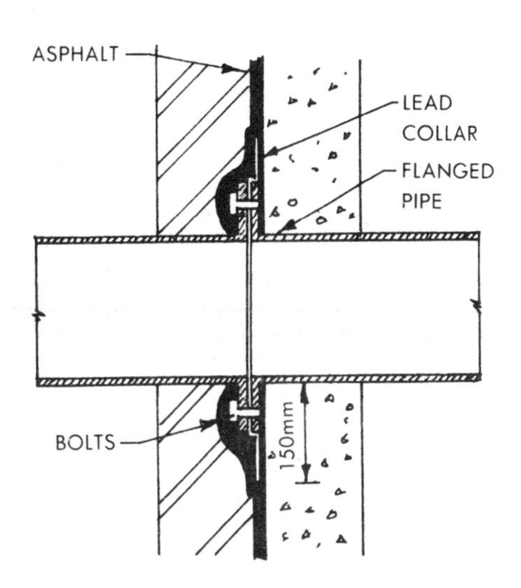

TREAMENT OF PIPES PASSING THROUGH TANKING

One method of dealing with a pipe passing through the tanking is illustrated above. An alternative method using flanged pipes with a sheet lead collar sandwiched between the flanges is shown on the left. This method may be used where ground water pressure is a problem.

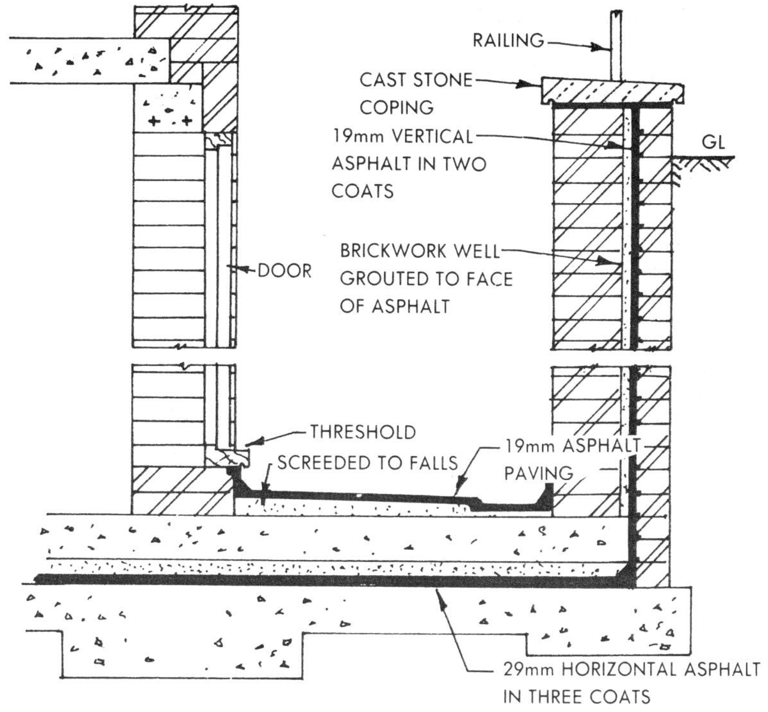

SECTION THROUGH OPEN AREA

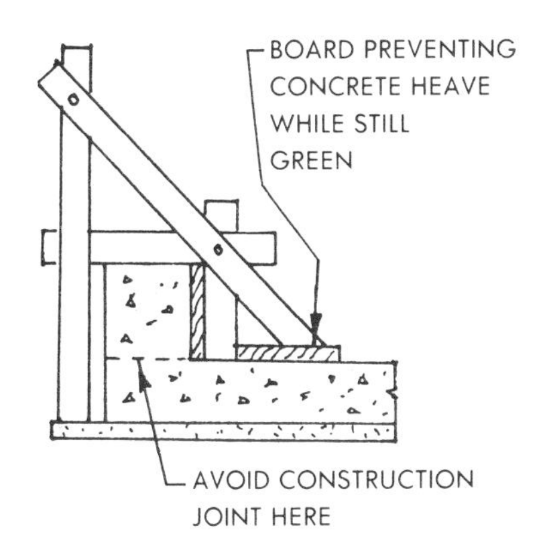

WATERTIGHT CONCRETE CONSTRUCTION

Dense, well compacted concrete is impervious, and basements may be constructed with or without a membrane. It is however difficult to achieve a completely watertight concrete under site conditions, and integral waterproofing additives, which fill the pores of the concrete may be used to facilitate the production of impervious concrete. These are added in liquid or powder form at the mixing stage.

If a completely watertight basement is to be achieved then certain precautions must be observed:-

1. Formwork must be well constructed, loose or ill fitting shuttering allows seepage of cement grout resulting in honeycomb, porous concrete.
2. All spaces where leakage might occur, e.g. where reinforcement projects must be sealed with putty etc. Any spaces in the oversite concrete through which grout might seep should be filled in before fixing the base slab reinforcement.
3. Care in handling and transporting concrete to see that no segregation takes place.
4. The mix must have good workability and the concrete must be thoroughly compacted, preferably by vibration, and compaction must be completed within 30 min of mixing.
5. Do not stop de-watering system until the concrete has matured for at least 24 hrs.
6. Joint between base slab and wall should be 230mm or so above the surface of the slab. A 'kicker' should be constructed (as shown left) in one pour with the base slab.

CONCRETE I

Concrete is composed of a mixture of fine aggregate, usually sand, coarse aggregate e.g. uncrushed gravel, a matrix or binding material, usually Portland cement, and water.

FINE AGGREGATE

May be natural sand dug from pits or dredged from river beds, or artificial sand obtained by crushing stone or gravel. Sand for concrete should be clean, well graded and not contain an excess of fine, dusty particles.

CLEANLINESS

Sand should not contain any chemical impurities which might adversely affect the cement, or contain excessive silt or clay particles, as these will prevent a good bond between the cement and the grains of sand and lead to shrinkage and cracking occurring. When the sand is rubbed between the fingers it should not stain the hands, and if on inspection there is doubt as to the cleanliness, a settling test should be carried out.

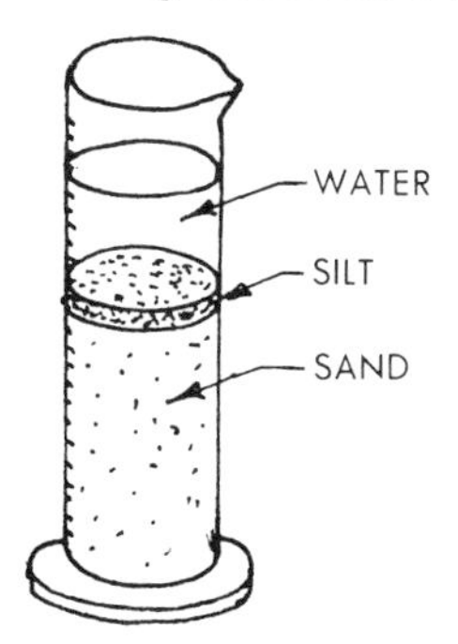

FIELD SETTLING TEST

About 50mls of a solution of common salt in water (l teaspoonful to 0.5 litre of water) are placed in a 250ml B.S. measuring cylinder. Sand is added gradually until the volume of sand is about 100mls. The volume is then made up to 150mls by adding more salt solution. The mixture is well shaken up and then the cylinder is placed on a level surface and gently tapped until the sand is level. After 3 hrs the silt will settle out on top of the sand and can be expressed as a percentage of the height of the sand itself. The amount of clay and silt should not exceed 8%. N.B. The presence of salt causes the colloidal particles, which would remain in suspension, to coagulate and deposit with the silt.

WELL GRADED SAND

UNIFORM SAND

GRADING

The size of the grains should vary from 4.7mm down, with a good proportion of larger grains. Such a sand is said to be well graded, the smaller grains fill the spaces between the larger grains and a strong, workable mix can be obtained with the minimum of cement.

The specific area (surface area) of sand is an important factor in deciding the proportions of a mix. The surfaces should all be coated with cement paste for maximum strength. The surface area per unit weight of volume is doubled when the particle size (linear) is halved. It follows that sand should not contain too high a proportion of fine particles. Research is being carried out to perfect a site method of measuring the specific area of sand, so that the sand content of a mix may be accurately calculated.

Sand having grains all one size is known as uniform sand and will tend to give poor workability and produce a weak, porous mix.

Sand may be tested for grading by passing samples through a nest of B.S. sieves, the amount (by weight) retained on each sieve being about equal for a well graded sand. A poorly graded sand may be improved by mixing with a different sand to obtain a better balance of particle size.

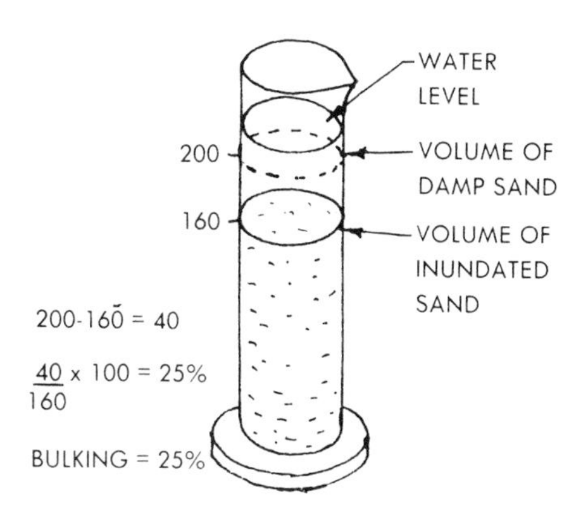

BULKING OF SAND

Dry sand and saturated sand have approximately the same volume. Damp sand however will show a considerable increase in volume. This is most marked when the water content is 5% to 10% of the dry weight, when the increase in volume may be as much as 30%. This phenomenon is known as 'bulking' and is more pronounced in fine sand. Allowance should be made for this factor when measuring out sand in order to avoid a deficiency of sand in a mix. This may be done by increasing the depth of the gauge box for the sand by the required amount. The amount of bulking may be found by measuring out a sample of sand, then saturating it with water and noting the decrease in volume.

STORAGE

Sand should be kept clean on site and not allowed to become contaminated with soil or rubbish. Stockpiles should be on a hard, clean surface, e.g. concrete or steel sheet, with the surface sloping outwards to facilitate drainage. Partition walls around the stockpile will ensure the sand is not mixed with different size aggregate.

CONCRETE 2

COARSE AGGREGATE
May be gravel, crushed rock or stone and hard, durable, and not contain any materials likely to decompose. It should be clean, free from any impurities and well graded from particles of 4.7mm upwards. It should be of the correct size for the job, for normal work up to 38mm dia, but in no case should the larger particles have a diameter greater than one-quarter the thickness of the finished concrete. For heavily reinforced work the maximum particle size should be 5mm less than the spacing between bars, or less than the thickness of the concrete cover to the reinforcement, whichever is the smaller.

ALL-IN AGGREGATE
A natural mixture of sand and gravel otherwise known as 'ballast' often used for mass concrete, but since the proportions of sand and gravel are variable it is not usually considered suitable for reinforced concrete. If 'all-in' aggregate is used for concrete a 1:6 mix is not the equivalent of a 1:2:4 cement, sand, coarse aggregate mix, since some of the fine aggregate will fill the voids in the coarse aggregate. A 1:5 mix is a closer equivalent.

STORAGE
Should be on a hard, clean surface sloping to facilitate drainage. If necessary the storage area should be covered with a 102mm thick slab of lean concrete. Aggregate should stand for 12 hrs before use so that any excess moisture can drain away.

CEMENT
The most widely used is Portland cement made from a mixture of chalk and clay, burnt in a rotary kiln, the resultant clinker being ground in a ball mill with a little gypsum added to control the set. When cement is mixed with water, chemical reaction (hydration) occurs, resulting in setting and hardening.

CEMENT STORAGE
Bags of cement should be stored clear of the ground in a weather-proof shed to guard against air setting. The bags should preferably not be stacked more than 1.500m high, and arranged so that they are used in the same order in which they are delivered, i.e. that new deliveries are not placed on top of existing stacks. Cement should not be kept too long, cement in bags shows marked loss of strength if stored longer than 4 to 6 weeks. In sealed drums cement keeps indefinitely. For jobs where large quantities of cement are used, bulk delivery using a storage silo is likely to be more economical.

MIXES
For small jobs volume batching may be used and the table below gives a selection of mixes based on either a 1 cwt or a 50kg bag of cement, and their suitability. The quantities are for dry sand and if the sand is moist allowance must be made for bulking.

MIX	CEMENT	SAND	COARSE AGG	USE
1:3:6	1 cwt 50kg	$3\frac{3}{4}ft^3$ $0.10m^3$	$7\frac{1}{2}ft^3$ $0.20m^3$	Mass concrete, foundations, solid ground floor
1:2:4	1 cwt 50 kg	$2\frac{1}{2}ft^3$ $0.07m^3$	$5ft^3$ $0.14m^3$	Concrete generally, walls, R.C. work
$1:1\frac{1}{2}:3$	1 cwt 50kg	$1\frac{7}{8}ft^3$ $0.05m^3$	$3\frac{3}{4}ft^3$ $0.10m^3$	Watertight and strong concrete

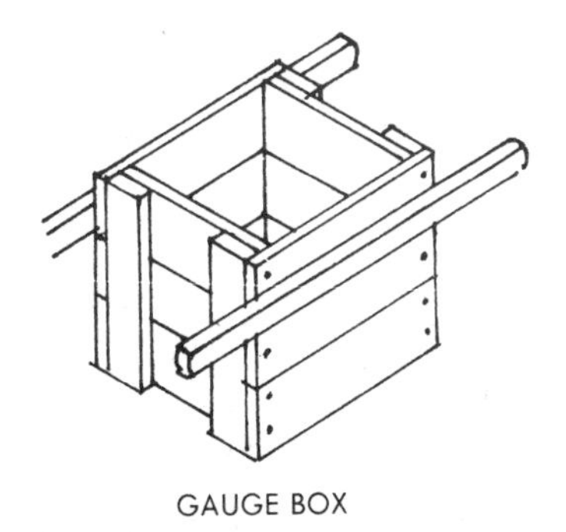

GAUGE BOX

VOLUME-BATCHING
It is desirable to base a mix on a 1cwt or 50kg bag of cement and avoid splitting a bag if possible. A gauge box should be used for measuring out the aggregates. This should be deep and narrow rather than wide and shallow, filled with aggregate, which should be screeded off with a straightedge to ensure accuracy. Gauge boxes should either hold the full amount for a batch of concrete or an exact fraction e.g. $\frac{1}{2}$ or $\frac{1}{3}$. Boxes larger than about $1\frac{1}{2}ft^3$ or $0.05m^3$ are difficult to handle. Dimensions can be worked out for required capacities, a box 12in x 12in x 18in deep will hold $1\frac{1}{2}ft^3$ of aggregate. A box 350mm x 350mm x 410mm deep will hold $0.05m^3$ of aggregate (approx).

CONCRETE 3

WATER CEMENT RATIO
A batch of concrete should contain sufficient water to produce a dense concrete of adequate workability. The quantity of water for a mix is expressed as a decimal fraction of the weight of cement e.g. a water/cement ratio of 0.60 means that for every 50kg of cement in the mix, 30kg (30 litres) of water are used (i.e. 50kg x 0.06).

The water/cement ratio is probably the most important factor influencing the strength of concrete. In general the drier the mix the stronger the concrete, provided it can be fully compacted. Hand compaction becomes difficult with a water/cement ratio of 0.50, and below this figure mechanical vibrators should be used. A wet mix may have better workability, but it tends to segregate and produces a weak concrete which may shrink and crack.

If an accurate water/cement ratio is to be maintained then the moisture content of the aggregate must be ascertained, and this should be measured twice daily, for each new load of aggregate delivered and after rain. One method used is to weigh a sample of the aggregate (WI) dry out the sample (by heating in a pan), and then re-weigh (W2). The percentage water content is then calculated as $\frac{W1 - W2}{W2} \times 100$

It is important that the water gauge on a machine is accurate and the gauge should be checked at the start of a job, at least once a week and whenever the machine is moved.

The water must be clean and should be potable. The normal mains supply should be used where possible.

WORKABILITY
The workability of batches of concrete should be consistent and control may be exercised by checking the water content of batches. The consistence of concrete may be measured by using the slump test or the compacting factor test, although the latter is a more sensitive method of measuring the workability of a mix, and is particularly suitable for mixes of low workability.

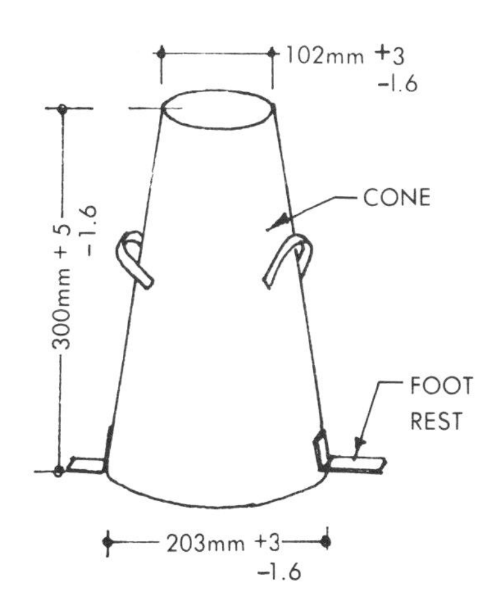

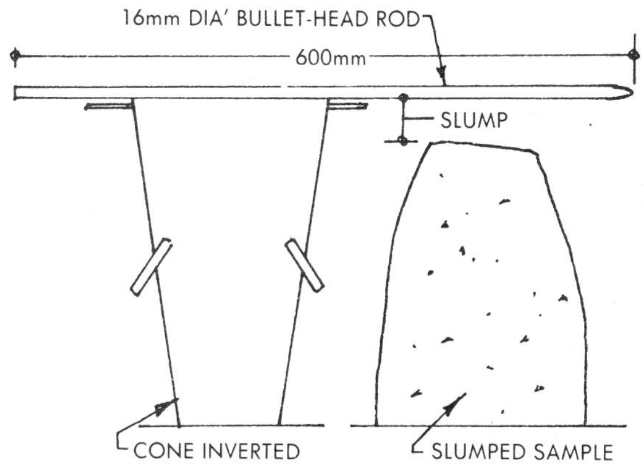

SLUMP TEST
The test is carried out using a standard hollow metal cone and a rod as shown. The cone is placed on a flat impervious surface and held down by the foot rests. It is then filled with concrete in four 76mm layers, each layer being given 25 strokes with the rod. The concrete is struck off level at the top and the cone carefully lifted clear of the concrete. The cone is then placed next to the slumped concrete and the slump measured as shown.

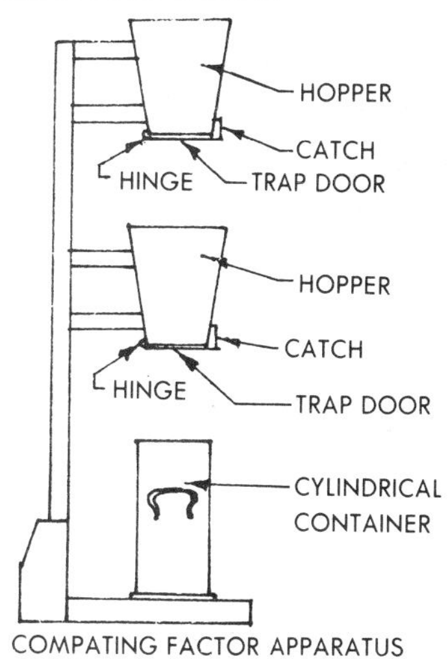

COMPATING FACTOR APPARATUS

COMPACTING FACTOR TEST
The apparatus consists of two conical hoppers mounted above a cylindrical container as shown. The top hopper is filled with concrete without compacting it. The hinged trapdoor at the bottom is then released and the concrete allowed to fall into the second hopper. The trapdoor of this hopper is then released and the concrete allowed to fall into the cylindrical container, the top surface being cut off level using two trowels from the outside to the centre. The cylinder and contents are then weighed and the weight of partially compacted concrete found.

It is now necessary to find the weight of fully compacted concrete, and this is done by refilling the cylinder in layers approximately 51mm deep, compacting each layer by hand-punning or by vibrating and striking the top off level. The compacting factor is the ratio of the partially compacted weight to the fully compacted weight. It is best to take the average of three tests.

CONCRETE 4

The table below indicates the slump and compacting factors for concrete for different purposes.

PURPOSE	COMPACTING FACTOR	SLUMP
Very high strength concrete for prestressed concrete sections compacted by heavy vibration	0.70-0.78	0
High strength concrete sections, pavings and mass concrete compacted by vibration	0.78-0.85	0-25mm
Normally reinforced concrete sections compacted by vibration hand-compacted mass concrete	0.85-0.92	25mm-50mm
Heavily reinforced sections compacted by vibration. Hand compacted concrete in normally reinforced slabs, beams, columns and walls	0.92.0.95	50mm-l00mm
Heavily reinforced concrete sections compacted without vibration, and work where compaction is particularly difficult	over 0.95	l00mm-l50mm

WEIGH-BATCHING
This method is more satisfactory than volume batching in producing good concrete. There are a number of machines available incorporating a weighing hopper. Care must be taken to see that the machine is set level and that the hopper is carefully loaded, uneven loading can give an incorrect reading of the weight. The plant should be checked for accuracy when beginning a job, whenever the machine is moved and at least once a week. First check that the indicator reads zero when the hopper is empty, then load with a known weight, e.g. three 50kg bags of cement, which weigh approximately l52kg (including the weight of the bags) and check that this weight is indicated. The machine should be accurate to within ± 2% of the indicated load.

HAND MIXING
Should be carried out on a clean, hard surface. The materials should be turned over and thoroughly mixed in a dry state, water added gradually, through a rose and mixing continued until the colour is uniform.

MACHINE MIXING
The mix should be turned over in the mixer for two minutes from adding the water, to ensure a uniform distribution of the materials. Mixing for insufficient time produces weak concrete. The first batch tends to be harsh because some of the cement will stick to the side and blades of the mixer. This may be offset by either using extra cement or half the amount of coarse aggregate. Alternatively use the first batch for less important ;work, e.g. in a trench bottom, for backfill etc.

COMPACTING
When placing concrete it must be well tamped and rodded to thoroughly compact the mix and ensure there are no voids or air pockets. If compacted by hand the concrete should be placed in thin layers not exceeding l52mm thick, each layer being thoroughly tamped before the next is placed.

VIBRATORS
The internal type of vibrator is inserted into the concrete at intervals of approximately 450mm and vibration continued until air bubbles practically cease to appear on the surface. In some situations external vibration may be necessary and this may be carried out by means of a clamp-on type of vibrator or by using a vibrating hammer against the shuttering. When external vibrators are used the formwork will need to be more robust than for internal vibration.

HANDLING
If concrete is transported for some distance the vibration, especially over rough ground may cause segregation, and runs should be kept as short as possible. For the same reason concrete should not be dropped from a height of more than lm. If the concrete cannot be placed nearer than this then a suitable chute, not steeper than necessary may be used. Care must also be taken to see that there is no loss of mortar (i.e. the sand/cement content) due to using vehicles with leaky bodywork. Contamination by oil or from dirty plant must be avoided, and the concrete should not be exposed to heavy rain.

CURING
Concrete should not be allowed to dry out too quickly but should be kept damp for 7 days after placing (3 days for rapid hardening cement). This may be achieved in a number of ways:-

l. By spraying the surface with a membrane curing compound which prevents excessive evaporation (not suitable where a screed has to be bonded to the surface).
2. By covering with waterproof paper or plastic sheet, weighted down all round.
3. By covering with hessian, matting, or a layer of sand which is kept damp.

CONCRETE 5

REINFORCED CONCRETE

Concrete is strong in compression but weak in tension, and where tension occurs steel bars are incorporated in the concrete to supply the tensile strength that the concrete lacks. A simple lintel or beam illustrates how two forces which attempt to bend it as shown, thus inducing compression in the top and tension in the bottom. Fig I. To combat the tension mild steel rods are placed 25mm up from the bottom of the lintel to give protective cover to the steel. Fig 2. The rods are hooked at the ends to ensure effective grip. Fig 3.

A useful rule of thumb for lintels up to l.829m span is to allow one l2mm diameter rod for each half-brick thickness of wall, the rods being positioned in the centre of each half-brick width. Fig 3.

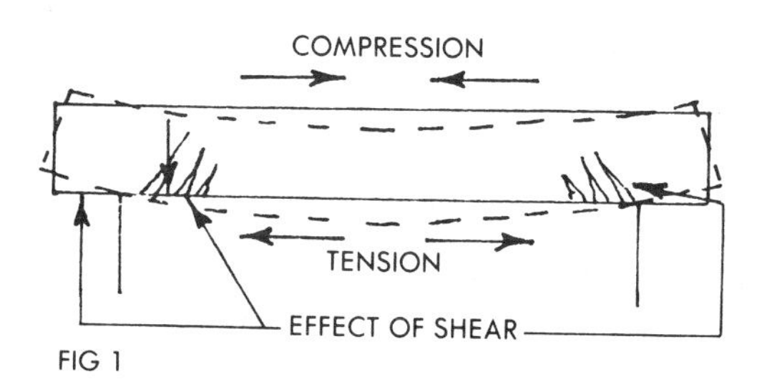

FIG 1

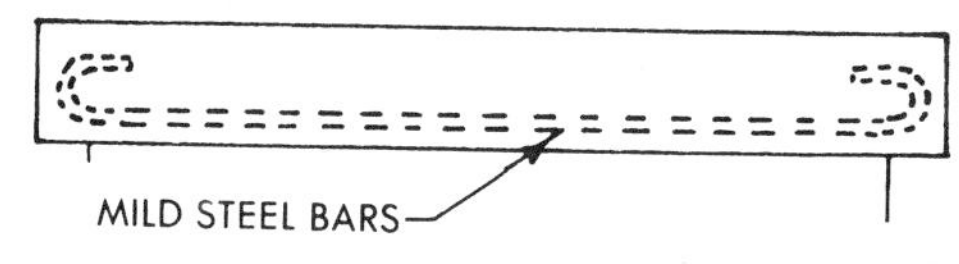

FIG 2

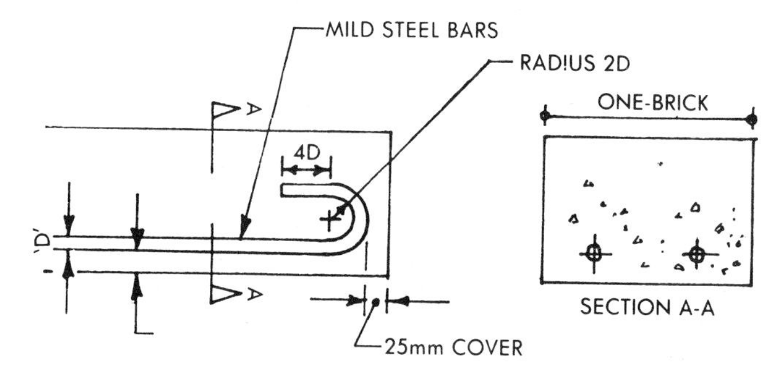

FIG 3 DETAIL OF REINFORCEMENT

SHEAR

Beams are subjected to shear stresses which are greatest adjacent to the supports. While the effect is not serious in small lintels and beams, larger beams over spans in excess of 2.438mm may fail as shown in Fig I unless reinforcement is incorporated to combat the shear stresses. This may be done in three ways:-

(i) By incorporating vertical steel stirrups, the stirrups being closer together near the supports.
(ii) By using cranked bars bent up near the supports.
(iii) By using both stirrups and cranked bars. An example is shown. Fig 4.

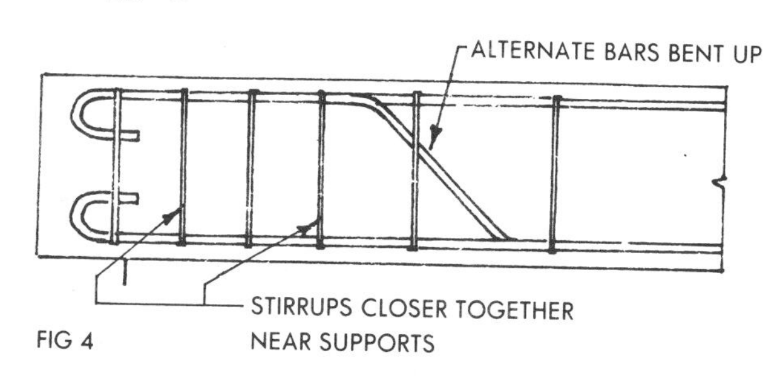

FIG 4

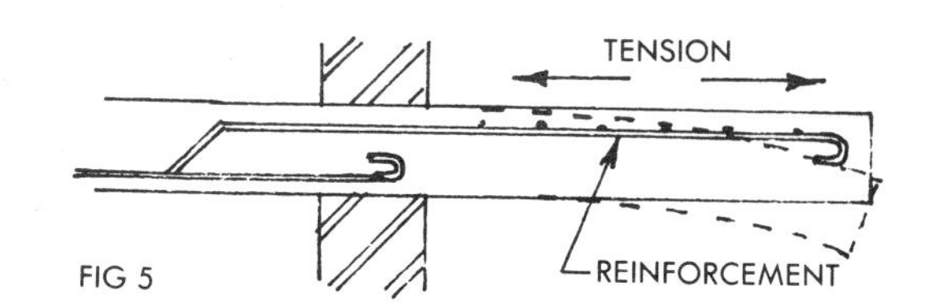

FIG 5

CANTILEVER

A cantilever e.g. a balcony slab, tends to bend as shown by the dotted line Fig 5. Thus tension is induced in the top of the slab and the reinforcement is required in the top as shown.

CONTINUOUS BEAMS

A continuous beam over a number of supports attempts to bend as shown in Fig 6. The positive bending between supports induces tension at the bottom of the beam, while the negative bending over the supports induces tension at the top of the beam. Thus reinforcement is required in the bottom of the beam (Fig 7) between the supports and in the top of the beam over the supports. Cranked bars are used as shown in Fig 7.

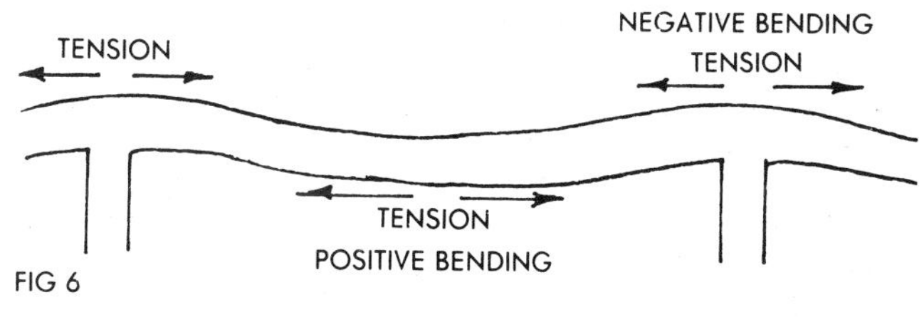

FIG 6

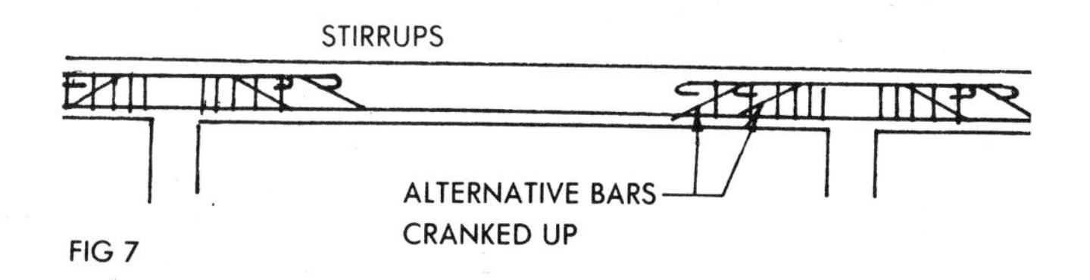

FIG 7

CONCRETE 6

COLUMNS

A concrete column under load has a tendency to bend as shown in Fig I. Thus the reinforcement must be spaced so that it will combat the tension on all sides. The vertical reinforcing bars are bound by steel links which prevent the bars from buckling outwards. Fig 2. Alternatively helical reinforcement may be used. Fig 3. The diameter of this transverse reinforcement should not be less than one-quarter the diameter of the main bars, and in no case less than 5mm. The pitch of the transverse reinforcement in a column should be not more than the least of the three dimensions:-

I. The least lateral dimension of the column.
2. Twelve times the diameter of the smallest longitudinal reinforcement in the column.
3. 300mm.

A column with helical reinforcement should have a minimum of six longitudinal bars within the reinforcement. Fig 3.

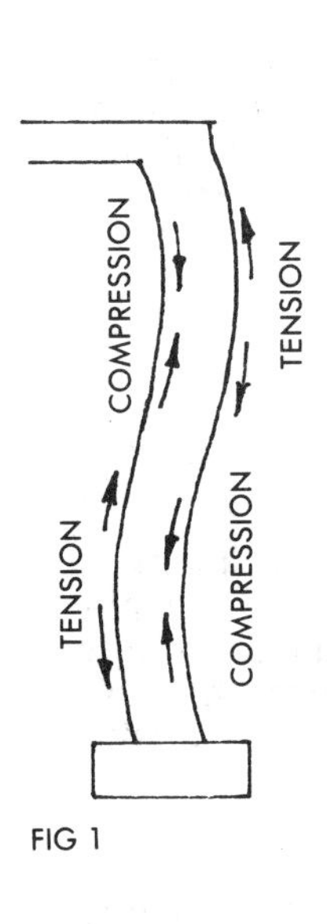

FIG 1

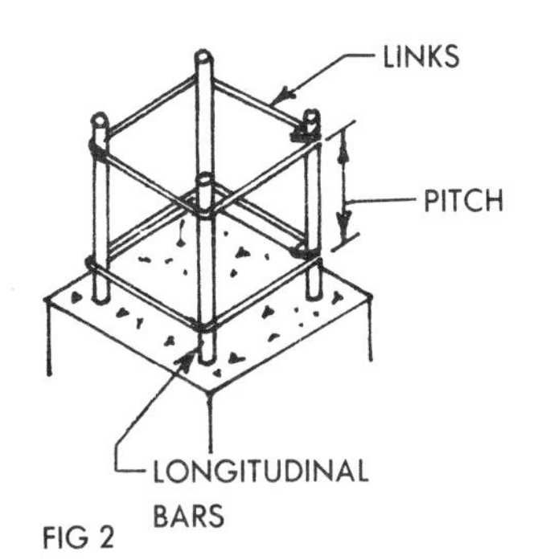

FIG 2

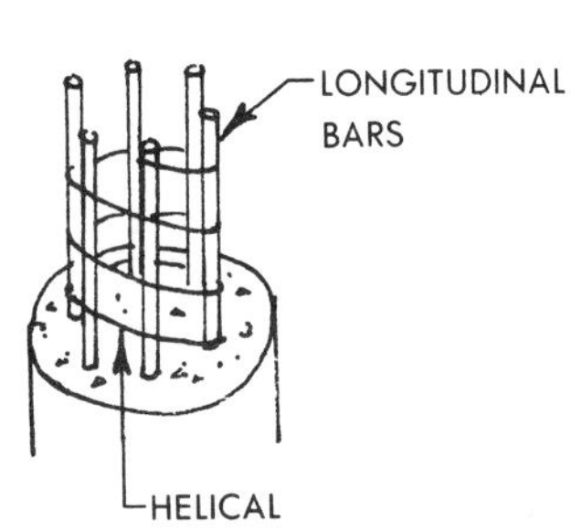

FIG 3

REINFORCEMENT

The steel must be free from loose mill scale, loose rust, grease, oil, paint, mud etc, which impairs the bonding of the steel to the concrete. Storage is important, bars should be stored on racks above ground level, and mesh reinforcement stacked flat on closely spaced timbers above ground level. The steel should be grouped according to size and clearly marked.

The most common method of reinforcing concrete is to use plain round bars, usually of mild steel. Medium tensile steel bars and high tensile steel bars are also available. Deformed bars which are twisted and ribbed, provide a better bond and greater frictional resistance than round bars and obviate the need for hooked ends to the bars. Three types of deformed bars are shown. Fig 4.

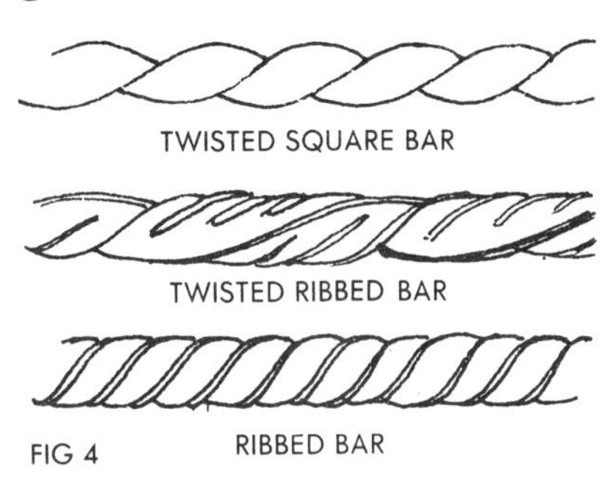

FIG 4

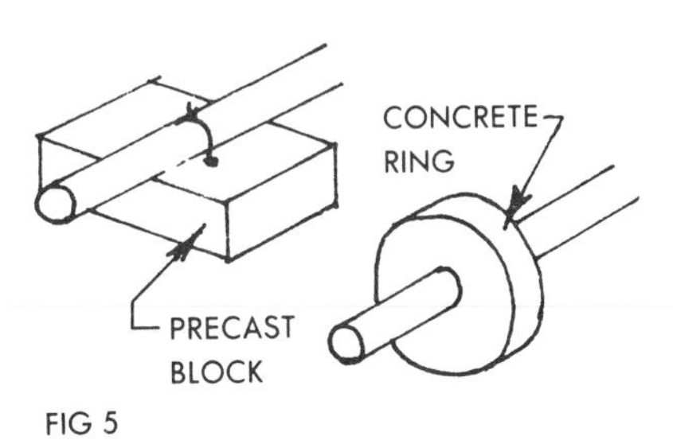

FIG 5

Fixing reinforcement:- It is important that the steel is fixed so that it is not displaced, and that correct cover is maintained when the concrete is placed. At intersections the steel bars should be securely tied together with I6 s.w.g. soft iron wire. Small precast spacing blocks (which should not be weaker than I part cement to 2 parts sand) may be wired to the bars, or small concrete rings threaded over the bars. Fig 5. Plastic spacers are also available which simply snap over the rods.

COVER

It is important to provide adequate cover to reinforcement to ensure there is no risk of corrosion. Minimum cover should be:-

I. At end of a bar not less than 25mm nor less than twice the diameter of the bar.
2. For a longitudinal bar in a column, not less than 40mm nor less than the diameter of the bar. For columns with a minimum dimension of 200mm or under, whose bars do not exceed I2mm diameter, 25mm cover may be used.
3. For a longitudinal bar in a beam not less than 25mm nor less than the diameter of the bar.
4. For tensile compressive, shear or other reinforcement in a slab not less than I5mm nor less than the diameter of the reinforcement.
5. For any other reinforcement not less than I5mm nor less the diameter of the reinforcement

For external work against earth faces and internal work in corrosive conditions, 40mm minimum cover, for all steel including stirrups, links etc; except where the face of the concrete is protected by suitable cladding or by a protective coating.

CONCRETE 7

SULPHATE ATTACK
When sulphate salts are in contact with Portland cement in damp conditions then sulphate attack may occur. The sulphates combine with the aluminate constituent of the cement to produce a new compound, resulting in considerable increase in volume. This expansion causes cracking and splitting of concrete or brickwork and failure of cement renderings. Sulphate attack may occur due to sulphates in soil and ground water and work below D.P.C. is particularly vulnerable. The extent of sulphate attack will depend on:-

1. The amount and nature of the sulphate salt present
2. The level of the water table and seasonal fluctuations
3. The type and quality of concrete and the form of construction.

Common salts which may be encountered are magnesium sulphate (Epsom salt), sodium sulphate (Glauber's salt) and calcium sulphate (Gypsum).

When sulphate salts are encountered ordinary Portland cement should not be used for mortar or concrete, but a sulphate resistant cement.

It is essential that concrete exposed to sulphate attack is of a dense impermeable quality. The water/cement ratio should not exceed 0.55 and the concrete should be weigh-batched and preferably vibrated.

SULPHATE-RESISTING PORTLAND CEMENT
The composition is adjusted to give a very low aluminate content and thus there is resistance to sulphate attack. When using this cement the concrete should not be leaner than a 1:2:4 mix with the minimum water/cement ratio.

SUPER-SULPHATE CEMENT
Composed essentially of granulated blastfurnace slag, calcium sulphate and a small amount of Portland cement or lime, resists attack by strongest concentrations of sulphate normally present in soil, and is useful where acid conditions exist either with or without sulphates present. Has low heat properties. May be necessary for use mixes richer than 1:2:4 in order to obtain adequate workability, increasing the ratio of sand to coarse aggregate will help produce concrete of low permeability. Thus 1: $2\frac{1}{4}$: $3\frac{3}{4}$ or 1:$2\frac{1}{2}$: $3\frac{1}{2}$ may be used.

Cold weather adversely affects the hardening of the cement, but after three days strength develops quicker than ordinary or sulphate-resisting PC. Special care is needed when using the cement, it will deteriorate rapidly if stored under damp conditions, generally it requires a longer mixing time, and the concrete must be carefully cured or a powdery surface may develop.

CONCRETE 8

SPECIAL CEMENTS

RAPID-HARDENING PORTLAND CEMENT
Setting time is the same as for Portland cement, but after setting rapid-hardening cement develops strength more rapidly, allowing formwork to be struck earlier and concrete slabs to be loaded and used sooner. Mixes are the same as for Portland cement.

EXTRA-RAPID-HARDENING CEMENT
Manufactured by adding an accelerator, e.g. calcium chloride, to rapid hardening Portland cement. Has quick initial set and concrete made with this cement should be placed and compacted within 30 minutes of mixing. Hardens much faster than rapid-hardening P.C. and its use speeds up the use of formwork and enables moulds to be re-used with a minimum of delay, also useful when running water is encountered. Particularly useful in cold weather, when its quicker evolution of heat allows concreting to be continued during low (not freezing) temperatures. Is also useful for marine work in inter-tidal conditions.

PORTLAND-BLASTFURNACE CEMENT
Made from a finely ground mixture of Portland cement and granulated blastfurnace slag. Slow rate of hardening in the early stages but at 28 days strength is equal to P.C. Because of slow hardening rate it requires a longer curing period than P.C, especially in cold weather. Has greater resistance than P.C. to soft ground water, dilute acids and sea-water. Evolves less heat than P.C. and this low heat of hydration makes it suitable for mass concrete.

LOW HEAT PORTLAND CEMENT
Setting and hardening times are longer than for P.C. and heat is evolved more slowly. The hardening may be considerably delayed in cold weather. Is used for large masses of concrete, e.g. in dams and retaining walls etc where the generated heat cannot be easily dissipated, and where the use of P.C. might give rise to expansion and resulting cracking. Not held on stock, is made up to special order.

WHITE CEMENT
By special manufacturing processes and selection of raw materials the amount of iron oxide in Portland cement is considerably reduced. Used for manufacture of precast concrete products, reconstructed stone, for renderings and in the production of cement paints.

COLOURED CEMENTS
Made by mixing chemically inert pigments with white or ordinary P.C. The strength of these

cements is slightly reduced but this may be compensated for by adding 10% or so more cement to a mix.

WATERPROOF & WATER-REPELLANT CEMENTS
Made from Portland cement blended with selected salts. Concrete made with these cements is more resistant to water penetration and to some oils, although for water resistance concrete must be dense and thoroughly compacted. May be used in renderings to prevent moisture penetration. In backing coats it helps provide uniform suction. Some types entrain air and time of mixing must be controlled to avoid entraining excess air.

HYDROPHOBIC CEMENT
Made by adding substances to Portland cement during the grinding process, which form a water-repellent film around each grain of cement. The cement can thus be stored under humid, damp conditions for a long time without deterioration. During the mixing process the film is rubbed off and normal hydration takes place.

MASONRY CONCRETE
Has plasticising and air entraining properties and when mixed with sand produces a "fatty" mortar having high workability. Suitable for blockwork, brickwork, internal plastering and external renderings. When mixing avoid too much water at the start as the mix becomes more fluid as air is entrained.

ACID & CHEMICAL RESISTANCE
Special cements, mastics and epoxide coatings are available for use where acids, oils and industrial chemicals present problems. Each case should be considered individually and manufacturers instructions strictly observed.

CONCRETE 9

LIGHTWEIGHT CONCRETE
Lightweight concrete may be produced by either using a lightweight aggregate, instead of the usual stone or gravel, or by forming air or gas bubbles in a plastic mix of cement, with or without sand or other fine aggregate. These concretes are used in construction where low density and weight reduction is important, where thermal insulation is required and for fire protection.

Normal concrete has a density of around 2250kg/m³ the density of lightweight concrete ranges from approximately 320 kg/m³ to 1600 kg/m³

It is possible to produce strong, high grade lightweight concrete (cube strength up to 55 MN/m², having a density up to 40% less than ordinary concrete.

Not all lightweight concretes are suitable for a particular situation, and properties, cost, strength etc, must be considered when choosing concrete for a particular purpose.

LIGHTWEIGHT AGGREGATES
For load bearing purposes the main lightweight aggregates used are furnace clinker, foamed slag, expanded clays and shales and sintered pulverised fuel ash.

Furnace clinker:- should conform to B.S. 1165 'Clinker aggregate for plain and precast concrete', which lays down limits for combustible content and sulphate content, and includes a test for soundness. Used mainly for concrete blocks, unsuitable for reinforced work.
Foamed (expanded) blastfurnace slag:- produced by bringing molten slag into contact with water. The resulting steam causing the slag to expand before it hardens producing particles containing voids. Should conform to B.S. 877 'Foamed blastfurnace slag for concrete aggregate'. Used for in-situ concrete, for thermal insulation of roofs and in lightweight concrete blocks, may be used for reinforced work.
Expanded clay & shale:- obtained by heating the material to a point where the generation of gases in the material expands it, producing a cellular structure. Used for blocks, cast-in-situ work, precast R.C. units and in prestressed concrete units.
Sintered pulverised-fuel ash:- Fly-ash is the residue from the burning of powdered coal, produced largely by power stations. The ash is moistened, made into pellets and fired at about 1200°C, the resulting nodules are crushed and graded. Used for block-making, lightweight screeds and for reinforced concrete.

For non-load bearing and insulating purposes exfoliated vermiculite and expanded perlite may be used.
Expanded vermiculite:- Vermiculite is a mineral resembling Mica the flakes of which when heated rapidly between 650° and 1000°C exfoliate or open out, producing a very lightweight material. Used for very lightweight concrete having low structural strength but high insulating properties. Used for in-situ screeds for insulation of flat roofs. Also used as an aggregate for plaster.
Expanded perlite:-Perlite is a glassy rock of volcanic origin, which when heated rapidly expands to produce a very lightweight cellular material. Has low strength to high insulation properties and is used in a similar manner to expanded vermiculite.

Both these aggregates are somewhat expensive.

Some lightweight aggregates have poor workability and it may be necessary to use a richer mix or add a proportion of sand to increase workability in order to obtain sufficient compaction for in-situ construction. Workability can also be improved by using air-entraining agents.
Fire resistance:- Lightweight aggregates have a very high fire resistance and are graded as Class I. Having low thermal conductivity lightweight concrete transmits less heat to reinforcing steel and this affords a higher measure of protection.

AERATED CONCRETE
Produced by forming minute bubbles of air or gas in a plastic mix of cement, with or without sand. The bubbles are formed either by chemical action with the cement before it sets, or by adding a foam or air-entraining agent to the mix. Some reduction in strength occurs, but this can be partly offset by reducing the proportion of fine aggregate in the mix, since the entrained air makes a mix more cohesive and workable and reduces the risk of segregation occurring. Aerated concrete has a high drying shrinkage and concrete blocks should be steam cured to reduce the shrinkage effect and achieve requisite strength.

CONCRETE 10

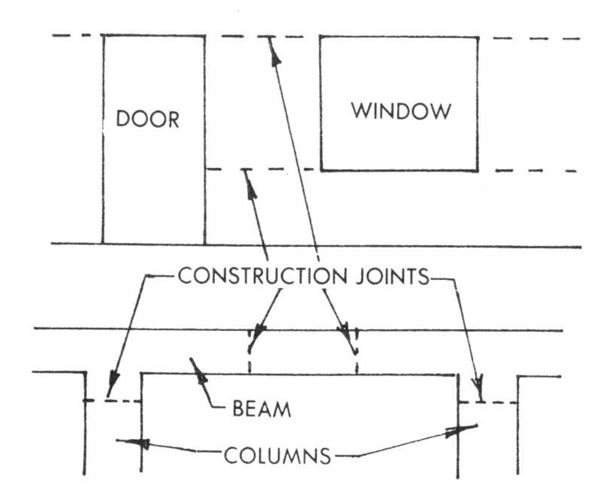

CONSTRUCTION JOINTS
Location:- Joints are a potential source of weakness and should be located and formed with care. Where possible joints should occur on lines logically fitting in with the structure. In walls at the top and bottom of openings, in beams and slabs at the centre or within the middle third of the span, and in columns as near as possible to the beam haunching. Vertical joints in walls should be kept to a minimum, and such joints formed against a stop board.

It is desirable to decide the position of joints before concreting begins.

Preparation:- Any laitance which has formed on the concrete of the previous lift, should be removed by spraying with water and brushing to expose the aggregate, preferably an hour or so after placing. If the concrete has been left overnight wire brushing may be necessary, and where the concrete has hardened, the surface should be hacked to provide an adequate key.

Making the joint:- The surface of the concrete should be brushed clean and all loose particles and dust removed. Methods of making the joint vary.

1. A layer of mortar composed of sand and cement in the same ratio as the cement and sand in the concrete mix, is spread over the concrete. The mortar should be freshly mixed, placed immediately before concreting is continued and the joint should be about 3 mm to 6 mm thick. Thick joints tend to be porous and unsightly.
2. A thin layer of cement grout is brushed over the concrete, being worked well into the surface. Concreting should be commenced immediately as the think layer of grout dries out quickly and a weak joint would result.
3. The surface of the concrete is well brushed and wetted sufficiently to take up most of the absorption of the existing concrete. No surplus water should be allowed to remain on the surface, and the fresh concrete mix should have good workability.

COLD WEATHER CONCRETING
In cold weather the rate of setting and hardening of cement is slowed down and practically ceases altogether as freezing point is approached. Special precautions must be taken if concreting is to continue in temperatures below 2°C (36°F) on a falling thermometer. The main thing is to ensure that the temperature of the concrete is at least 4°C (40°F) when placed and that it does not fall below 2°C (36°F), until the concrete has thoroughly hardened. A list of possible precautions is given below and those selected will depend upon prevailing circumstances, severity of conditions and facilities available.

A thermometer should be used and a record of temperatures kept.

1. Aggregates and mixing plant should be under cover.
2. Cover exposed surfaces with insulating material, e.g. straw or mineral wool blankets, as soon as concreting is completed. N.B. keep insulation dry, most materials lose their effectiveness when staturated.
3. Timber formwork, if thick enough, gives good insulation, but steel formwork should be insulated by covering with suitable material.
4. Use a richer mix, increasing the cement content helps.
5. Use rapid hardening or extra rapid hardening cement and retain formwork in position longer.
6. The method of curing should be one which retains the water within the mix, e.g. covering with tarpaulins, plastic sheet or waterproof paper.
7. Frozen aggregates should never be used, and concrete should not be placed on frozen soil or in frozen formwork.
8. Heat water and aggregate. (i) the water should not be heated above 80°C (176°F) or a flash set may occur. The aggregate and water may be turned over in the mixer and the cement added when the temperature is not above 32°C (90°F). (ii) small quantities of aggregate may be heated by spreading on sheets of corrugated iron over a fire. For larger jobs steam pipes may be used.

9. Concrete should be placed quickly and insulated immediately, continuous heating being provided. N.B. Various heating appliances are available and on some sites industrial electric blankets have been used with good effect.

10. Ready mix concrete supplied at temperatures approaching 10°C (50°F) may be used.

An accelerator, e.g. calcium chloride may be added to the ordinary P.C. or R.H.P.C. but not to extra R.H.P.C or sulphate resisting cement. Not more than 2% must be used or corrosion of reinforcing steel may result.

BRICKS I

STANDARD SIZES

The format (i.e. brick dimensions plus joints) for imperial bricks is laid down in B.S. 3921: 1974. Mortar joints are taken as ⅜ in for imperial work and 10 mm for metric.

Imperial bricks will be met with for some time to come, particularly in alteration and maintenance work, and so imperial sizes are included in this sheet.

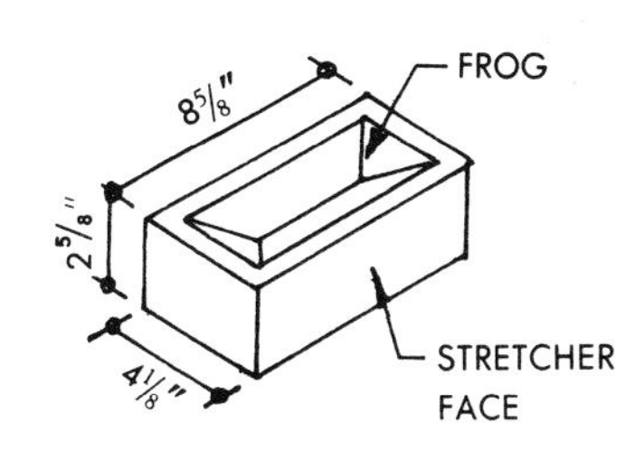

IMPERIAL BRICK

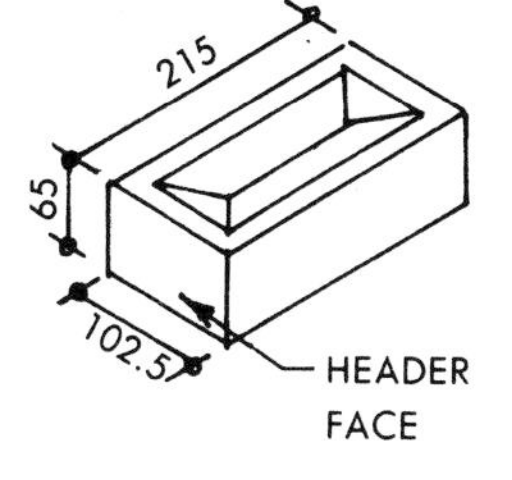

METRIC BRICK

TOLERANCES

When checking dimensions 24 bricks are placed in contact in a straight line on a level surface and the overall measurements then taken using a steel tape.

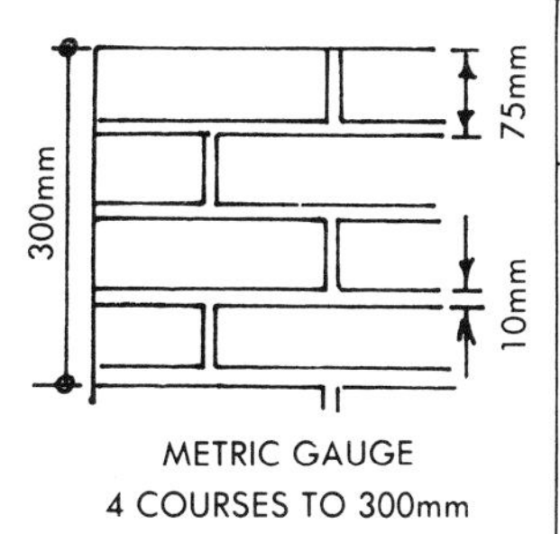

METRIC GAUGE
4 COURSES TO 300mm

FORMAT	DIMENSIONS	24 BRICKS IN CONTACT	
		OVERALL MEASUREMENTS	TOLERANCES
3in	2⅝in	63 in	± 1¾in
4½in	4⅛in	99in	± 1¾in
9in	8⅝in	207in	± 3 in
75mm	65mm	1560mm	+ 60mm - 30mm
112.5mm	102.5mm	2460mm	± 45mm
225mm	215mm	5160mm	± 75mm

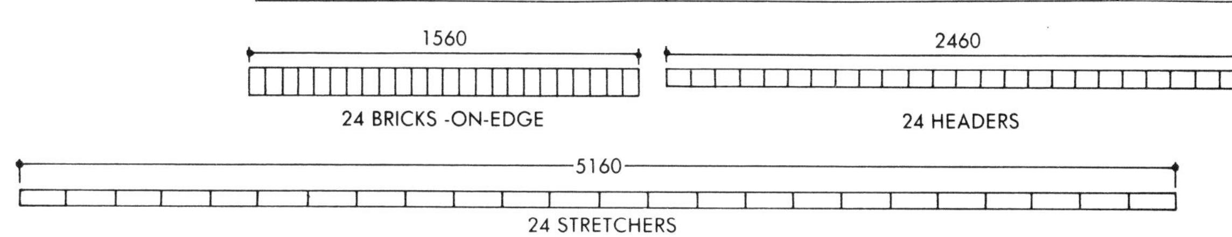

'Phorpres' metric bricks:- The London Brick Company is producing their 'PHORPRES' bricks to a work size of 215mm x 102.5mm x 66mm. This is within the tolerances of B.S. 3921 as shown above, will give a bed joint of 9mm in a 75mm coursing height and facilitate the use of the bricks, for maintenance and repair work in conjunction with imperial sizes.

B.S. 3921 classifies bricks under three headings:-

VARIETIES

Commons:- Suitable for general building work but having no special claims to give an attractive appearance.

Facings:- Specially made or selected to give an attractive appearance.

Engineering:- Having a dense and strong semi-vitreous body conforming to defined limits for absorption and strength.

QUALITY

Internal quality:- Bricks for internal use only.

Ordinary quality:- Less durable than special quality but normally durable in the external face of a wall. Some types are unsuitable for exposed positions.

Special quality:- Durable even when used in situations of extreme exposure where the structure may become saturated and be frozen, e.g. retaining walls, pavings, etc. N.B. Engineering bricks normally attain this standard of quality.

TYPES

Solid:- In which small holes passing through or nearly through a brick do not exceed 25% of its volume. (A small hole is less than 20mm wide or less than 500mm^2 in area). Up to three larger holes not exceeding 3250mm^2 each, may be incorporated as aids to handling, within the 25%.

Perforated:- Small holes passing through a brick exceed 25% of its volume. Up to three larger holes not exceeding 3250mm^2 may be incorporated for handling.

Hollow:- Holes passing through a brick exceed 25% of its volume and the holes are not small, as defined above.

Cellular:- Holes closed at one end exceed 20% of the volume of the brick.

Type 3 brick:- This designation formerly referred to bricks having a height of 2⅞in, but these are no longer recognised. However some manufacturers are producing, for the time being bricks 216mm x 102.5mm x 73mm (actual size) for working to the old type 3 size, particularly for maintenance and repair work.

BRICKS 2

CLAY BRICKS
Made from clays composed mainly of silica and alumina with small amounts of lime, iron, manganese, etc. Several different types of clay are used for brickmaking, producing a variety of bricks with a wide range of colours, textures and strength. A few bricks are still hand moulded, but the majority are machine pressed or wire cut.
Hand-made bricks:- Are irregular in size and shape with uneven arrises. Produced mainly as facings.
Machine-pressed bricks:-The clay or shale is fed into steel moulds and shaped under heavy pressure. May be single or double frogged. The bricks are uniform in size and shape, with sharp arrises and even surfaces. The surface of the bricks may be treated in different ways, e.g. by sanding, to produce a variety of facings.
Wire-cut bricks:- The clay is extruded from a pugmill in a continuous band and then cut into individual bricks by means of wires attached to a frame. Wire-cuts have no frogs and wire marks can be seen on the beds.
Burning:- A few bricks are still burnt in a large stack or 'clamp', but most bricks are kiln burnt. There are a number of types of kiln, but the majority of bricks are burnt either in a Hoffman kiln, in which the fire is made to pass through a series of chambers, or in a tunnel kiln in which the bricks pass through a long chamber with the firing zone in the centre.

CALCIUM SILICATE BRICKS (Sandlime and flintlime)
Made from sand and lime (or crushed flint and lime) moulded under heavy pressure and then subjected to steam pressure in an autoclave.

B.S. 187: 1978 grades calcium silicate bricks into eight classes as shown in the table on the right.

Each class number gives the strength in thousands of Lbf/in^2. Guidance is given on the choice of bricks for various purposes and also recommendations for suitable mortar mixes.

In general classes 7, 5, 4, 3A & 3B are suitable for use where the particular minimum strengths given are required, and may be used for external work, in exposed positions, and work below D.P.C.

Class 2A and 2B are suitable for the strength requirement given and for external and internal work, but not in very exposed conditions or work below D.P.C. Class I is only suitable for internal non-facing work above D.P.C.

The difference between Type 3A and 3B, and 2A and 2B is one of drying shrinkage, Type 'A' having the lower shrinkage. A weaker mortar should be used with class 'B' bricks.

When facings are required they should be specified as such, so as to avoid wide variations in colour and to ensure careful handling, avoiding damage to faces and arrises of bricks.

CONCRETE BRICKS
Made by compacting the concrete under heavy pressure in a mould.
Aggregates:- (i) Natural sand, gravel or stone, whole or crushed or a combination there of conforming to B.S.882: 1983; (ii) Air cooled blast furnace slag conforming to BS. 1407: 1970. (iii)Foamed or expanded blast furnace slag conforming to B.S. 877: 1977; (iv) Well burnt furnace clinker to B.S. 1165: 1977. Pigments may be incorporated in the concrete to produce various colours.
Cement:- May be Portland, Portland blast furnace or high alumina. BS 6073: Part I grades concrete bricks into three classes:-
Special purpose:- Suitable for exposed positions and below D.P.C. Strength 2500 Lbf/in^2 (17.5N/mm^2).
Class A(i):- For general external face work. Strength 1750 Lbf/in^2 (12.25 N/mm^2)
Clase A(ii):- For external face work in mortars other than strong mortars N.B. Class A(ii) bricks have the same strength as Class A(i) but the drying shrinkage is higher.
Class (B):- For internal work in mortars other than strong mortars. Strength 1000 Lbf/in2 7N/mm2).

FIRE BRICKS
Made from refractory clay having a high fusing point, and laid in refractory mortar with tight joints.

Class of brick	Compressive Lbf/in^2	Strength N/mm^2	Drying shrinkage (% of wet length)
7	7000	48.0	0.25
5	5000	34.5	0.25
4	4000	27.5	0.25
3A	3000	20.5	0.25
3B	3000	20.5	0.35
2A	2000	14.0	0.25
2B	2000	14.0	0.35
I	1000	7.0	–

BRICKS 3

STANDARD SHAPES

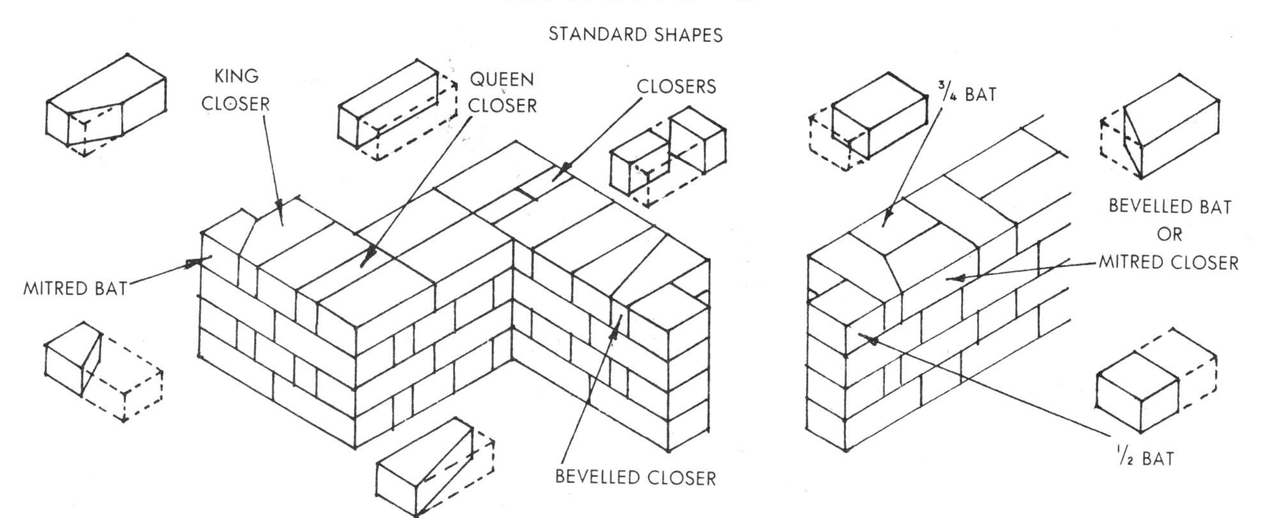

EXAMPLES OF THE USE OF STANDARD SHAPES OR CUTS

STANDARD SPECIALS AND PURPOSE MADE BRICKS

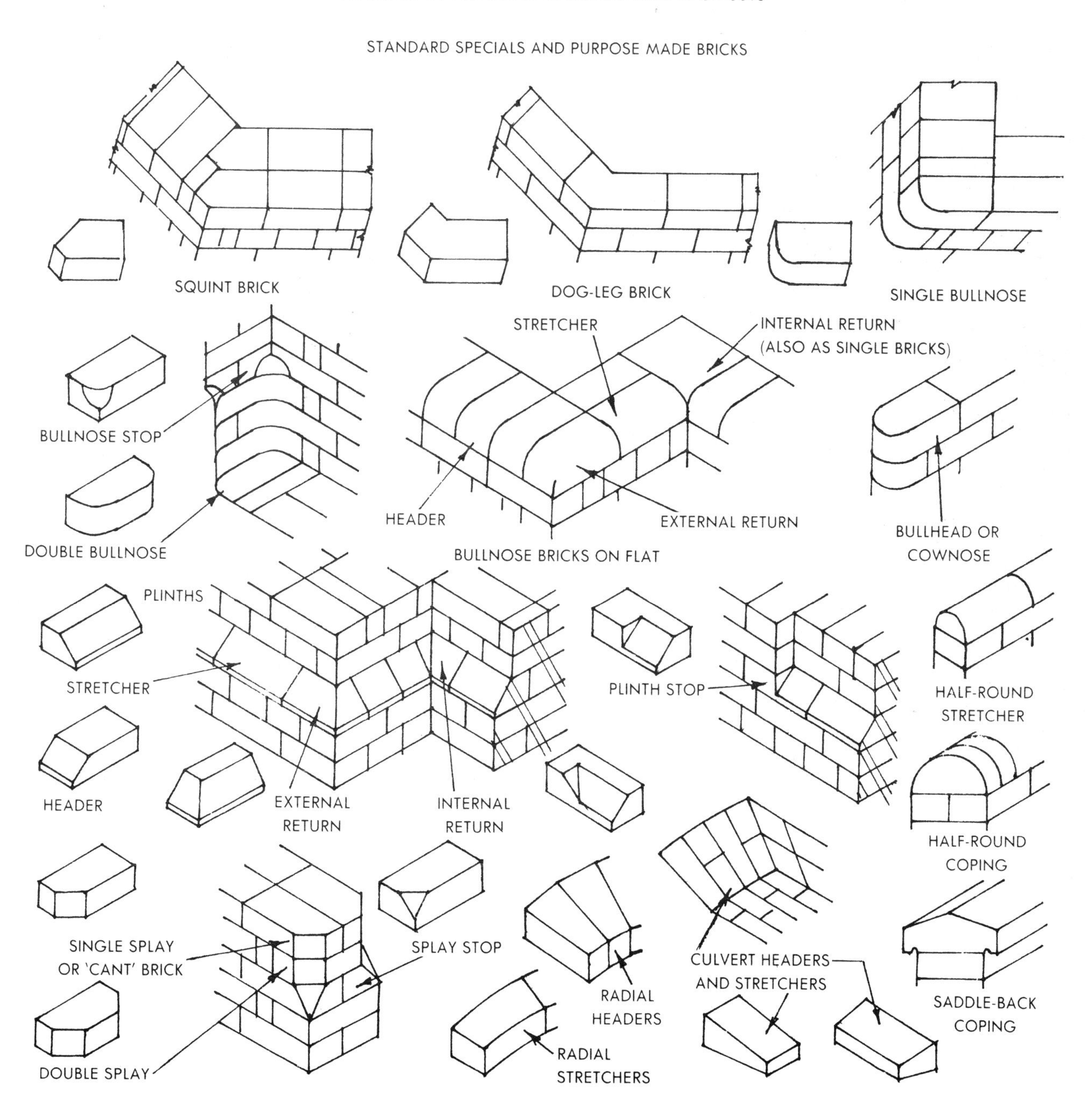

BRICKS 4

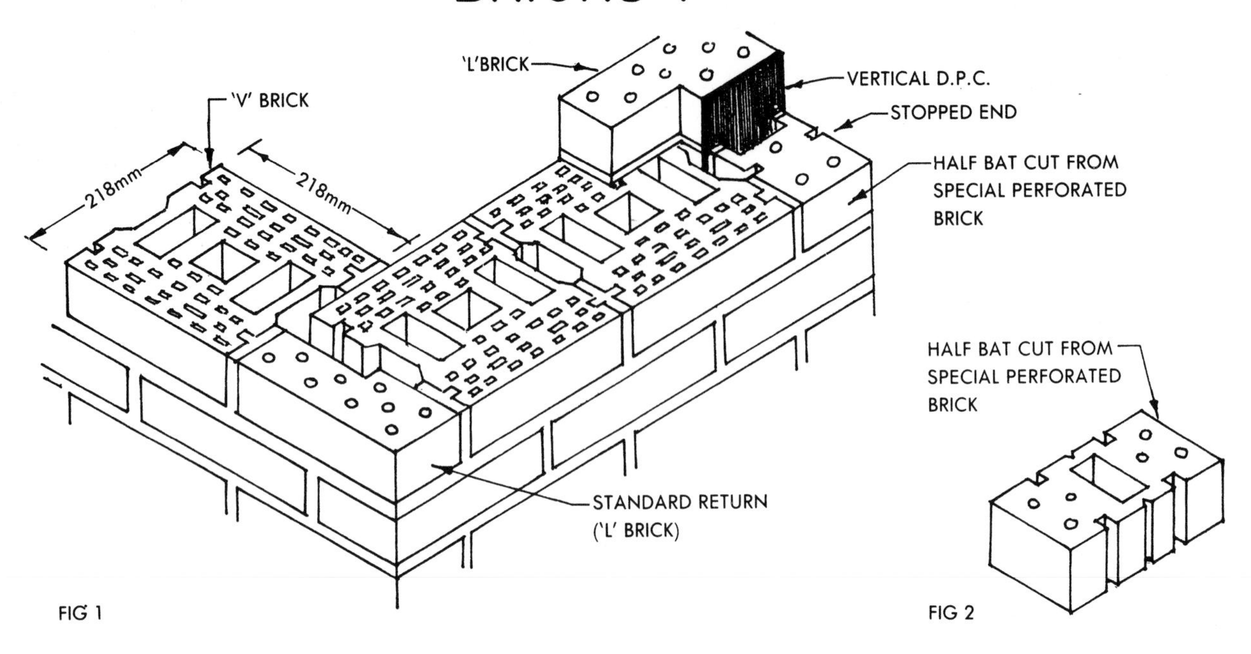

FIG 1

FIG 2

'V' BRICKS

The vertically perforated 'V' bricks Fig I were designed by the Building Research Station to provide at lower cost, the equivalent of traditional cavity walling by a single leaf wall. Advantages claimed compared to a cavity wall having brick inner and outer leaves are:- (i) increased output of approximately 30%. (ii) saving in mortar of approximately 20%. (iii) a considerable saving in weight.

The bricks are bedded on two strips of mortar, back and front, the centre of the wall being kept clear of mortar. Fig 3.

'V' bricks are difficult to cut and it is advisable to keep walls and piers to brick lengths. Where cutting is unavoidable it is an advantage to use special perforated bricks which are available. Fig 2. Standard return bricks ('L' bricks) are available for bonding quoins and stopped ends. Fig I.

'V' bricks are produced as facings and it is economical to use ordinary bricks below ground level. The D.P.C. should not be laid as a continuous sheet across the wall at one level. Either separate strips should be laid at the front and back of the wall as the mortar beds, or a flexible D.P.C. should be stepped down across the cavity as shown. Fig 4. This necessitates cutting bricks longitudinally, alternatively matching perforated bricks may be used on the face, commons on the back and the D.P.C. taken vertically between them.

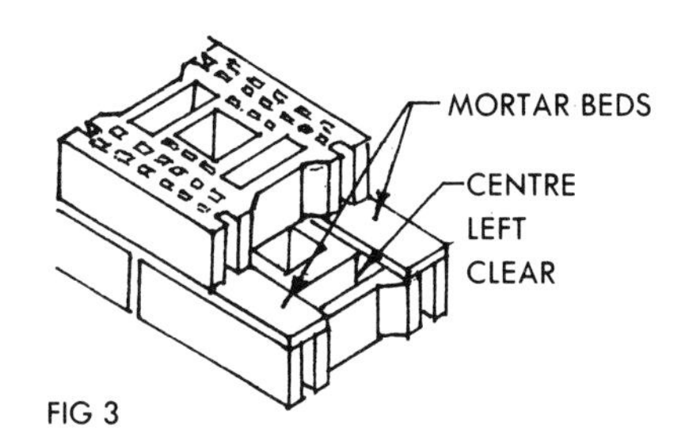

FIG 3

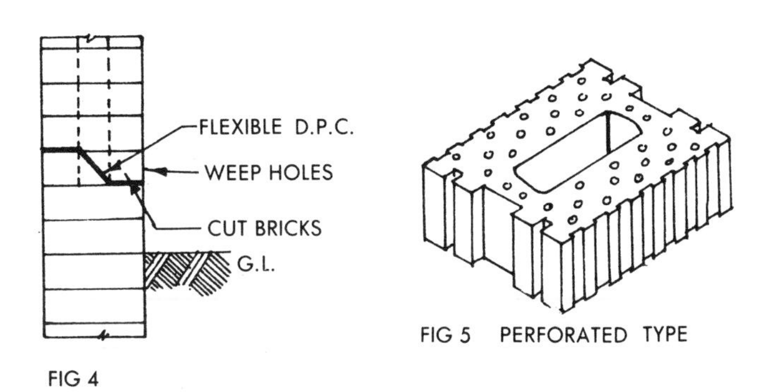

FIG 4

FIG 5 PERFORATED TYPE

CALCULON BRICKS

A clay brick designed for highly stressed brickwork. Figs 5 & 6. Can be used for internal load bearing walls and is particularly suitable for crosswall construction.

There are three grades as shown in the table. Quarter, half, and three-quarter bricks are available to avoid wasteful cutting.

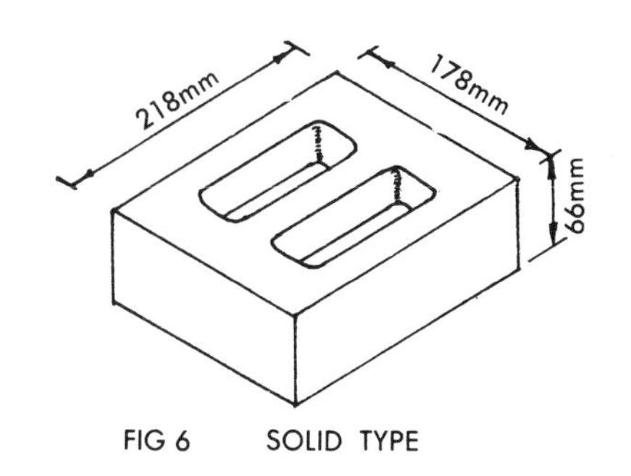

FIG 6 SOLID TYPE

MORTAR

The type of mortar is dependent on loads, stresses, and brick strength. Reference should be made to C.P. 111: 1970 Table 3.

There is a saving of approximately 40% in mortar required compared with a one-brick solid wall.

GRADE	TYPE	Compressive Strength Lbf/in²	N/mm²
A10	Perforated	10000	69.0
B75	Perforated	7500	51.7
C5	Solid	5000	34.5

MODULAR BRICKS

Metric modular bricks are available to the following formats:- 300 x 100 x 100, 200 x 100x 100, 300 x 100 x 75 and 200 x 100 x 75, although the latter two are not strictly modular. Two types are illustrated Fig 7 bed joints and cross joints are 10mm.

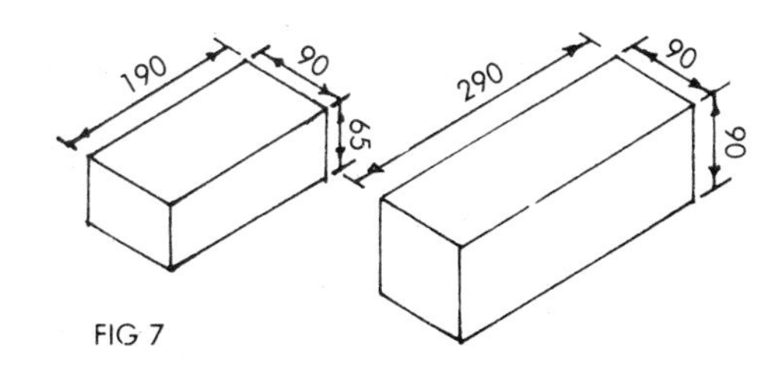

FIG 7

BONDING BRICKWORK I

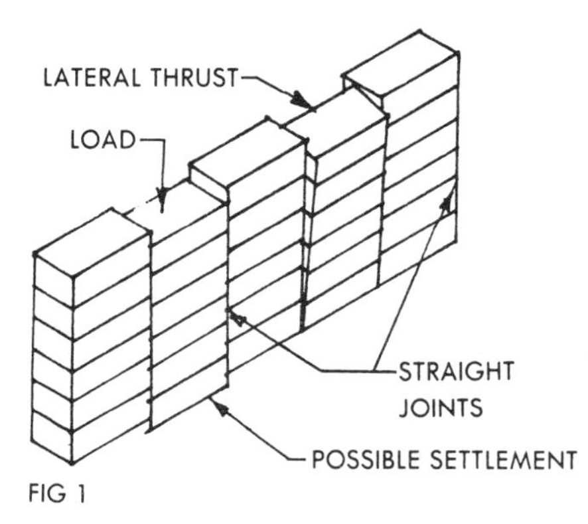

FIG 1

Bond:- The arrangement of bricks in a wall to give a required pattern whilst maintaining adequate lap.

Purposes of bond:-

1. To strengthen the wall and ensure that any load carried is distributed over the whole wall.
2. To ensure lateral stability and resistance to side thrust.
3. A number of bonds are used primarily to give a pleasing appearance to the face of the wall.
4. Some bonds require fewer facings than others. These bonds are sometimes used for economy when expensive facings are required (See table).

An unbonded wall is comparatively weak and liable to fail under a load or lateral thrust as shown in Fig I. When a wall is bonded any load is distributed over the whole wall and there is greater resistance to side thrust. Fig 2.

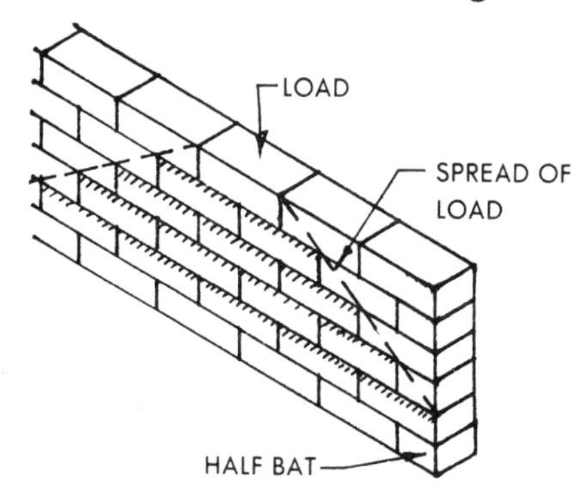

FIG 2 BONDED WALL

A square metre of one-brick walling requires approximately 125 bricks, allowing 5% for cutting and waste. The table on the right gives the approximate number of facings required per m^2 allowing 5% waste.

BOND	Facings/m^2
English	94
English Cross	94
Dutch	94
Flemish	84
English Garden Wall	78
Flemish Garden Wall	78
Stretcher	68

RULES OF BONDING

When working out bond the bricklayer applies a number of basic rules. These rules which are given below, are taken as a guide when solving a problem, but other relevant factors such as economical use of bricks, avoidance of wasteful cutting, strength requirements etc; are also considered. There may be more than one solution to a particular problem and the bricklayer will use the solution applicable to the particular circumstances.

1. The bond should be set out along the face of the wall, working from each end to the centre, with the end bricks in each course corresponding. If the wall is of a length such that the bonding pattern does not work out, or a cut is necessary, then we have what is known as 'broken bond'. If rule I is implemented, it follows that any broken bond will be as near the centre of the wall as possible. N.B. No cut brick should be less than header width other than the closers at the end of the wall.

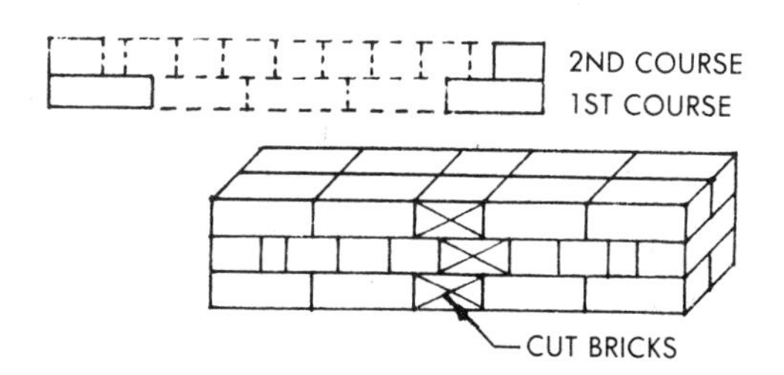

EXAMPLE OF BROKEN BOND IN ENGLISH BOND

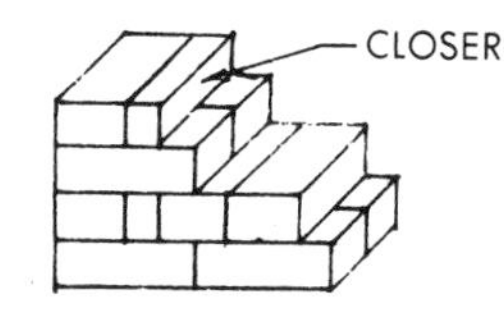

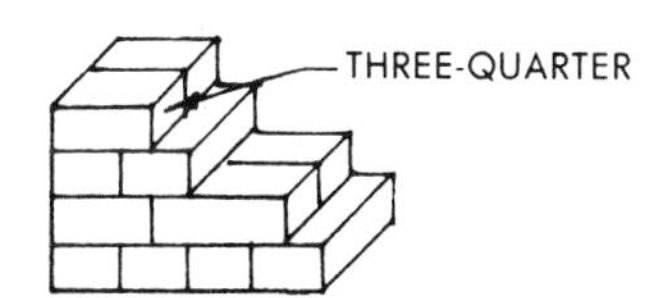

2. Whe half bond is used a half bat is used at a stopped end. Fig 2. When quarter bond is used a closer is placed next to the header at the quoin or stopped end. Alternatively in some bonds a three-quarter brick may be used to achieve quarter bond.
3. All transverse joints should continue unbroken across the width of the wall unless stopped by the centre of a stretcher.

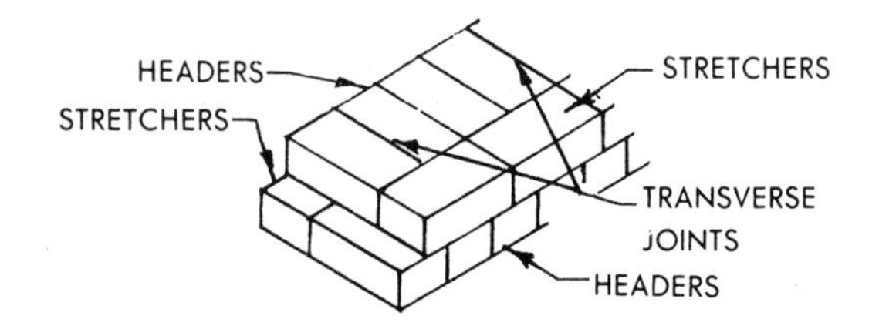

4. In English Bond where the wall is of odd half-brick thickness, when stretchers are shown on the face headers are shown on the back and vice versa.

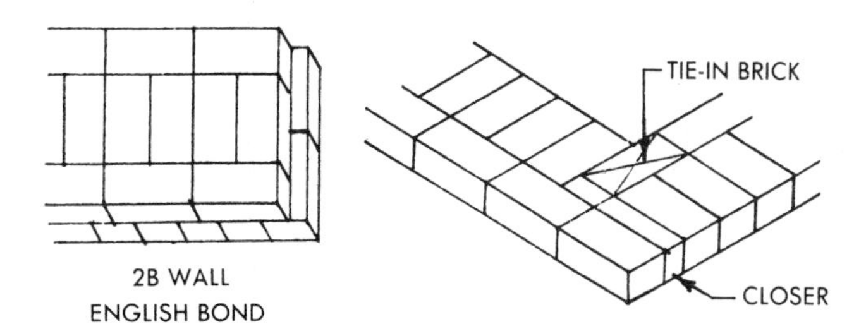

5. The bricks in the interior of thick walls are laid headerwise.
6. The tie-in brick at a corner is opposite the closer.

Reverse bond:- Occasionally used for reasons of economy where appearance is not important. The end bricks in each course do not correspond. On certain lengths it is possible to avoid a central cut in this way.

The use of reverse bond is less obvious in Flemish bond than in English bond.

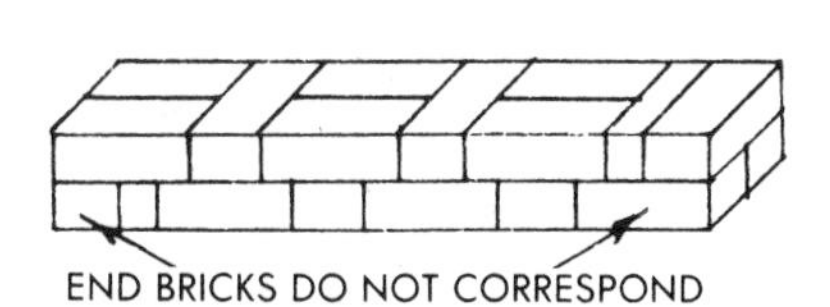

EXAMPLE OF REVERSE BOND

BONDING BRICKWORK 2

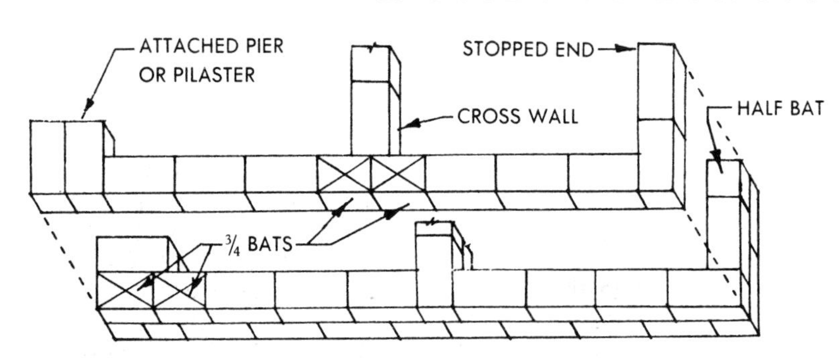

STRETCHER BOND

Used for half-brick walls, all the bricks are laid as stretchers and half bond should be kept as far as possible. A half-bat is used to maintain the bond at a stopped end. Three-quarter bats may be necessary at junctions of cross walls or where pilasters occur.

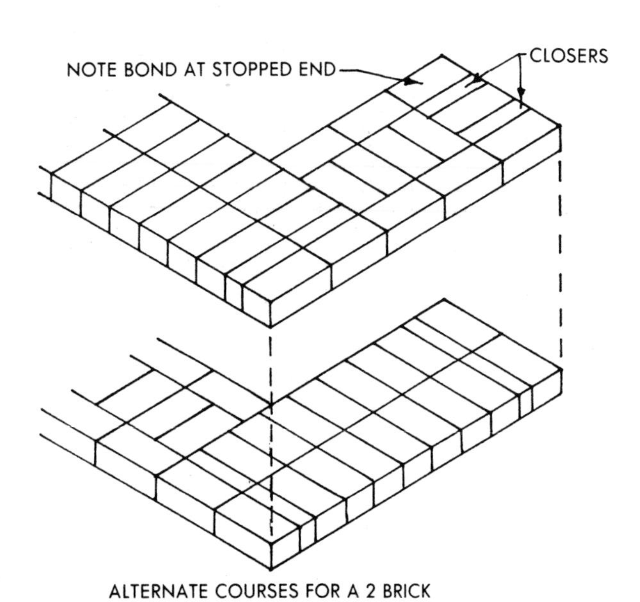

ALTERNATE COURSES FOR A 2 BRICK QUOIN WITH STOPPED END

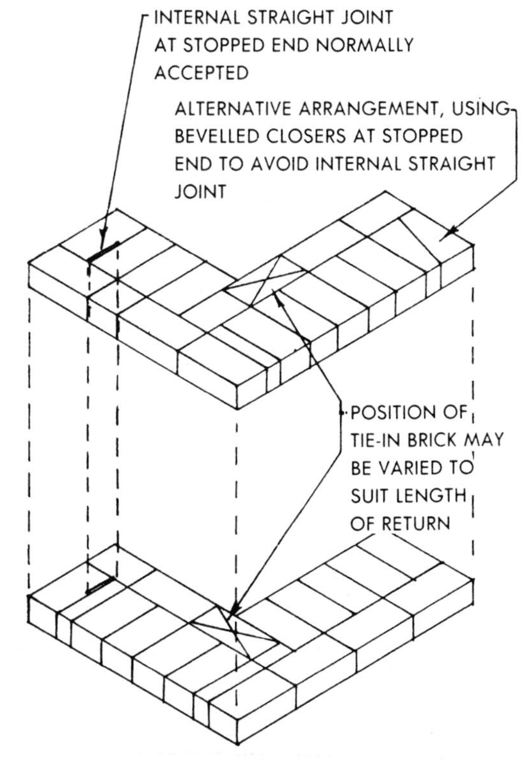

ALTERNATE COURSES FOR A 1½ BRICK QUOIN WITH STOPPED ENDS

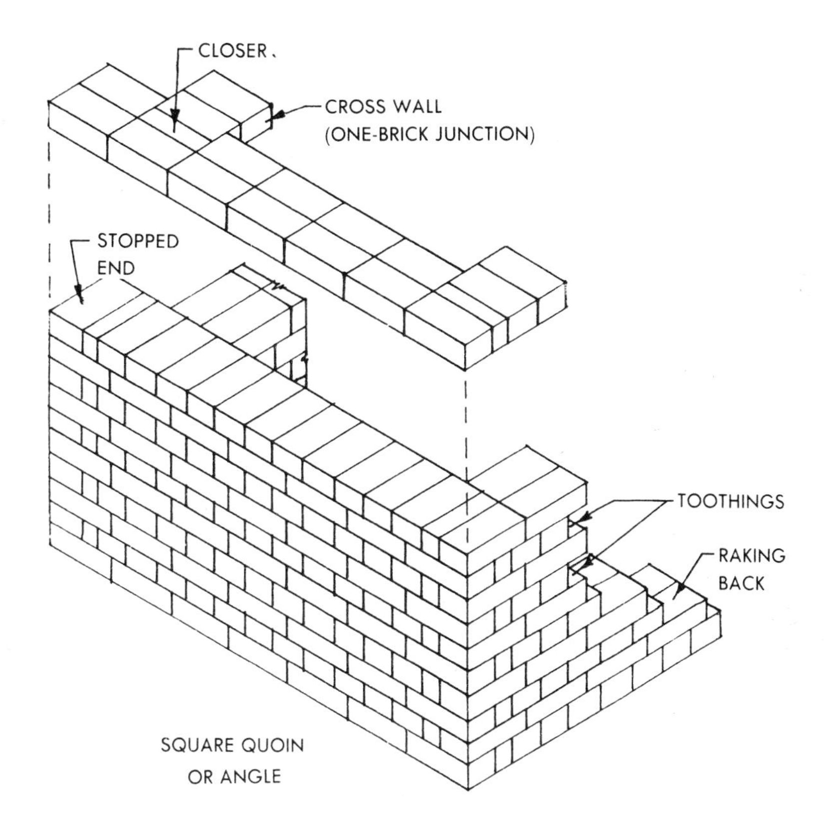

ENGLISH BOND
Consists of alternate courses of headers and stretchers. To obtain the quarter bond, a closer is placed next to the quoin header or the header at a stopped end.

One of the strongest bonds, because it is free of any internal straight joints.

Toothings:- When raising a corner the bricklayer will often tooth one or two bricks as shown in order to avoid raking out too far. Toothings should be avoided as far as possible as they are a source of possible weakness. If they have to be left, e.g. where a wall is to be continued at a later date, then particular care must be taken in 'chopping in' the mortar and filling the joints when building on the subsequent wall. If this is not done properly, cracking of the brickwork may result.

BONDING BRICKWORK 3

FLEMISH BOND
Consists of headers and stretchers laid alternatively along each course, the headers being central over the stretchers in the course below. The bond is really only suitable for walls of one-brick thickness or more, but is occasionally used in the external leaf of cavity walls, using snap headers. This is not to be recommended.

Flemish bond is said to present a more pleasing appearance than English bond, there is also a saving in facings. English bond requires approximately 90 bricks/m^2 of elevation and Flemish bond 80 bricks/m^2.

Flemish bond is weaker than English bond, because of the internal straight joints which occur either side of the headers, as shown. In walls of odd half-brick thickness the larger number of bats required to maintain the bond is also a weakness.

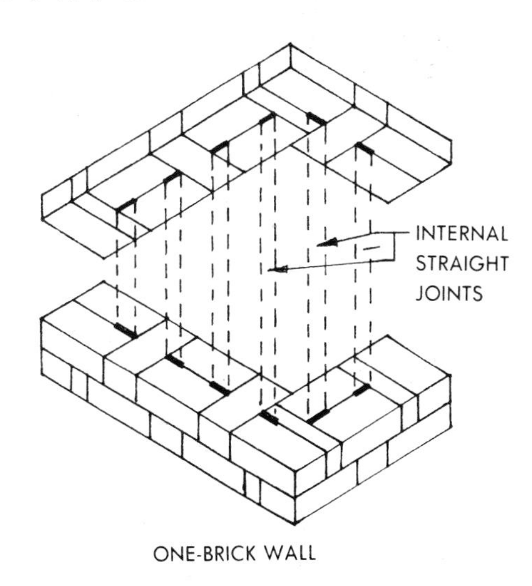

ONE-BRICK WALL

DOUBLE FLEMISH BOND
Refers to the method of bonding walls over one-brick thick, to show Flemish bond on both faces.

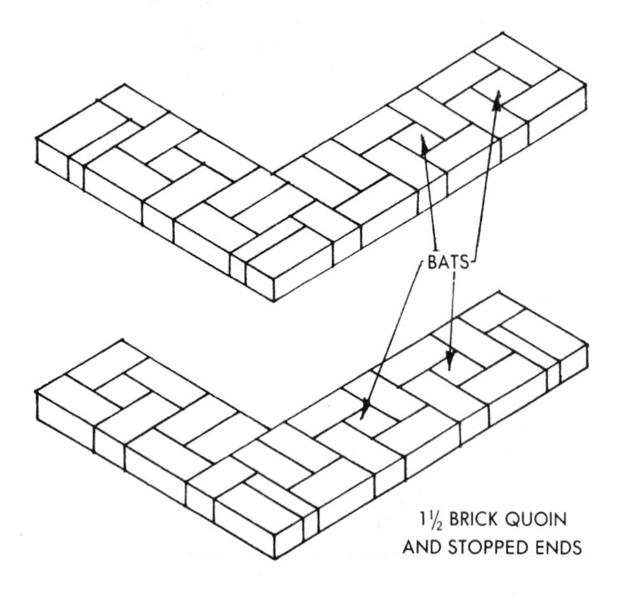

1½ BRICK QUOIN AND STOPPED ENDS

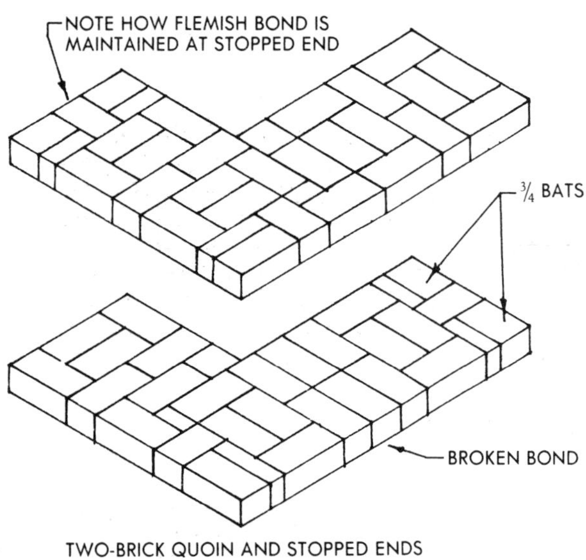

TWO-BRICK QUOIN AND STOPPED ENDS WITH EXAMPLE OF BROKEN BOND

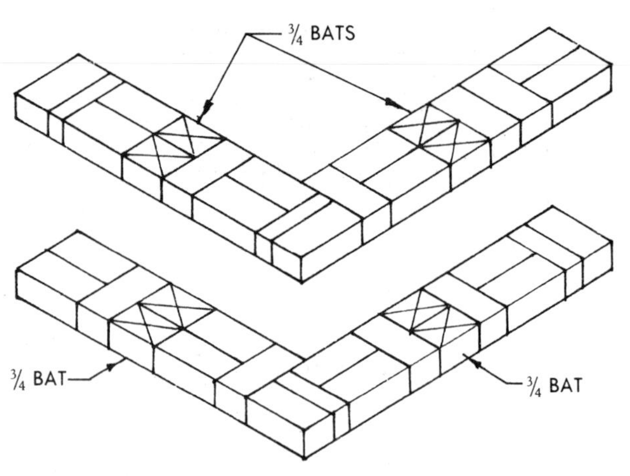

ONE-BRICK QUOIN AND STOPPED ENDS SHOWING EXAMPLES OF BROKEN BOND

SINGLE FLEMISH BOND
In walls over one brick thick Flemish bond is sometimes shown on the face side only, the remainder being built in English bond. This is known as single Flemish bond and is achieved by using snap headers in alternative courses.

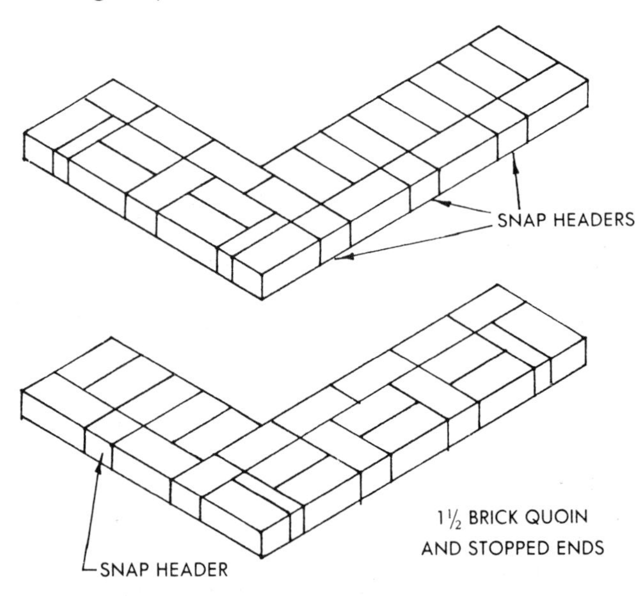

1½ BRICK QUOIN AND STOPPED ENDS

BONDING BRICKWORK 4

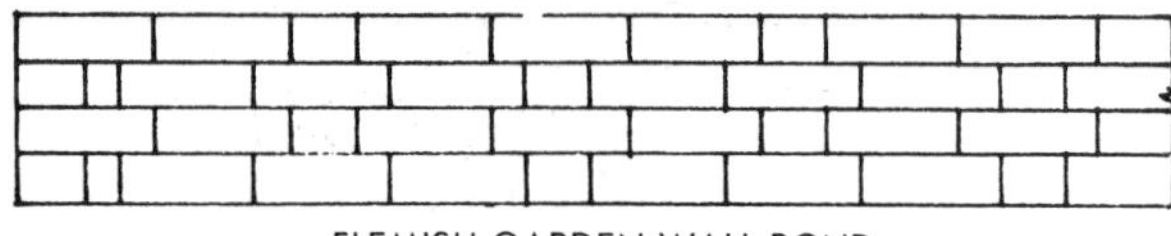

FLEMISH GARDEN WALL BOND

FLEMISH GARDEN WALL BOND
Also known as Sussex bond. Consists of three stretchers and a header alternately along each course, the headers coming over the centre stretchers. The object of the bond is to enable a fair face to be kept on both sides of a one-brick wall. As the proportion of stretchers to headers is higher than in English or Flemish bonds, the wall joints between stretchers can be adjusted to allow for variations in the lengths of headers.

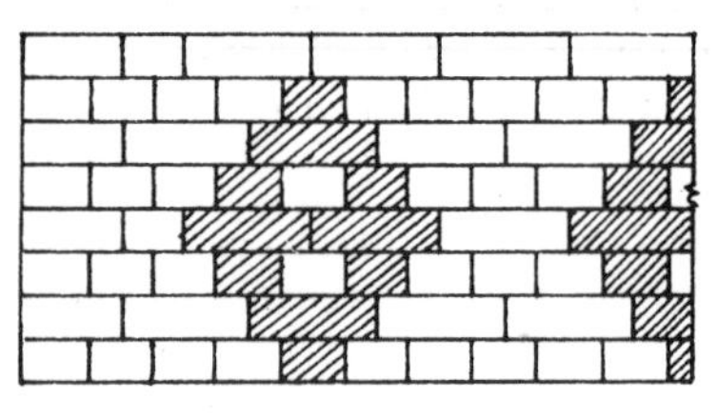

DUTCH BOND

DUTCH BOND
The same as English Cross bond in the centre of the wall, but there are no closers, ¾ bricks are used at the ends of the wall. Patterns or diapers can be picked out as shown.

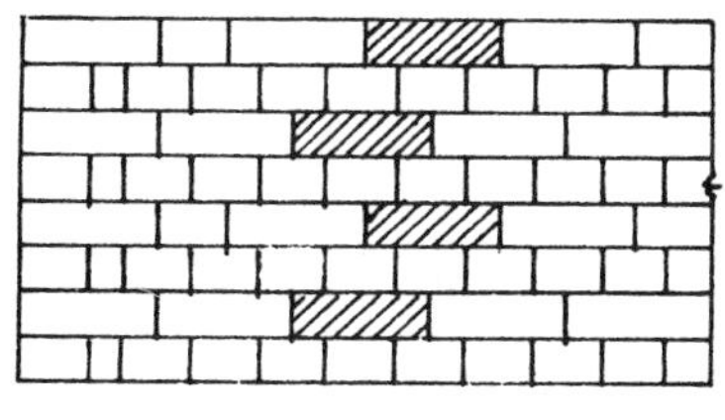

ENGLISH CROSS BOND

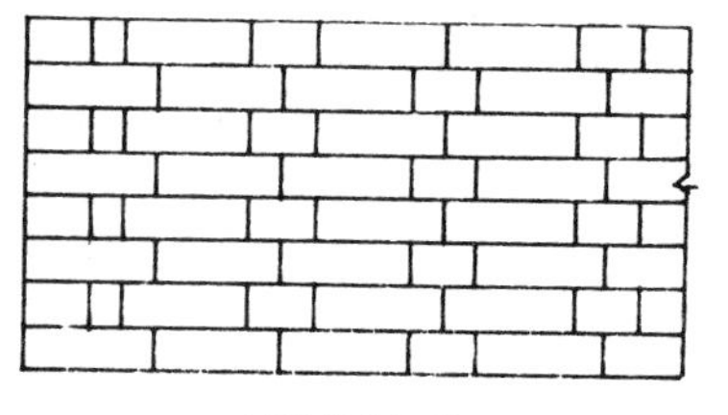

MONK BOND

ENGLISH CROSS BOND
Similar to English bond but a bat is placed next to the end stretcher every other course. Thus the stretchers are 'staggered' as shown.

MONK BOND
The basis of this bond is two stretchers and a header, with the headers over the joints between the stretchers. Variations of the bond are met with.

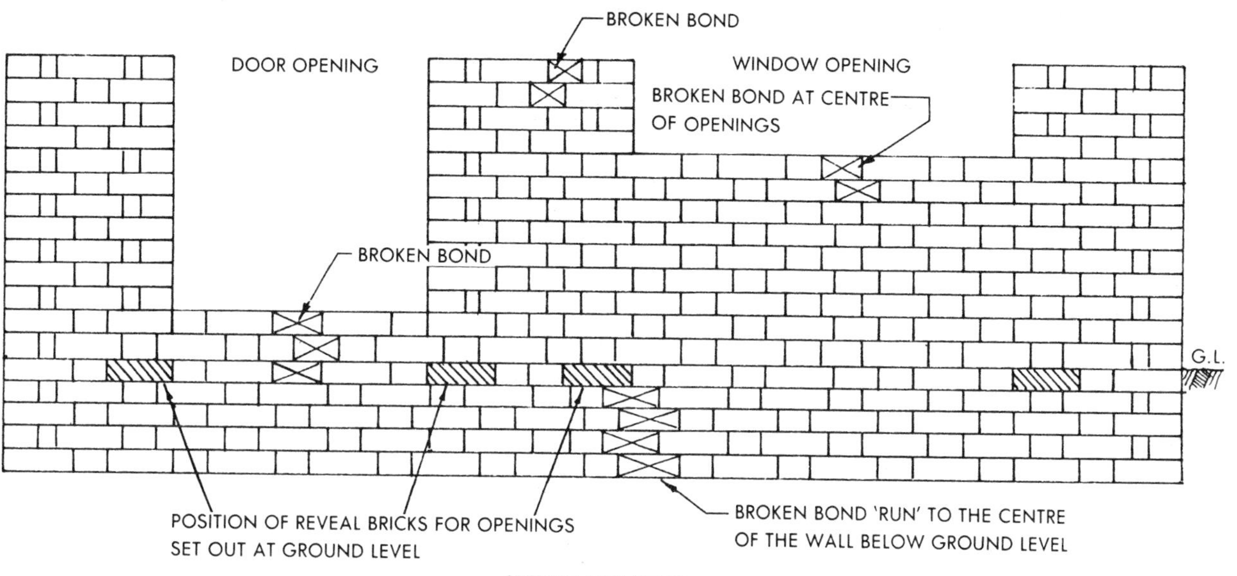

SETTING OUT BOND

SETTING OUT BOND
Below ground level the brickwork is 'run' from each end of the wall, any broken bond occurring at the centre of the length of wall. At ground level the positions of any openings occuring are marked along the wall, and the bond set out, any broken bond being kept to the centre of piers and openings.

Although in the example above a cut brick is shown at the centre of the wall below ground level, in practice the bricklayer would probably 'lose' this broken bond by very slightly adjusting his cross joints so that a cut is avoided.

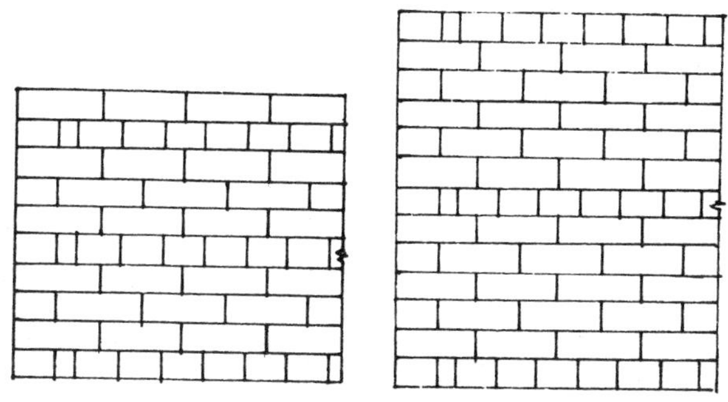

ENGLISH GARDEN WALL BOND

ENGLISH GARDEN WALL BOND
Consists of three courses of stretchers and one course of headers repeating for the height of the wall, as shown left. The stretching courses being half bond.

A variation of the bond has five courses of stretchers to each header course as shown right.

As English Garden Wall bond gives a quick lateral spread of a load it is usually adopted for the construction of tall chimneys.

MORTARS FOR BRICKWORK

MIX	USE
1:3 Hydraulic Lime Sand 1:3:10-12 Portland Cement Lime Sand 1:7 Masonry Cement Sand 1:8 Portland Cement sand with plasticizer	Internal walls and partitions. Clay, sandlime and concrete bricks. Concrete partition Blocks having high drying shrinkage
1:2 Hydraulic Lime sand 1:2:8-9 Portland Cement Lime Sand 1:6 Masonry cement sand 1:7-8 Portland Cement sand with plasticizer.	Clay, sandlime and concrete bricks or blocks in sheltered positions between eaves and D.P.C. for internal walls of sandlime and concrete bricks or blocks when there is a possibility of frost.
1:5:5-6 Portland cement Lime Sand 1:4½ Masonry cement sand 1:5-6 Portland cement sand with plasticizer	Clay bricks, sandlime or concrete bricks A(i) Class in exposed conditions below D.P.C., free standing walls, parapets etc. Between eaves and D.P.C. in conditions of severe exposure.
1:½:4-4½ Portland Cement Lime Sand	Clay bricks in conditions of severe exposure
1:0-¼: 3 Portland cement Lime Sand	Retaining walls. Heavy engineering work with appropriate bricks.

REQUIREMENTS

1. Good workability
2. Retain plasticity long enough for bricks to be laid and adjusted, but stiffen within a reasonable time.
3. Early development of strength
4. Mature strength should be adequate but no greater than is needed for the design. (Mortar for a wall need be no stronger than the bricks)
5. Bond well to the bricks
6. Adequate durability.

The strength of the mortar has less influence on the ultimate strength of a wall than might be supposed. An excessively strong mortar tends to concentrate the effects of differential movement and produce fewer and wider cracks. Weaker mortar will take up small movements and cracks will tend to show as hair cracks in the joints. Thus mortar should contain only sufficient cement to give adequate strength in the walling, unless there is good reason for a richer mix, e.g. in cold weather, when a richer mixes develops strength quickly enough to resist the effects of frost.

The following table indicates the most suitable mixes for various situations. Where a range of sand contents is given, e.g. 10-12 the higher figure is for well graded sand and the lower figure is for coarse or uniformly fine sand.

N.B. In table, lime refers to non-hydraulic or semi-hydraulic lime.

Non-hydraulic lime is obtained by burning pure chalk or limestone in a kiln, the result being quicklime. When water is added to quicklime the resultant reaction liberates considerable heat and a large expansion occurs. This action is known as 'slaking' and lime must be slaked before use. The slaking may be carried out on site on a clean mixing stage or in a suitable container, but most of the lime used today is supplied ready slaked as 'hydrated lime'. Non-hydraulic lime will not set under water. Semi-hydraulic lime is obtained from beds of greyish chalk or limestone and will harden under water in a few weeks. Hydraulic lime is obtained principally from the 'lias' beds. Will set fully under water, it should not be mixed with cement.

MORTAR PLASTICIZERS

These have the effect of entraining micro bubbles of air and breaking down surface tension, resulting in increased workability.

A 1:6 cement : sand mix with plasticizer is an alternative to a 1:1:6 cement lime: sand mix. A 1:8 cement : sand, plasticized mix is an alternative to a 1:2:9 mix. A plasticed mix weaker than 1:8 is not recommended.

Plasticizers may be used with sulphate resisting Portland cements and with high alumina cement, and there is evidence that aerated cement : sand mortars have greater resistance to sulphate attack than cement : lime: sand mortars of equal strength.

Indications are that mortar plasticizers improve the resistance of freshly laid mortar to frost.

Aerated mortar should not be used as an alternative to a strong (eg 1:0-¼: 3) cement : lime : sand mix.

Roller type mortar mills are not suitable for mixing plasticized mortar, as they fail to entrain sufficient air. Revolving drum mixers are satisfactory, but prolonged mixing leads to excessive air entrainment and consequent weakening of the mortar.

POINTING & JOINTING

The faces of the joints in face brickwork may be finished either by 'jointing' i.e. finishing the joint as the wall is built, or by 'pointing' i.e. raking out the joints while still soft to a depth of at least 13mm and then filling the joints with different mortar at a later stage.

JOINTING

The bricklayer finishes the face of the joints as he builds the wall, using his brick trowel, a pointing trowel or some form of jointing tool. (Sometimes a piece of bent steel rod or an old bucket handle). This method has the advantages that it is quicker and cheaper than pointing and, the surface finish being part of the bedding mortar there is less chance of face joints failing by frost action or falling out due to insufficient adhesion. Disadvantages are that it may not look as neat as a pointed finish and it is difficult to keep the work clean and maintain uniform colour in the joints because of variations in the sand and possibly slight variations in the proportions of the mix.

POINTING

Preparation is important, the joints must be well raked out and brushed down to remove any loose mortar. Before commencing pointing, the wall should be wetted sufficiently to obviate the initial suction, so that the wall is damp but not saturated. This will ensure good adhesion. Advantages of pointing are that it is easier to maintain a neat and attractive appearance, to keep the face of the work clean, maintain uniform colour in the joints and to use coloured mortars.

JOINT FINISHES

Struck joint:- may be used for internal fair face work but is unsuitable for external work as water tends to collect on the upper arrises of the bricks and may accelerate frost action.
Flush joint:- has a roughish texture and can present an attractive appearance particularly when used with sand-faced bricks.
Weather struck joint:- suitable for external work as it provides good protection against rain penetration. When pointing, the cross joints are neatly cut off using a pointing trowel and the bed joints trimmed with a frenchman. Thus any slight unevenness in the joints can be corrected and this type of pointing presents a very neat appearance.
Tooled joint:- also known as a keyed joint or bucket handle joint. A well compressed finish particularly suitable for jointing.
Recessed joint:- to be effective the bricks should have sharp straight arrises and the joint should be even. If used externally the bricks should be hard, dense, and well burnt.
Tuck pointing:- not often used today. Main use is for repointing old brickwork where the original joints have decayed and the arrises of the bricks have spalled. The joints are well raked out, brushed down, and then stopped with a coloured mortar to match the bricks. Quite often the whole wall is given a colour wash. A narrow groove is then incised in the centre of the joints, filled with a mix of lime putty and silver sand and then cut off accurately with a frenchman.

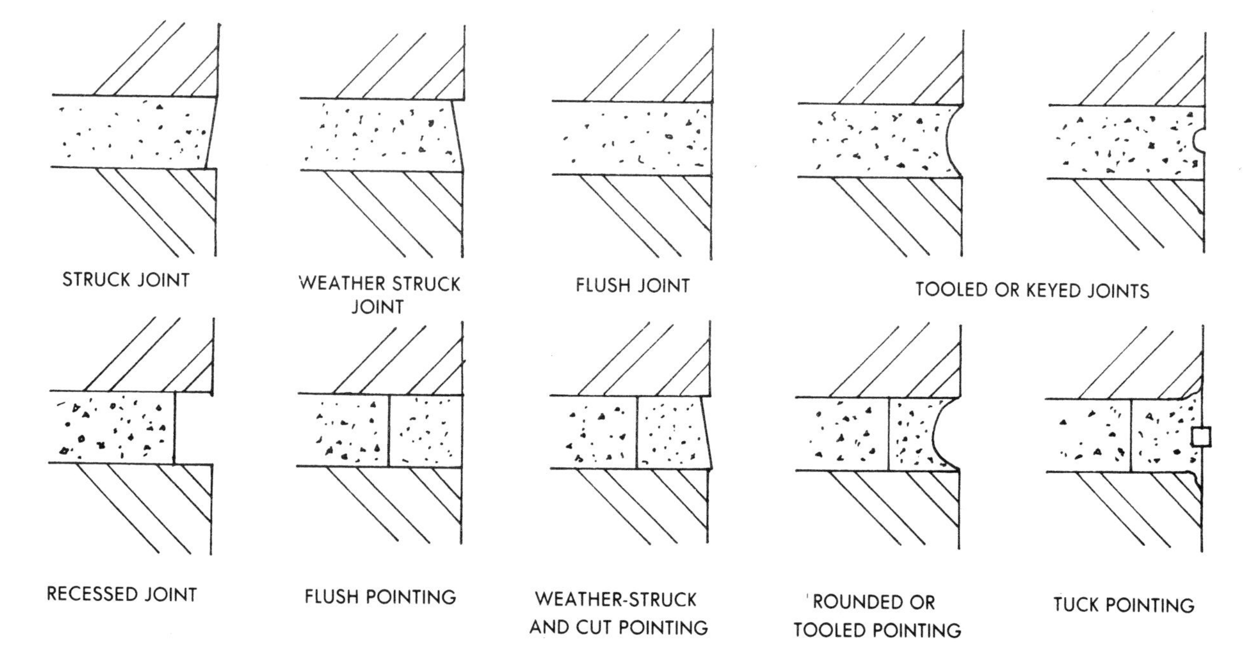

MORTAR

Mortar for pointing should match the type of brick and in general not be stronger than the bricks. Strong dense mortar for engineering bricks, weaker mortar for soft bricks.

DAMP-PROOF COURSES

Damp proof courses are required to prevent the penetration of moisture into a building. D.P.Cs are required below ground floors, at the base of walls, around openings in cavity walls, in chimney stacks and parapet walls. Constructional details of these will be given in later sheets.

D.P.Cs may be classified as flexible, semi-rigid. Sheet materials should be lapped at least 100mm at all joints.

FLEXIBLE MATERIALS

Bitumen D.P.C.:- consist of bitumen with a base of hessian, fibre or asbestos, with or without lead. Should comply with B.S. 743. Supplied in rolls to suit various wall widths, standard length 7.3m (24ft). Should be bedded on an even bed of mortar, free of stones or lumps which might prevent downward seepage of moisture, e.g. in parapets, the joints should be sealed with bitumen compound. Care must be taken when unrolling this type of D.P.C. It is advisable in winter to warm the material first to prevent cracking.

Polythene:- Black, low density polythene is used, supplied in rolls. Laps should be at least equal to the width.

Pitch polymer:- a pitch based plastic reinforced with fibres. Should be lapped at all joints and lapped and sealed against downward seepage. Will not creep under load.

Sheet lead:- should comply with B.S. 1178 and weigh at least 19.5 kg/m². Is liable to corrosion in contact with lime or Portland cement mortar, and to prevent this should be given a coating of bitumen as extra protection. Joints should be lapped and if used to prevent downward seepage the joints should be welted.

Copper:- should comply with B.S. 1569 and be grade 'A' annealed condition. Against downard seepage joints should be welted. In some cases joints may be welded, in this case phosphorus-deoxidised copper to B.S. 1172 should be used. Copper may cause staining of external surfaces, especially stone.

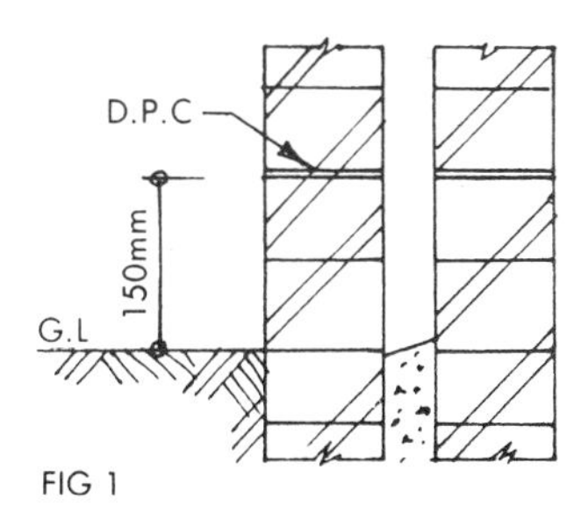

FIG 1

SEMI-RIGID MATERIALS

Mastic asphalt:- a mixture of suitably graded aggregates (natural asphalt rock or limestone) and asphaltic cement (Bitumen, lake asphalt etc). Should comply with B.S. 1097 or 1418. Normally laid by specialists, is jointless and particularly useful for vertical work and tanking. The surface of the asphalt should be scored whilst still warm or have grit beaten into it to provide a good key for the mortar bed which covers it.

RIGID MATERIALS

Slate:- consists of two courses of stout slates free from any flaws, laid breaking joint in 1:3 cement/sand mortar. The slates should be at least 225mm long. Not suitable against downward seepage. A brittle D.P.C. which may crack if any unequal settlement occurs.

Engineering blocks:- should conform to the requirements for bricks for D.P.Cs as specified in B.S. 3921. Each brick should have a max. water absorption of 4.5% and contain no holes or indentations other than frogs. The D.P.C. should consist of at least two courses of bricks, laid to break joints and bedded in 1:3 cement/sand mortar.

Epoxy resin/sand:- the resin content should be approximately 15% and the appropriate hardener should be used. The D.P.C. should be at least 7mm (0.25in) thick.

A D.P.C. is required at the foot of a wall and should be at least 150mm above ground level. If the D.P.C. is to be effective care must be taken to see that soil (particularly where flower beds are adjacent to buildings) is not heaped against the wall.

The D.P.C. should extend across the full width of the wall and not be kept back to allow for pointing. Fig 1.

Where a wall is rendered externally it is good practice to stop the rendering at the D.P.C. (Fig 2) and not carry the rendering down to ground level.

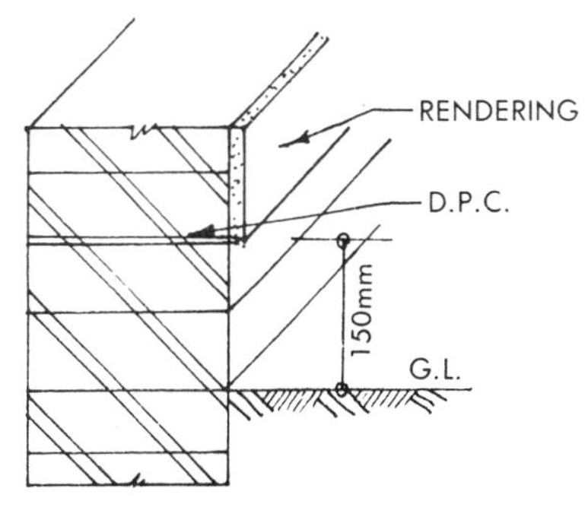

FIG 2

GROUND FLOORS I

Approved Document C of the Building Regulations 1985 lays down the requirements for ground floors as follows:

Part C4 — Resistance to weather and ground moisture states that the walls, floors and roof of the building shall adequately resist the passage of moisture to the inside of the building.

Section 1 — Floors next to the ground.

1.1) This section describes provisions for three kinds of ground floors.

1.2) A floor next to the ground should:

A) prevent ground moisture from reaching the upper surface of the floor. This provision does not apply to a building to be used wholly for:

i) storing goods or accommodating plant or machinery, provided that any persons who are habitually employed in the building are engaged only in storing, caring for or removing the goods, plant or machinery, or

ii) a purpose such that provision would not serve to increase protection to the health or safety of any persons habitually employed in the building, and

B) not be damaged by moisture from the ground.

SUSPENDED TIMBER FLOORS

Section 1.9 of Approved Document C of the Building Regulations 1985 states that any suspended timber floor next to the ground will meet the performance if:

a) the ground is covered so as to resist moisture and prevent plant growth, and

b) there is a ventilated air space between the ground covering and the timber, and

c) there are damp-proof courses between the timber and any material which can carry moisture from the ground.

A suspended timber floor next to the ground may be built as follows.

A) ground covering either:

i) concrete at least 100mm thick, composed of 50kg of cement to not more than 0.13m³ of fine aggregate and 0.18m³ of course aggregate, or BS5328, mix C7.5P if no embedded steel. The concrete should be laid on a hardcore bed of clean broken brick or any other inert material free from materials including water soluble sulphates in quantities which could damage the concrete, or

ii) concrete at least 50mm thick composed as described above, and laid on at least a 1000 gauge polythene sheet with the joints sealed on a bed of material which will not damage the sheet. To prevent water collecting on the covering, either the top should be entirely above the highest level of the adjoining ground or the covering should be laid to fall to a drainage outlet above the lowest level of the adjoining ground.

B) Ventilated air space. Measuring at least 75mm from the concrete to the underside of any wall plates and at least 125mm to the under side of the suspended floor. Each external wall should have ventilation openings placed so that the ventilating air will have a free path between opposite sides and to all parts. The openings should be large enough to give an actual opening

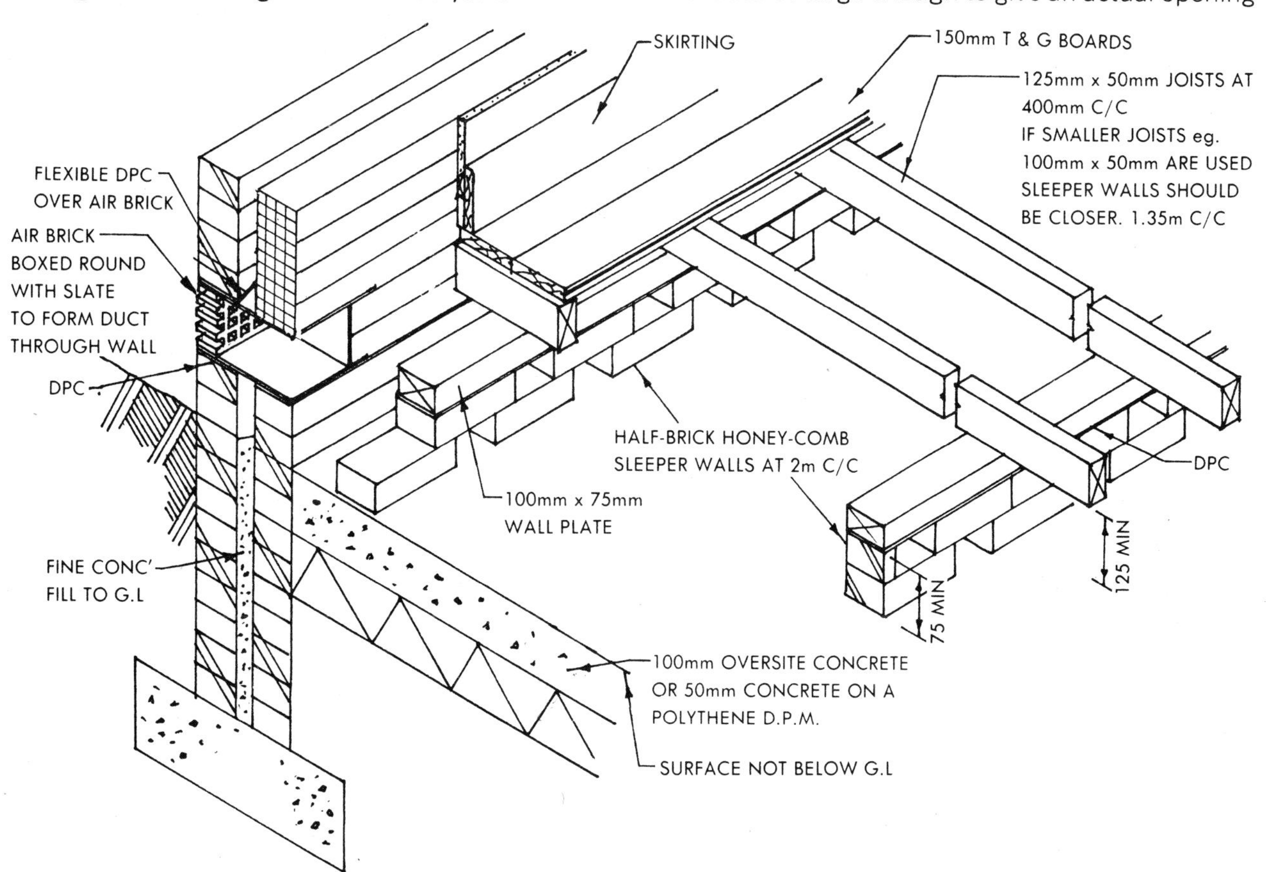

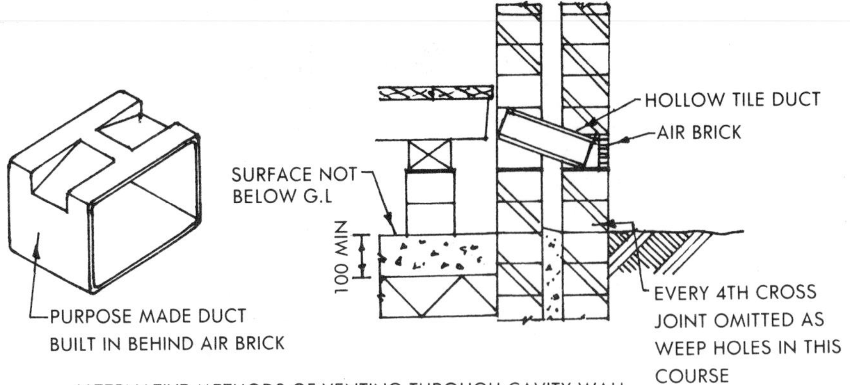

ALTERNATIVE METHODS OF VENTING THROUGH CAVITY WALL

of at least equivalent to 3000mm² for each metre run of wall. Any pipes needed to carry ventilating air should have a diameter of at least 100mm.
C) Damp proof courses of impervious sheet material, engineering brick or slates in cement mortar or other material which will prevent the passage of moisture.

It should be noted that the performance can also be met by following the relevant recommendations of clause 11 of CP102:1983 – Protection of buildings against water from the ground.

There should be through ventilation below the floor, and to ensure this air bricks should be built into external walls, sleeper walls, built honeycomb and vent holes left in any partition walls which might obstruct the air flow. Ducts are formed behind air bricks as shown, to prevent air being dissipated into the cavity.

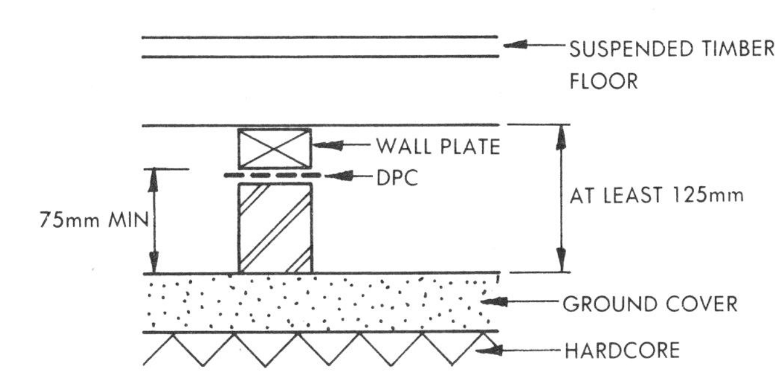

GROUND FLOORS 2

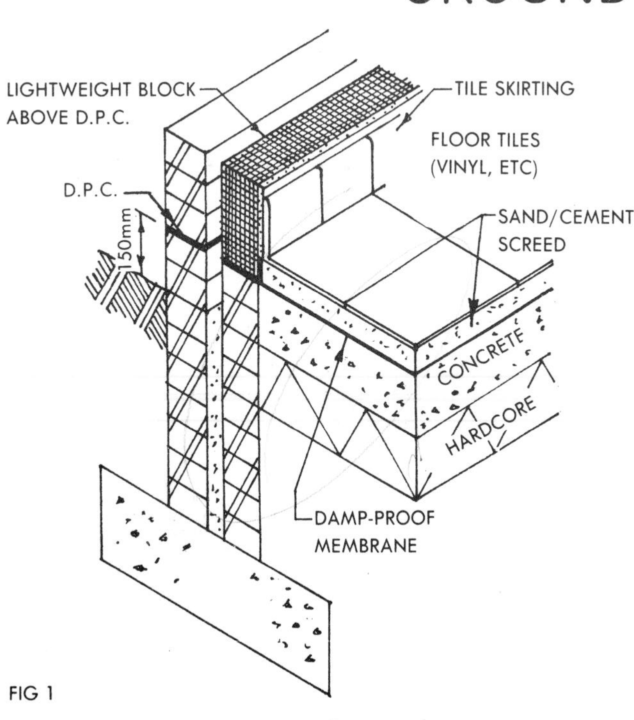

FIG 1

SOLID FLOOR WITHOUT TIMBER
(Any floor which is next to the ground shall be so constructed as to prevent any part of the floor being adversely affected by moisture or water vapour from the ground).

SOLID FLOORS
Approved Document C4, diagram 5, of the Building Regulations 1985, illustrates a concrete ground supported floor that will meet the performance if built as follows (unless it is subjected to water pressure) (See Fig 4):-

A. concrete at least 100mm thick. 50kg cement mixed with 0.11m³ of fine aggregate and 0.16m³ of coarse aggregate, or BS5328, mix C10P if steel is embedded within.
B. hardcore bed – as specified for suspended floor.
C. damp-proof membrane – above or below concrete, and continuous with the damp-proof courses in the walls, piers, etc.

A membrane laid below the concrete should be at least 100 gauge polythene
A membrane laid above the concrete should be either polythene sheet or three coats of cold applied bitumen solution or similar moisture and vapour resistant material. In each case it should be protected by a screed or floor finish, unless the membrane is pitch mastic or similar material which will also serve as a floor finish.

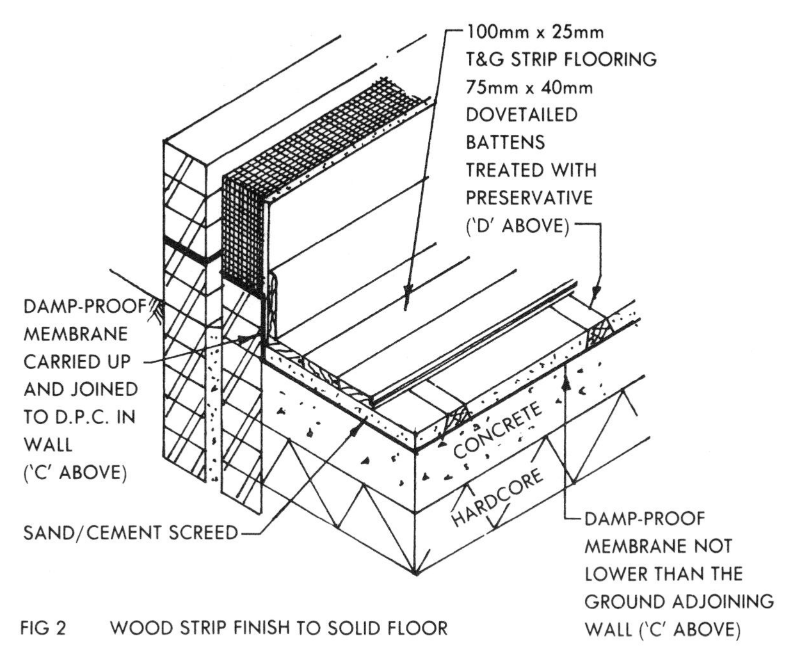

FIG 2 WOOD STRIP FINISH TO SOLID FLOOR

A timber floor finish laid directly on concrete may be bedded in a material which may also serve as a damp-proof membrane. Timber fillets laid in the concrete as a fixing for a floor finish should be treated with an effective preservative unless they are above the damp-proof membrane (see Figs 5 and 6).

Floor clips may be fixed directly to the oversite concrete but this means piercing the damp-proof membrane. The method shown on right avoids this, but it is important that the screed is thoroughly dried out before the boards are fixed.

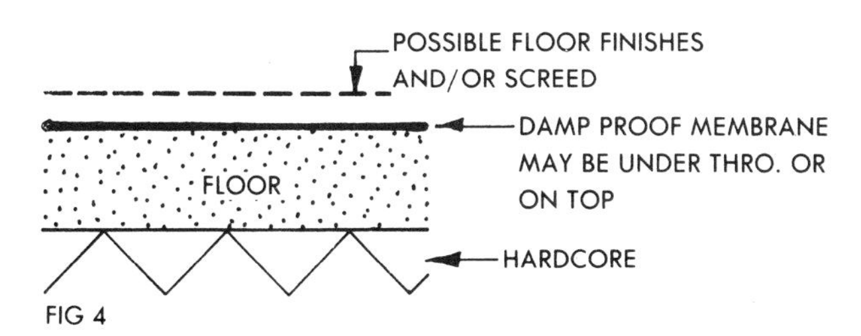

FIG 4

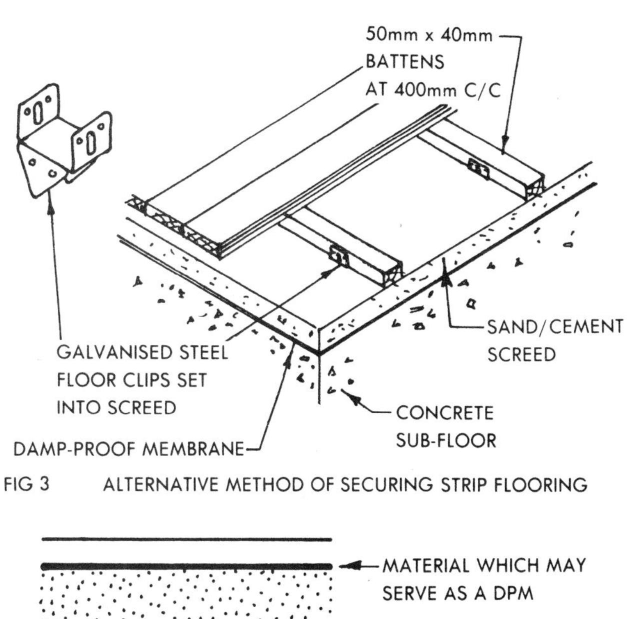

FIG 3 ALTERNATIVE METHOD OF SECURING STRIP FLOORING

FIG 5 TYPE 1

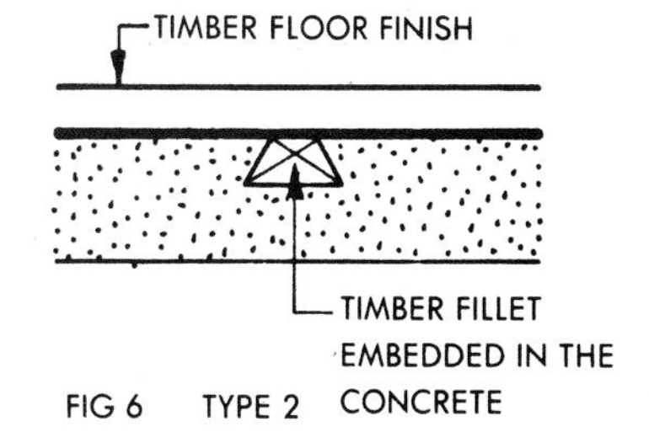

FIG 6 TYPE 2

GROUND FLOORS 3

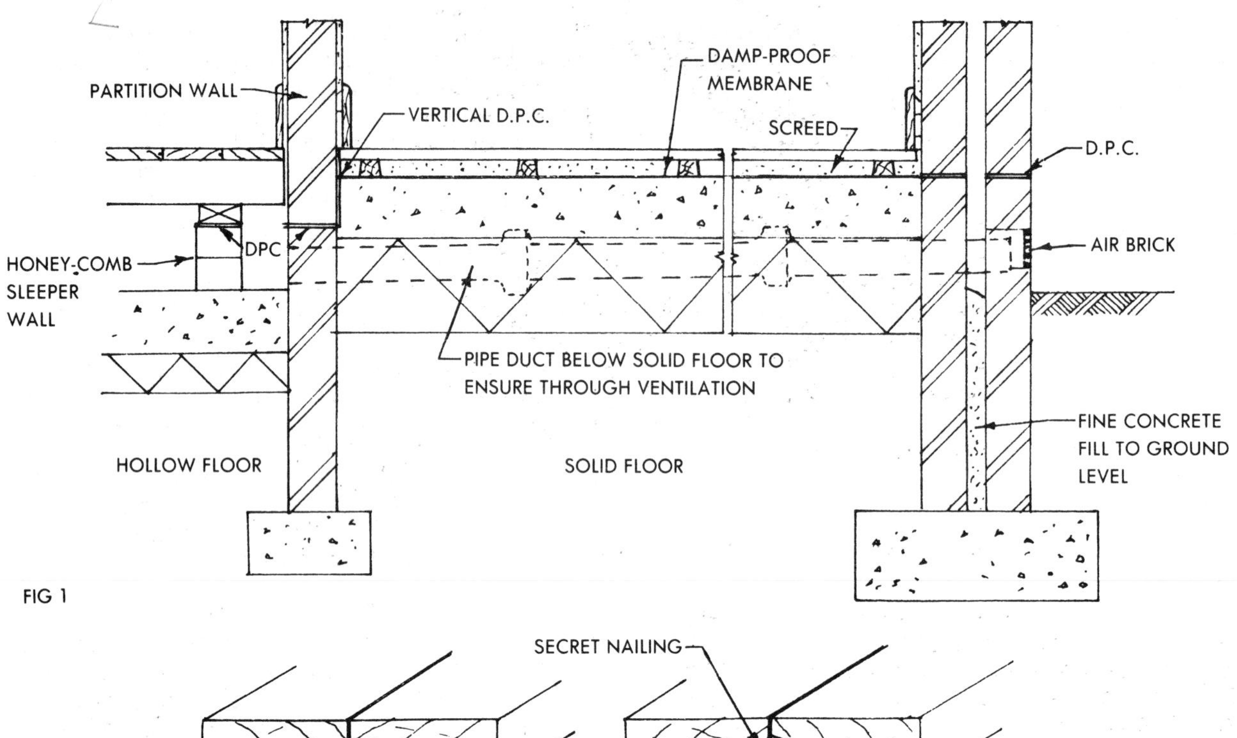

FIG 1

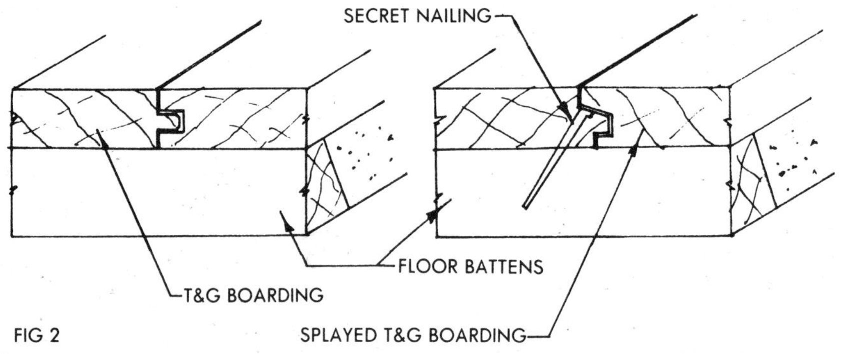

FIG 2

JUNCTION OF SOLID AND SUSPENDED FLOORS
It is important to provide through ventilation beneath a suspended timber floor, and where such a floor meets a solid floor as in Fig I, pipes should be laid beneath the solid floor, connecting with air bricks in the external wall as shown. The damp-proof membrane should be linked with the D.P.C. in the crosswall by a vertical D.P.C.

TONGUED & GROOVED BOARDING
When a boarded floor is required T&G boards (Fig 2) are recommended for both suspended floors and for solid floors. In the case of a suspended timber ground floor, if the under floor ventilation is adequate, draughts may pass through the joints between plain edged boards and the use of T&G boards will prevent this. For solid floors the use of T&G boards will ensure that spilt liquid or washing down water does not easily penetrate through the joints. Water below the boards would dry out very slowly and the resulting dampness may encourage the development of dry rot. It is important to protect the timber from any dampness which might arise from the sub-floor and in addition to treating the battens with preservative it is advisable to give the underside of boards a brush coat of preservative.

For strip flooring, and where the boarding provides the floor finish, secret nailing may be used so that the nails do not show on the surface. In this case splayed tongued and grooved boarding is preferable as there is less chance of the tongue splitting when the board is nailed.

WOOD BLOCKS
Fig 3 illustrates the requirements of the Building Regulations in respect of wood block floors.

The blocks are normally laid in the bonded pattern shown or may be laid in a herring-bone or basket weave pattern. Figs 4 & 5.

The blocks are usually dovetailed on the bottom (Fig 6) to ensure a good grip in the bitumen.

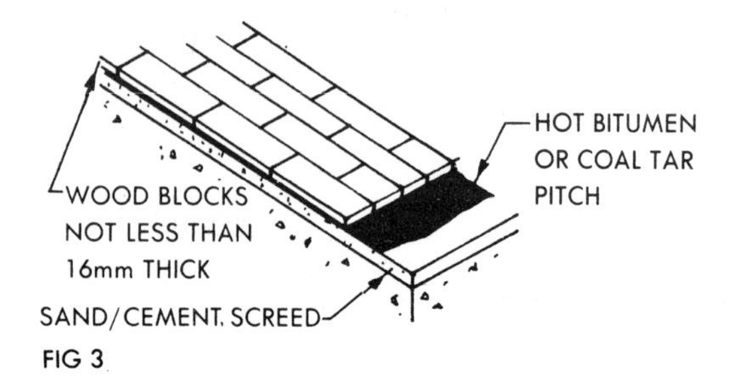

FIG 3

FIG 4 HERRING BONE

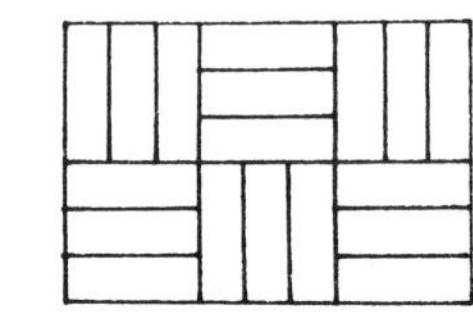

FIG 5 BASKET-WEAVE

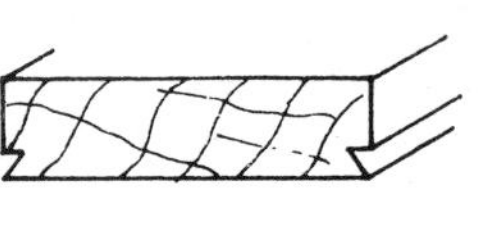

FIG 6

FLOOR SCREEDS

Sand:cement screeds are not usually considered suitable as wearing surfaces and if an in-situ wearing surface is required it should be in plain concrete, or a granolithic or terrazzo topping.

MIX

Where the floor finish is to consist of a thin, flexible covering e.g. rubber, vinyl tiles, asbestos tiles etc, a strong screed 1:3 pc:sand is recommended. This will resist pressure and impact from legs of chairs, tables and heavy furniture. Allowance must be made for the bulking of sand, and the water/cement ratio kept as low as possible (not above 0.45), although full compaction must be obtainable. A simple test is to squeeze a sample of the mix in the hand, when it should 'ball' together, but it should not be possible to squeeze out excess water.

For strong rigid floor finishes e.g. terrazzo, quarry tiles etc, a 1:4 pc:sand mix may be used, as the floor finish itself will resit impact and pressure from furniture etc.

A weaker mix 1:5 pc:sand may be used to enable a suitable floor finish e.g. cork tiles to be pinned to the screed.

THICKNESS OF SCREED

The table indicates minimum thickness of screeds recommended for various conditions.

CONDITION	Min' thickness in mm
Screed fully bonded to subfloor	20
Screed partially bonded to subfloor	40
Screed on in-situ concrete with damp-proof membrane between, or on concrete impregnated with oil or grease, or on concrete containing an integral water proofer, or on a weak concrete base.	50
Floating screed on compressible material	65
Screed over subfloor containing heating elements, or floating screed containing heating elements.	75

PREPARATION OF SUBFLOOR

Where the screed is laid on in-situ concrete before the concrete has set (i.e. within three hours of placing) the surface of the concrete can be prepared by spraying with water and brushing, removing any laitance and exposing the aggregate. This method ensures strong, monolithic construction reducing the possibility of cracking and curling of the screed.

It is however, more likely that the screed will be laid some weeks after the subfloor has been completed. In this case if the screed is required to be fully bonded to the base, the subfloor should be thoroughly hacked by mechanical means. The surface should be thoroughly cleaned, well wetted (preferably overnight) any surplus water removed before laying commences, and a brush coat of cement grout applied immediately before placing the screed. Alternatively a bonding agent may be used instead of the cement grout.

Partially bonded screed:- If the surface of the subfloor is well brushed before setting to remove any laitance and form a key, the screed may be laid at a later date without resorting to hacking the surface. In this case however, the possibility of some break in the bond between screed and subfloor must be accepted.

In cases where the screed cannot be bonded to the subfloor, e.g. where there is a damp proof membrane between, then a minimum thickness of 50mm should be maintained.

LAYING SCREEDS

The traditional method of screeding a large area is to divide the floor up into bays not exceeding 3.500m in width. The bays are laid alternately working to screeding battens which have been carefully levelled and aligned, and firmly bedded throughout their length. (Not on mortar pats). The first bays are left for 24hrs, then the battens removed and the remaining bays completed. It is difficult to ensure that there is no curling or undulations at the edges of the bays, and this would be very noticeable if a thin floor finish is used, e.g. vinyl tiles. The floor may be screeded in one operation, but in this case random cracks may develop as the screed dries out. These may be acceptable and are fairly easy to repair. As an alternative to battens a fillet of the screed material about 80mm wide may be laid and levelled around the edges of the floor and the floor then screeded to this fillet.

If heating cables are incorporated in the screed bays should be used, taking care to avoid long narrow bays and keeping sides of bays in the ratio of 1:1½.

SCREED FINISH

The floor covering largely decides the method of finishing the surface of the screed. A surface left from the screedboard or from a wooden float provides a good key and is recommended for clay tiles, concrete tiles, pitch mastic, terrazzo, wood block, asphalt tiles, magnesium oxychloride and cement emulsions. The smoother finish obtained with a steel trowel is recommended for cork tiles, hardbaord, linoleum, rubber, thermoplastic tiles, vinyl tiles and flexible PVC flooring. N.B. Do not overtrowel or cement and water will be drawn to the surface.

CURING

A screed should be cured for at least 7 days after laying. This may be done by covering with polythene sheet or waterproof paper.

FLOOR FINISHES I

In situations where appearance is relatively unimportant and where cheapness is a consideration, e.g. in garages, workshops etc, the surface of the oversite concrete may serve as the floor finish, or a cement and sand screed may be laid over the concrete as described.
A floor finished in this way however does not present a particularly attractive appearance, tends to stain easily and is liable to 'dust up' on the surface.

LIGHTWEIGHT AGGREGATE CONCRETE SCREEDS
Are not used as the floor finish, but are laid in a similar manner to the cement and sand screeds previously described.

Because of their low density these screeds can be laid in sufficient thickness to accommodate services without imposing excessive loading on structural floors. Lightweight screeds provide good thermal insulation, but care must be taken when laying not to over compact the screed, thus increasing density and reducing insulation. The screed should be at lest 40mm thick and if the proposed floor finish is incapable of spreading point loads, then a topping of cement and sand 1:4, and at least 10mm thick should be laid monolithically with the screed. Because of their insulating properties lightweight screeds are not suitable where under floor heating is used.

Lightweight screeds also provide good insulation against airborne sound, but a soft floor finish would be required to reduce the effect of impact noise.

DRYING
A screed must be allowed to thoroughly dry out before any floor finish sensitive to moisture is laid. This is a slow process, a useful rule of thumb being to allow one month for each 25mm thickness of the screed. The slower the drying out the least risk of cracking or curling of the screed. At least a month should elapse after laying before any attempt to speed up the drying process by the use of heaters is resorted to.

GRANOLITHIC CONCRETE
Widely used for industrial flooring, in factories, garages etc, where hardwearing qualities are essential. The usual mix is 2:5 Portland cement : granite chippings, the aggregate graded from 6.4mm down; or 1 part cement : 1 part dry fine aggregate : 2 parts dry coarse aggregate, by weight. The aggregate to comply with BS. 1201 'Aggregates for granolithic concrete floors'. An excess of fine dusty material must be avoided and there should not be more than 20% passing a 200 mesh sieve to B.S. The mix should contain sufficient water to permit thorough compaction, but too much water results in high shrinkage and a surface liable to dusting. The water cement ratio should not exceed 0.45.

The most reliable finish is obtained by laying the granolithic topping within 3 hours of laying the base, thus obtaining monolithic construction. In this case the topping may be from 15mm to 25mm thick. If the topping must be laid at a later stage and monolithic construction cannot be achieved, then the topping should be at least 40mm thick and it is important to ensure the best possible bond with the base. Where the granolithic is laid on a membrane, precast units, or where ever the bond with the base is likely to be weak, there is an increased risk of curling of the topping. In such situations the topping should be at least 76mm thick and laid in two 38mm layers, to ensure thorough compaction. The top layer should be laid within 1 hour of the bottom layer.

Trowelling:- When the granolithic has been spread and well compacted it should be trowelled just enough to ensure a level surface. About 2 hours later the surface should be re-trowelled to close any surface pores. Final trowelling should only be carried out when considerable pressure is required to make any impression, and a hard compact surface can be obtained without producing laitance.

Surface treatment:- The surface of granolithic concrete may be treated to reduce any tendency to produce dust. This may be done by the application of sodium silicate, magnesium or zinc silico-fluoride, or by using a sealer e.g. polyurethane.

It is often specified that carborundum shall be sprinkled over the surface, (1.35kg/m^2) and well trowelled in before the granolithic has set in order to obtain a non-slip finish.

Curing:-If high quality granolithic concrete is to be achieved it is important that it is properly cured and a minimum period of 7 days is desirable.

CONCRETE TILES
Some tiles have a wearing layer 7mm or so thick with a backing of fine concrete, others, particularly granolithic tiles are of one mix throughout. For industrial use hard wearing surfaces incorporating natural or metallic aggregates are preferred, while for non-industrial use, fine pigmented surfaces giving an attractive appearance are more common. The tiles are usually laid on a 10mm bed of mortar, either 1:3 cement : sand, or a 1:1½ : 4 or 5 cement : lime: sand mix. The mortar bed should be separated from the base by a layer of building paper or plastic sheet to allow for any shrinkage.

FLOOR FINISHES 2

TERRAZZO FLOORING

A decorative form of concrete usually made from white cement and crushed marble aggregate, but other suitable stone is sometimes used. For fine mixes aggregate of size 3mm to 5mm should be mixed with an equal or greater proportion of 5mm to 6mm aggregate. Aggregate up to 25mm is available. Mixes should not be richer than 1:2 cement : aggregate by volume.

Two methods of placing in-situ terrazzo are commonly used:-

1. All the aggregate is incorporated in the mix, which is spread to the required thickness. Mixes with up to 10mm aggregate should be spread at least 15mm thick; for larger aggregates the thickness is increased.
2. A mix incorporating only the fine aggregate is spread in position and the larger aggregate then sprinkled on the surface and beaten and trowelled in.

Monolithic construction:- The base concrete should be levelled off 12mm below finished floor level, the surface brushed with a stiff broom to remove any surplus water or laitance, and the terrazzo immediately laid followed by normal rolling and trowelling.

Separate construction:- An isolating membrane is laid over the base concrete, the screed laid on top and then the terrazzo placed while the screed is still green. If this is not possible the terrazzo may be laid over the screed immediately before laying.

Cracking of floor finish:- Unless care is taken there is a risk of cracking and crazing of the terrazzo finish. Precautions to obviate cracks are:-

1. The floor should be divided up into panels not exceeding $1m^2$ in area, with sides in the ratio of 3:1, and separated by dividing strips of metal, ebonite or plastic.
2. The water cement ratio should be as low as possible in order to reduce drying shrinkage.
3. Aggregates smaller than 3mm should not be used. A larger aggregate reduces the risk of crazing.
4. The terrazzo should be allowed to dry out slowly and if possible the building should not be heated for 6 to 8 weeks after the floor finish has been laid. N.B. There is less risk of curling if monolithic construction is used.

Surface finish:- The material must be thoroughly compacted by tamping, trowelling and rolling and further trowelling carried out at intervals, (depending on rate of setting of the mix) to produce a dense, smooth surface, free from laitance. After this surface has hardened sufficiently (usually after 4 days) it is ground, usually by machine and finally polished using a fine abrasive stone to produce a 'fine grit' finish.

MAGNESITE FLOOR FINISH (Magnesium Oxychloride)

A mixture of calcined magnesite, fillers such as sawdust, wood flour, ground silica, talc or powered asbestos, to which a solution of magnesium chloride is added. Needs to be laid by specialists and should comply with B.S. 776 'Magnesium oxychloride (magnesite) flooring'. Available in various colours and in mottled or grained effects. Suitable for industrial, commercial or domestic use depending on the type of fillers. A damp proof membrane should be provided as the material tends to soften and disintegrate in wet conditions.

The thickness may vary from 10mm for single coat work to 50mm for two or three coat work. If laid on timber, galvanised wire netting should be firmly nailed to the base to provide a key. When laid on a cement and sand screed the slightly rough surface left by a screeding board provides a good key.

Rapid drying of the floor finish should be avoided for at least 24 hours after laying and it should be left for at least 3 days before being used.

Metal is liable to corrode in contact with magnesium oxychloride and service pipes should be protected with bitumen or coal tar composition.

MASTIC ASPHALT

Should comply with one of the following:- BS 1162, 1418, 1410: 1973 Mastic Asphalt for Building (Natural rock asphalt aggregate). Three types of asphalt for building i) roofing and tanking, ii) damp proof coursing, iii) flooring. BS 988, 1076, 1097, 1451: 1973 Mastic Asphalt for building (lime stone aggregate). Other grades are available for special purposes e.g. acid resisting construction.

Mastic asphalt provides a jointless floor which is dustless and impervious to moisture. It has good durability, but concentrated loads which may cause indentation should be avoided. Suitable for a wide range of conditions from domestic to heavy industry use, in the latter case it may be necessary to incorporate metal armouring. For light duty a thickness of 15mm to 20mm may be used, but for heavy duty a thickness of 25mm or more is recommended.

Asphalt floors over 20mm thick may be laid directly on new concrete, but an isolating membrane of black sheathing felt to BS 747 is usually provided. Where the thickness is less glass fibre sheeting may be used as an alternative in special circumstances. A membrane should always be provided:-

1. On a timber base
2. On a porous base, e.g. lightweight concrete
3. Where the concrete surface shows fine cracks
4. Where the surface of the concrete has been treated, e.g. with sodium silicate
5. Where a coloured mastic asphalt floor finish has a polished surface.

N.B. Asphalt surface is liable to soften by prolonged contact with grease, fats, oils and unsuitable polishes.

FLOOR FINISHES 3

PITCH MASTIC

Should comply to BS 5902 (which covers both black and coloured pitch mastic flooring). Treatment of the base and conditions generally are similar to those for mastic asphalt. Suitable for a wide range of conditions from light domestic to heavy duty industrial flooring. Is resistant to attack by mineral oils and greases at normal temperatures, but vulnerable to attack by animal and vegetable oils and greases.

It is important to ensure that the pitch mastic does not become over-heated (the temperature when melting on site should never exceed 160°C) as this will cause defects to occur which will diminish the wearing qualities. The surface may be given a matt finish or a polished finish.

CEMENT RUBBER-LATEX

A mixture of Portland cement, aggregate, fillers and pigment, gauged on site with a stablized aqueous emulsion of rubber latex. Usually laid to give a thickness of 6mm. Not affected by dampness, but should not be regarded as a damp-proof membrane. Aggregates of vulcanised rubber and wood chips provide a resilient finish, while for harder grades providing good abrasive resistance, crushed marble, granite etc is used.

Provides a firm finish which has a high slip resistance even when wet and is quieter to the tread than hard finishes such as concrete.

A range of colours is available and the surface may be trowelled or ground to expose decorative aggregate.

The concrete base should have a wood float finish and be primed with a thin coat of latex and cement, which should be allowed to dry before laying the floor.

May be laid on a timber base only if it is strong and rigid. Any movement in the timbers will result in cracking.

Regular maintenance is required to keep the floor in good condition. Should be washed with warm water containing a neutral detergent. Scrubbing should be avoided. Aqueous emulsion polish may be used, but polishes containing solvents should not be used.

CLAY TILE FLOORING

Two types of tile are specified in BS 1286: Clay tiles for flooring. Type 'A' floor quarries and Type 'B' Semi-vitreous and vitreous floor tiles. Type 'B' are manufactured to closer dimensional tolerances than Type 'A', have a fine, smooth texture, are less absorbent, offer a wider colour range and may be manufactured from fine ceramic bodies as well as ordinary clay. Special non-slip tiles are available having a ribbed or a textured surface.

Bedding tiles:- A separating layer of building paper or polythene sheet should be laid over the base Fig I. The purpose of this is to ensure that there is no bond between the tile bedding and the base, thus allowing for any relative movement between them. The separating material should be lapped at all joins to prevent any contact between bed and base. Tiles may be bedded directly on to a damp-proof membrane, or a surface which has been treated, e.g. with a curing compound, wherever there is no strong bond between tile bed and the base. The tiles are soaked in clean water and then drained until almost dry (to reduce excessive suction) and then bedded in mortar. A cement, lime, sand mix is suitable, say 1:½:4 or 5, in any case the mix should not be richer than 1:3 cement : sand. A thickness of 13mm is common.

Thick bed method:- This method requires no separating layer and has been gaining popularity of late. A semi-dry bed of cement and sand, not richer than 1:4 is spread and consolidated about 40mm thick. The thickness should not be less than 20mm. A slurry of cement : sand 1:1 is immediately trowelled into the bed, and the tiles placed in position and tapped down firm and straight. Tiles are usually laid dry.

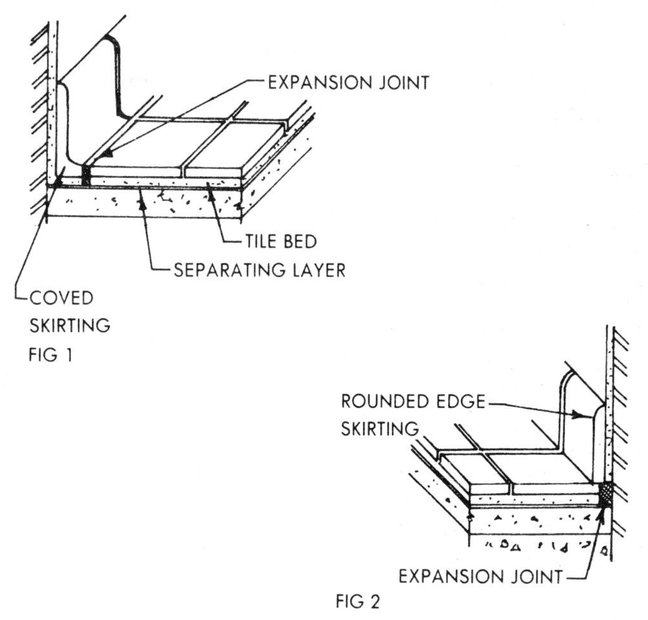

FIG 1

FIG 2

Adhesives:- These should conform to BS 5385. Bed should not exceed 5mm tiles are bonded to base, but slight movement can be accommodated.

Bitumen bed:- Used where high temperatures are likely, e.g. boiler houses. The tiles are bedded in a mix of bitumen emulsion/sand 1:2½. The base surface should be primed with a coat of bitumen emulsion first.

Expansion joint:- Should be provided around the perimeter of tiled floors, either between the tiles and coved skirting Fig I, or between tiles and wall Fig 2, if a superimposed skirting is used. For large areas expansion joints may be provided at intervals of about 7.5m.

FLOOR FINISHES 4

CORK FLOORING
Wear resistance is related to density, cork tiles are available in thicknesses of 3mm upwards and densities range from low, 450 kg/m^3 to extra heavy density of over 580 $kg/^3$.

Tiles should be exposed for 48 hours in the area where they are to be laid so as to ensure dimensional stability. They are bedded in an appropriate adhesive and may be secured with headless steel pins. May be wax polished or treated with a suitable sealer when dry. They provide a firm non-slip surface, (unless polished over a sealer) and give a significant reduction to impact sound transmission. Cork tiles may be laid over any level sub-floor, but it must be dry and a damp-proof membrane is required. When laid on a timber sub-floor a hardboard underlay is recommended.

CORK CARPET
An open textured material made with coarsely granulated cork. Is similar to linoleum but is more absorbent, and its surface should be sealed or polished. Has good sound insulating properties, is resistant to water and weak acids. But vulnerable to alkalis, will withstand heavy foot traffic but is liable to indentation from heavy furniture etc.

LINOLEUM
Should comply with BS 810 'Sheet linoleum, cork carpet & and linoleum tiles', or BS 1863 'Felt backed linoleum', the latter type having greater moisture stability. Thicker gauge is recommended where traffic is heavy, and hardened linoleum is available for use where indentation is expected.

Material should be kept at room temperature of not less than 18°C (65°F) for at least 24hrs before laying. The adhesive should be spread evenly using a serrated edged trowel, and after laying the material should be well rolled with a 68kg floor roller.

The concrete sub-base must be dry if linoleum is to be bonded to it, a damp-proof membrane is therefore essential. Where linoleum is fixed to a timber floor, the best method is to use hardboard or plywood as an underlay. Linoleum bonded to a sub-floor has a longer life than when laid loose.

Is resistant to oils and greases and provides a reasonably quiet resiliant floor, which is not slippery unless highly polished. The use of a polish with reduced tendency to slipperiness is recommended.

RUBBER FLOORING
Should comply with BS 1711 'Solid rubber flooring'. May be in sheet form or as tiles. Natural rubber is vulnerable to fat, grease, oil and petrol, but various synthetic rubber floorings are available which are resistant to these substances.

Concrete sub-floors should be finished with a cement/sand screed not less than 1:3 having a steel trowel finish. Wood floors must be smooth and even, or plywood may be used as an underlay.

Flooring is usually fixed with rubber solution and in this case it is essential that there is a damp-proof membrane in the sub-floor. Alternatively tiles are available having dovetail ribs on the underside, which will key-in to a wet cement/sand screed and provide a mechanical fixing, thus obviating the need for a damp-proof membrane. Heavy duty tiles of this type are available up to 20mm thick.

Rubber flooring may become slippery when wet, but varieties havinig a ribbed surface are available to provide a slip resistant finish. One type of flooring has a thin rubber surface mounted on a sponge rubber backing, and this gives a significant reduction in impact sound transmission.

THERMOPLASTIC TILES (ASPHALT TILES)
Should comply with BS 2592 'Thermoplastic flooring tiles'. The tiles are fixed with a bituminous adhesive, and the sub-floor or screed should be given a steel trowel finish. They are likely to deteriorate if in contact with grease or oil, but otherwise may be used in all normal situation. *Thermoplastic tiles — PVC modified*:- These have greater resistance to the action of grease and oil and are more flexible. A damp-proof membrane is only essential if there is any possibility of water pressure.

VINYL ASBESTOS TILES
Should comply with BS 3260 'PVC (Vinyl) asbestos floor tiles'. Manufactured from a mixture of PVC resin, chrysolite asbestos fibre, mineral fillers and pigments. Have reasonable resistance to oil and grease, but liable to be marked by cigarette or cinder burns. (May be removed with steel wool, unless serious). Warmed when laying, so that they become flexible enough to 'give' to any slight uneveness in the sub-floor. A damp-proof membrane is usually incorporated in the sub-floor.

FLEXIBLE PVC (VINYL) FLOOR TILE AND SHEET
Should comply with BS 3261 'Flexible PVC flooring'. Sheets can be welded at the joints using a rod of PVC and a hot air gun, to provide an impervious surface, used where spillage is likely. This type of flooring is resistant to grease, oils and most chemicals, and is liable to damage by cigarette burns. The 'finish' of the sub-floor is most important as these floor finishes are sometimes very thin (1.5mm) and any irregularities will be apparent. Varieties are available backed with cork composition, needled fibres etc, which provide good impact sound insulation.

UPPER FLOORS I

TIMBER FLOORS
Joists span from wall to wall and may be supported in various ways:-

1. Ends of joists bear directly on the wall. Fig I. This may involve awkward cutting and packing to ensure that the tops of the joists are level.
2. Joists rest on a wall plate built into the wall Fig 2.
3. Wall plates are not really suitable where joists are supported by an external wall, and in this case a mild steel or wrot iron bar may be used. Fig 3.
4. The joists may rest on a wall plate supported by wrot iron or MS corbels Fig 4. The corbels are nromally spaced at 750mm C/C
5. The joists are supported on joist hangers. This method avoids building the ends of the joists into the wall, and is useful where the joists are supported by a separating wall. Fig. 5.

Where joist from either side meet on a load-bearing partition they are usually arranged side by side and spiked together Fig. 6.

Herring-bone strutting Figs 7 & 8 are inserted between joists as shown to ensure rigidity. Folding wedges are insereted between the end joists and the wall as shown.

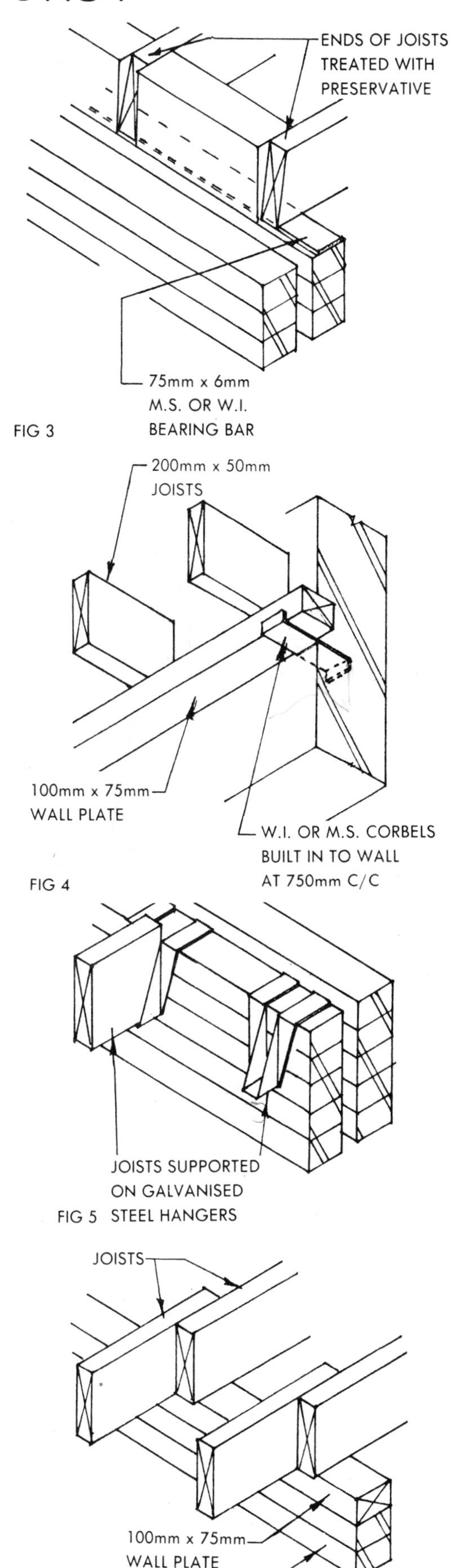

FIG 3

FIG 4

FIG 5

FIG 6

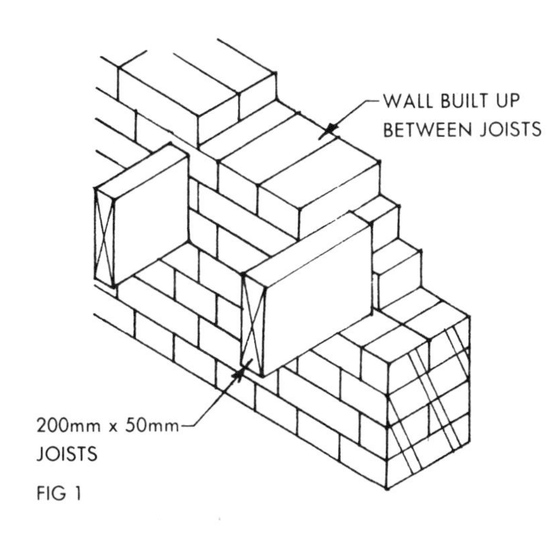

FIG 1

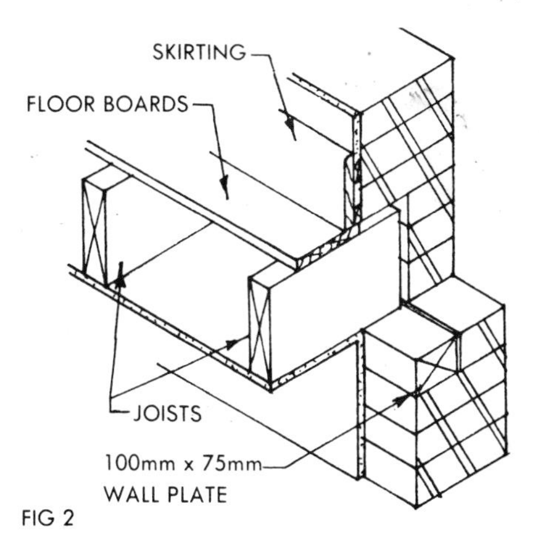

FIG 2

Joists spanning from wall to wall are known as bridging joists Fig 7 and are 50mm thick. Where openings occur e.g. for stair wells and around fireplaces, trimmer and trimming joists are used Fig 7. These are 75mm thick, and connected by a tusk tenon Figs 9 & 10. The trimmed joist Fig 7, are housed to the trimmer Fig 12. Alternatively a joist hanger Fig 11 may be used. A rule of thumb for calculating depth of joist is :-

$\frac{1}{24}$ of span (in mm) + 50 = depth in mm

e.g. 4m span :- $\frac{4000}{24} + 50 = 217$mm, or say 225mm (standard size)

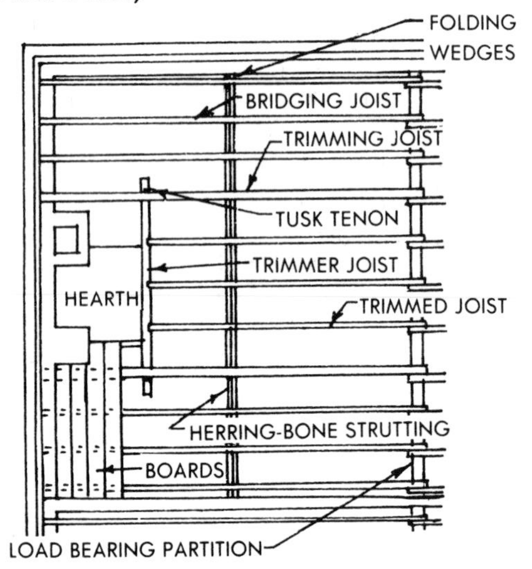

FIG 7 PLAN 1st FLOOR

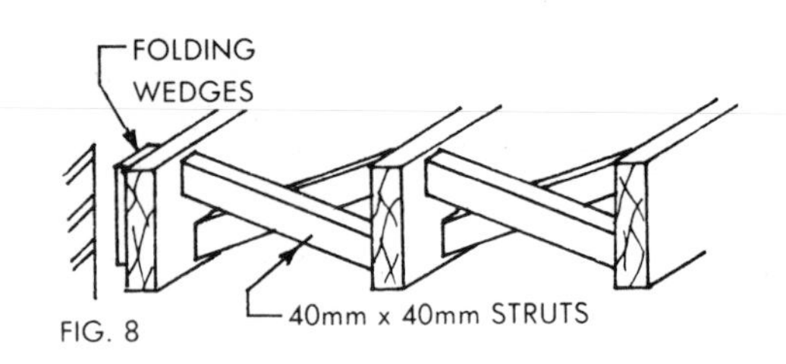

FIG. 8

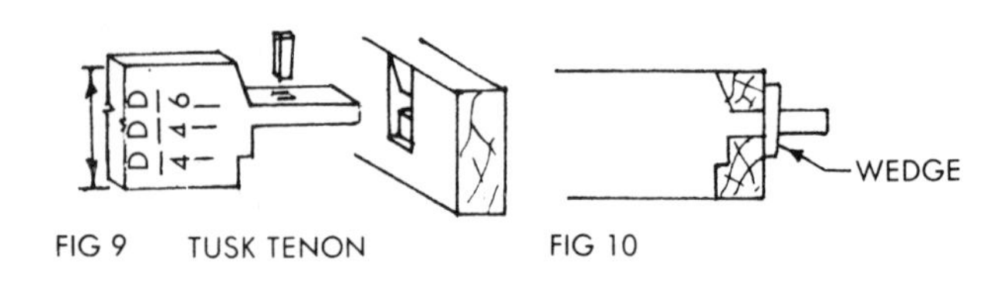

FIG 9 TUSK TENON

FIG 10

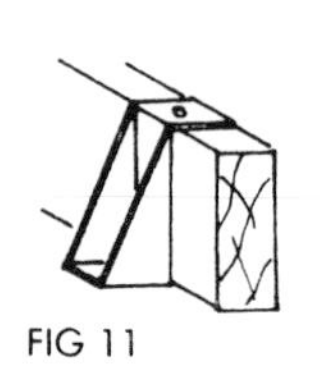

FIG 11

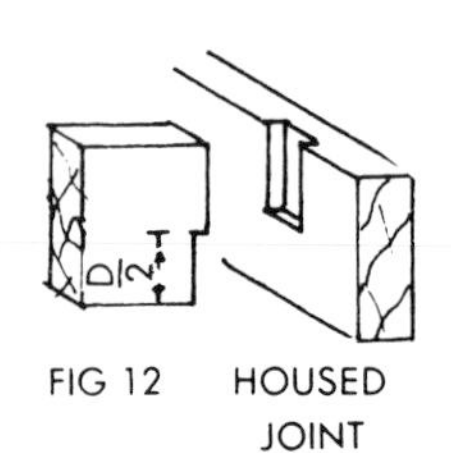

FIG 12 HOUSED JOINT

UPPER FLOORS 2

HOLLOW POT FLOOR

This type of floor has cast-in-situ reinforced concrete beams with hollow clay or terra cotta in-filling blocks cast in between the beams. Advantages are:- saving in weight, saving in quantity of concrete, good resistance to damage by fire, and it is possible to accommodate conduits within the topping screed. The thickness of the floor and amount of reinforcement will depend upon the span and the superimposed load. In some types of floor the concrete beams or ribs extend the full depth of the floor, but in others a clay slip tile is used at the bottom of the ribs. This reduces the possibility of pattern staining occurring, gives better resistance to fire and provides a uniform key for plaster.

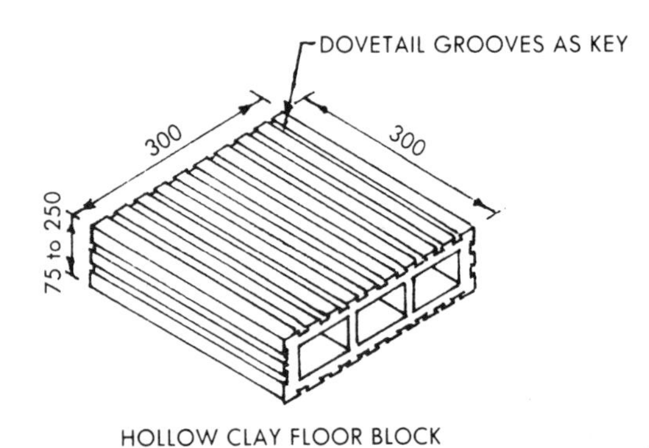

HOLLOW CLAY FLOOR BLOCK

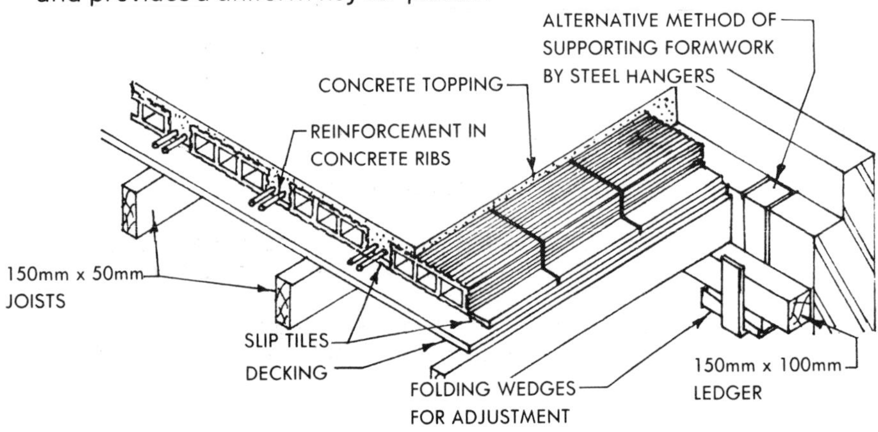

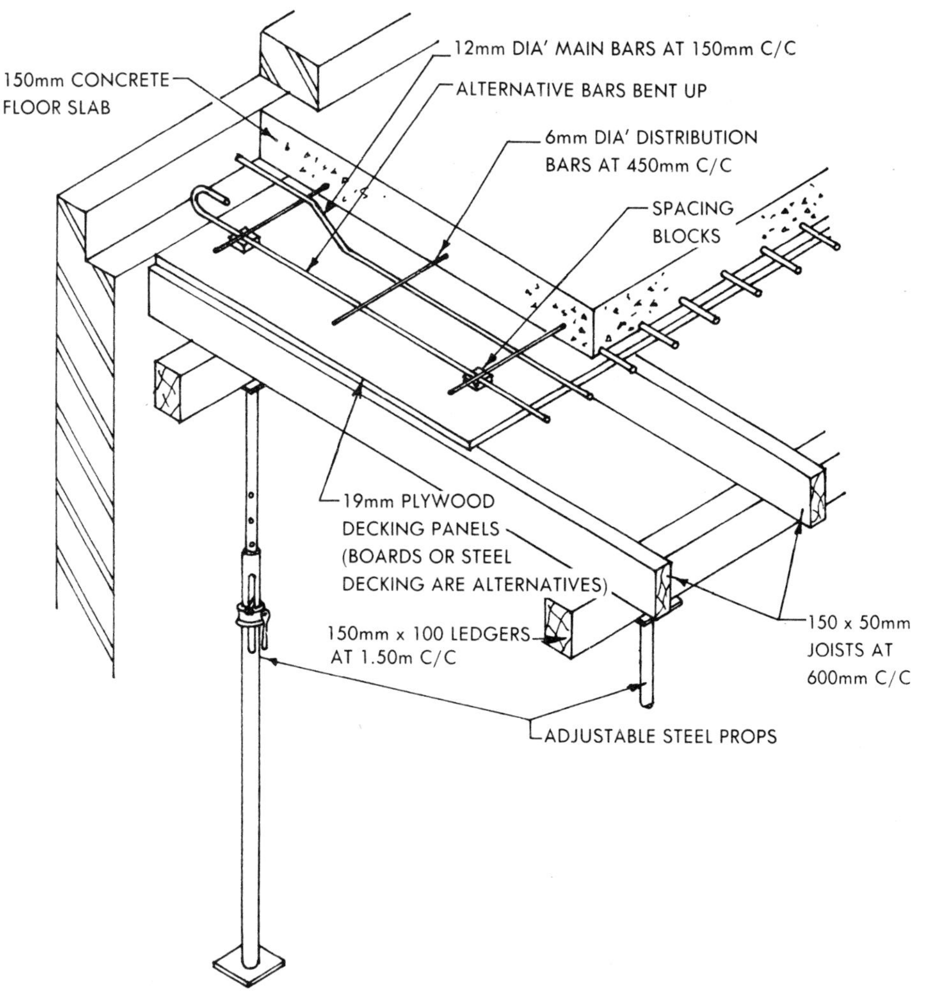

SOLID IN-SITU CONCRETE FLOOR

The floor slab is normally designed to span across the least width of a room, the reinforcement consisting either of mild steel main bars with distribution bars at right angles wired to them, welded steel fabric reinforcment, expanded steel mesh or ribbed steel lathing.

This type of floor is economic over spans up to about 5m, but for larger spans it is usually more economical to introduce secondary beams.

A floor of this type will carry heavier loads than a timber floor of equal thickness, provide good lateral rigidity, is highly fire resistant, and because of its mass provides good insulation against airborne sound.

Disadvantages are that it is necessary to wait several days until the concrete has set and hardened before the floor can be loaded, and the formwork and props required to support the floor slab during this period, usually about a week, delays the finishing operations.

The drawing shows an example of an R.C. floor slab and the type of centering required.

Where the slab is built into the wall a bending moment occurs at the top of the slab and alternate bars may be cranked up as shown to allow for this.

UPPER FLOORS 3

TELESCOPIC FLOORS

There are a number of types of these and they can be adjusted to suit various spans. Bearing plates at the ends of the centres rest in pockets in the walls or beams, and may be retracted when striking the formwork. They may be used to support timber or steel decking, or in some cases the floor may be supported directly on the centres as shown.

Advantages are that the centres may be erected quickly without expert labour and the clear space below enables follow-up work to proceed without interferences, and allows space for storage.

Fig I shows the 'Smiths fireproof floor' which has two-way reinforcement. This ensures safe distribution of point loads, permits the use of long spans and helps to obviate surface and ceiling cracking.

Reinforcement is placed in the channels formed by the projecting flanges of the blocks, which are blanked at one end and laid with open ends together to form a series of boxes. It is an easy matter to provide openings for lifts and staircase wells, as the lateral beams serve as trimmers for the longitudinal ones.

SELF-CENTERING FLOORS

Are of two types:-

1. Hollow pre-cast R.C. floor beams, spanning between walls or walls and beams. Fig 2.
2. Precast reinforced concrete floor beams or ribs which support some form of hollow blocks or lightweight concrete slab between them. The 'Armocrete floor' Fig 3 is an example of this type.

When the floor is supported by a beam, Fig 2, continuity bars are required in the top of the slab, or between the precast units as shown.

The hollow beams and blocks have comparatively thin walls and a constructional concrete topping is laid so as to spread the weight of superimposed loads.

These types of floor have the advantages of speed of erection, saving in weight, and do not obstruct the space below.

Many of these patent floor beams are prestressed. This increases the economic span of the floor and permits the thickness of the floor to be reduced.

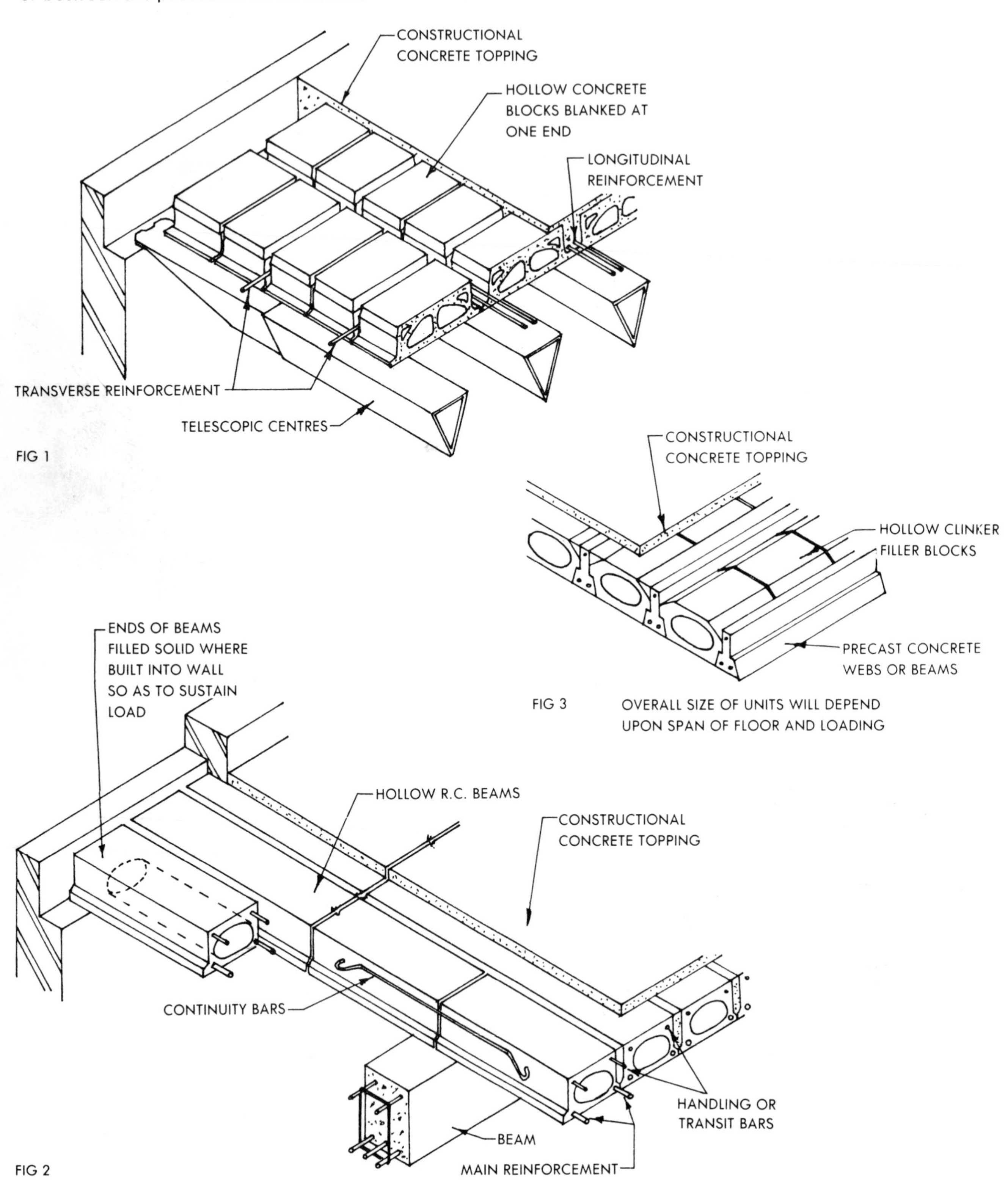

FIG 1

FIG 3

FIG 2

CAVITY WALLS I

TWISTED STEEL TIE
(GALVANISED)

BUTTERFLY WIRE TIE

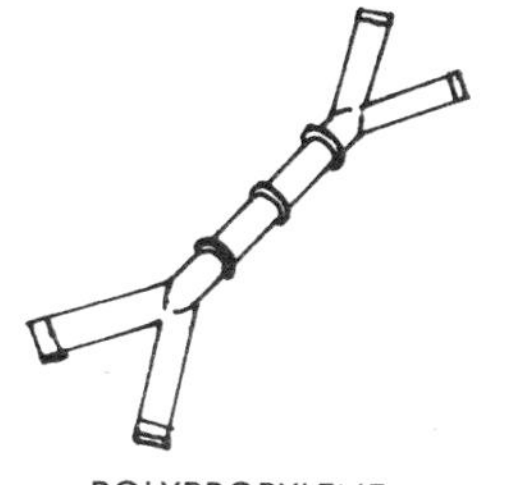

POLYPROPYLENE
PLASTIC TIE (KAVI-TIE)

FIG 1

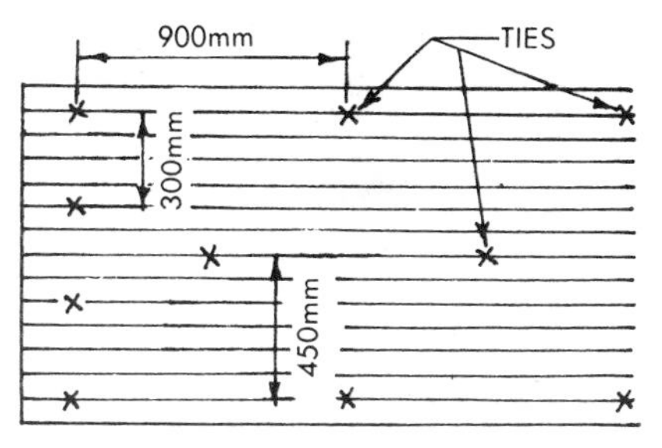

FIG 2 SPACING OF WALL TIES

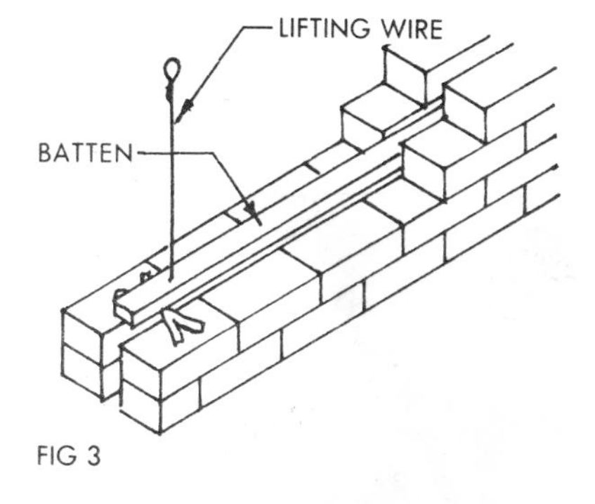

FIG 3

The purpose of cavity walls is to resist rain penetration and to ensure a dry inner leaf, in addition cavity walling provides good thermal insulation, and this may be further improved by using lightweight concrete blocks for the inner leaf of the wall. Concrete blocks are graded and only type 'A' blocks having a density of not less than 1500 kg/m^3 are suitable for general use below D.P.C. However Type 'B' blocks, of lower density may be used below ground level D.P.C. for internal walls and the inner leaf of cavity walls.

WALL TIES
The Building Regulations require that the cavity shall be not less than 50mm wide, or more than 100mm wide, and the leaves must be tied together with suitable ties. Spaced as shown in Table C3 of Approved Document A of the Building Regulations 1985 (See Fig 1). In addition at the sides of openings the vertical spacing should not exceed 300mm. Fig 2.

For loadbearing walls the galvanised steel tie is preferable. Where sound insulation is a consideration he butterfly tie is more effective.

Where walls are very exposed ties of non-ferrous metal or plastic are preferable.

Wall ties are designed so that water cannot pass across from the outer leaf to the inner leaf, and it is important that the cavity is kept clean, and no mortar droppings allowed to collect on the ties and so act as a bridge for moisture to cross. It is good practice when building to use cavity battens (Fig 3). These are laid on the wall ties and drawn up when the level of the next course of ties is reached, thus removing any droppings.

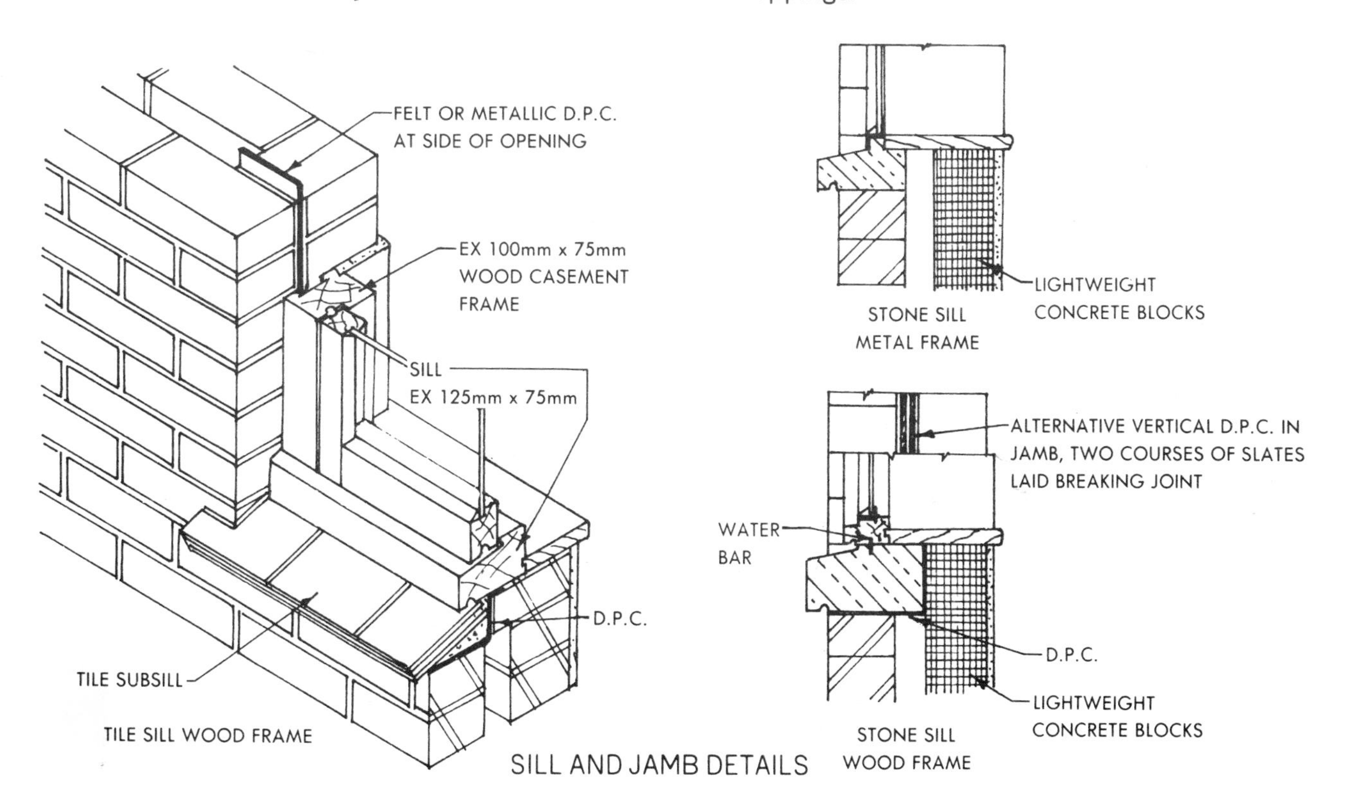

SILL AND JAMB DETAILS

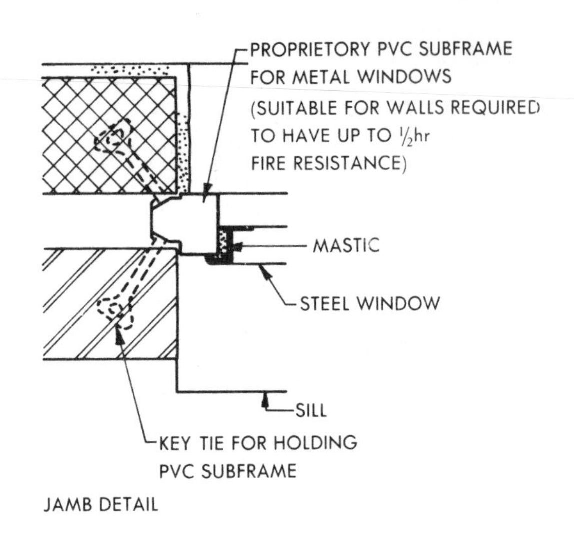

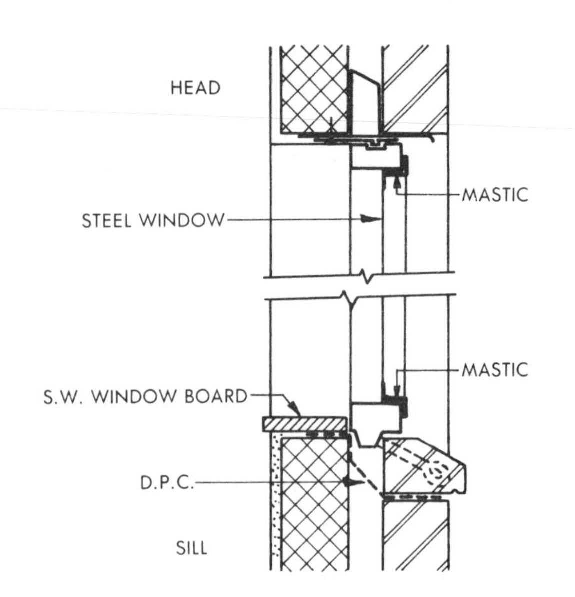

CAVITY WALLS 2

SILL AND HEAD DETAILS

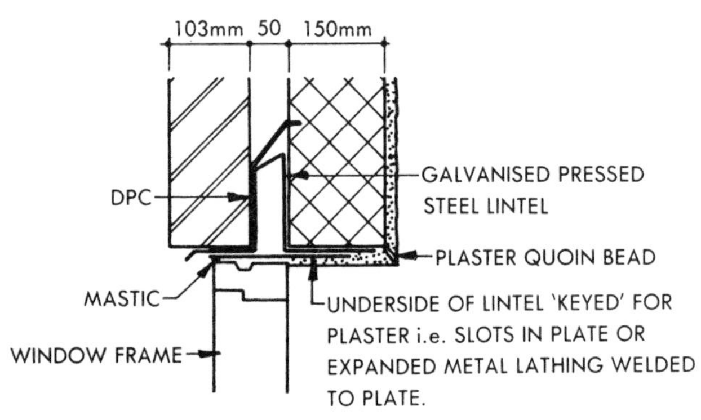

INNER LEAF THICKNESS CAN VARY; SUITABLE LINTELS AVAILABLE TO SUIT 'U' VALUE OF 0.6 USUALLY DECIDING FACTOR

TYPE A

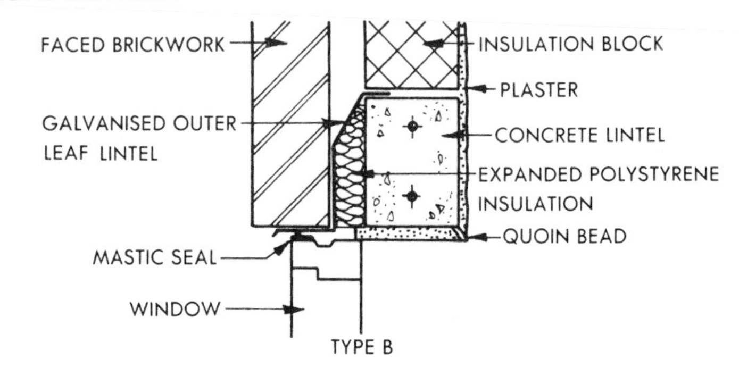

TYPE B

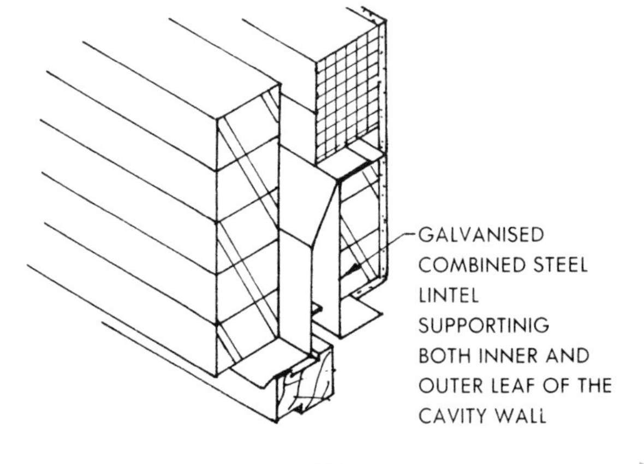

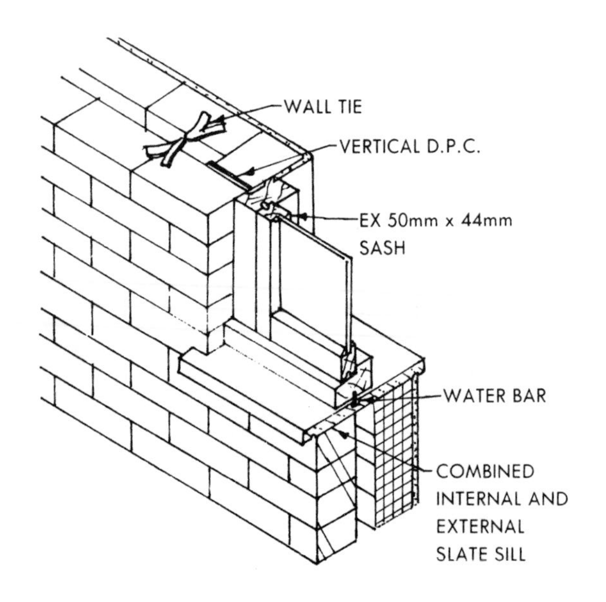

EAVES DETAILS

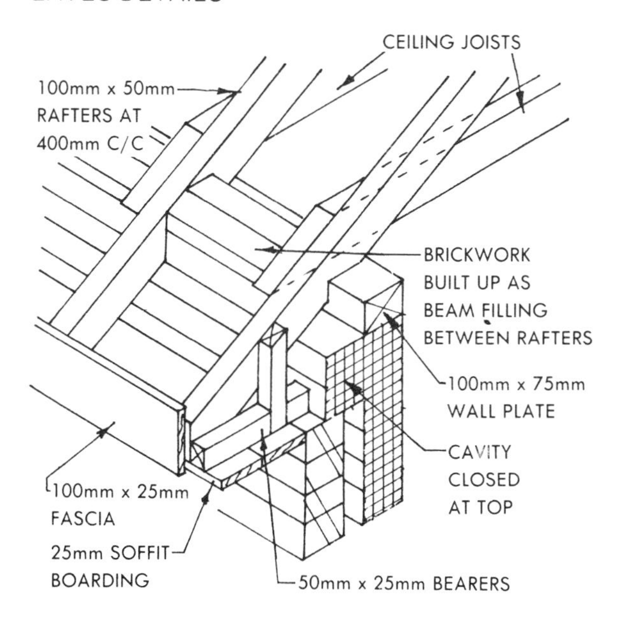

TRADITIONAL EAVES

PARAPET DETAIL

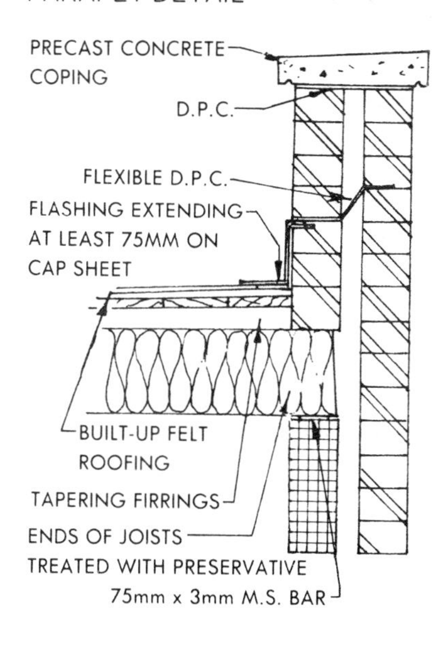

PITCHED ROOFS I

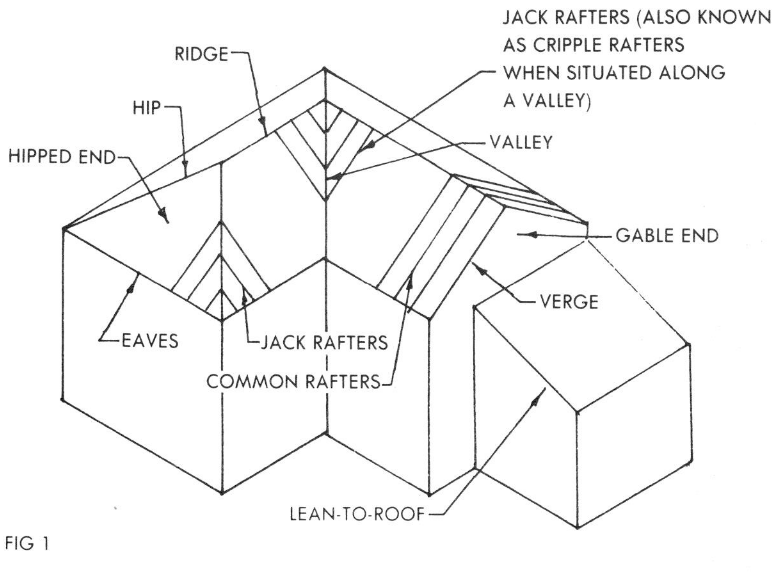

FIG 1

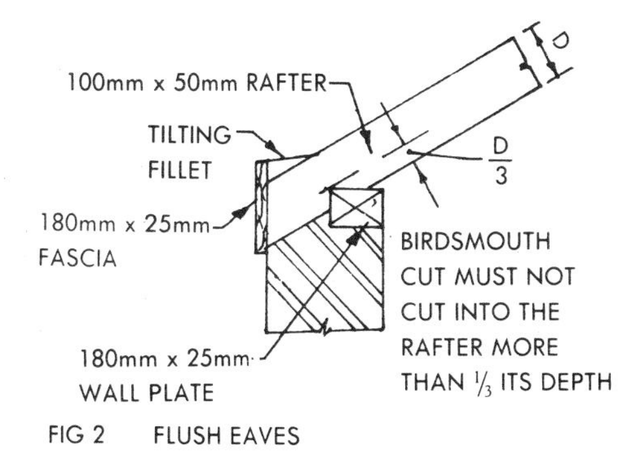

FIG 2 FLUSH EAVES

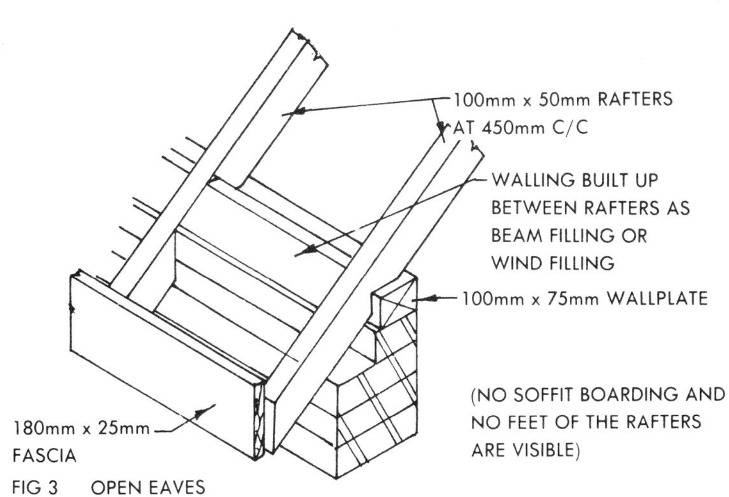

FIG 3 OPEN EAVES

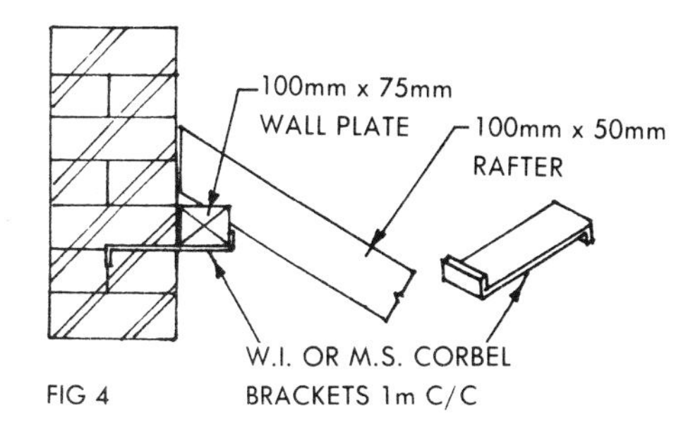

FIG 4

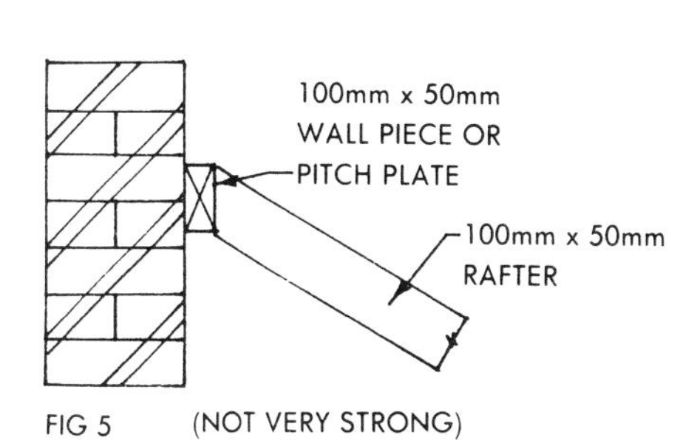

FIG 5 (NOT VERY STRONG)

The terms used in connection with pitched roofs are illustrated in the line diagram. Fig I.

LEAN-TO ROOF
As shown in Fig I is one of the simplest types of roof and suitable for spans up to about 2.5m. Sometimes known as a 'pent' roof.

The construction at the eaves may be as shown in Figs 2 or 3. At the top of the lean-to the heads of the rafters may be birdsmouthed over a wallplate Fig 4 or alternatively fixed to a wall piece or pitch plate plugged to the wall. Fig 5. The wall plate may be supported on brick corbels Fig 6, although this method is not much used today.

OPEN EAVES
(No soffit boarding and the feet of the rafters are visible).

SINGLE ROOFS
Roofs in which the rafters are supported at the head and foot. These include the lean-to roof, the double lean-to, 'pent' or 'vee' roof Fig 7; couple roof Fig 8; couple-close roof Fig 9; and collar roof Fig I0.

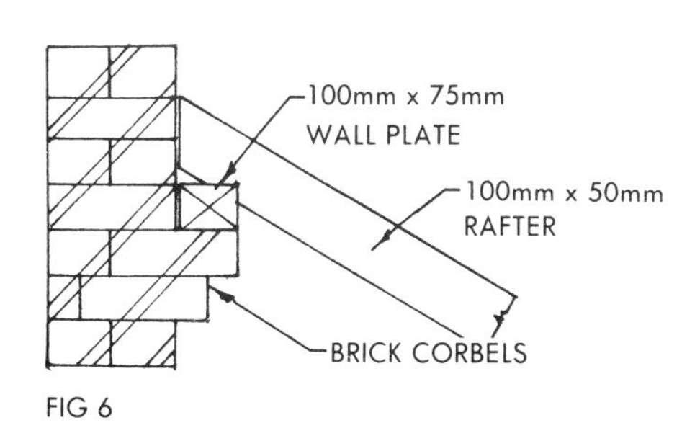

FIG 6

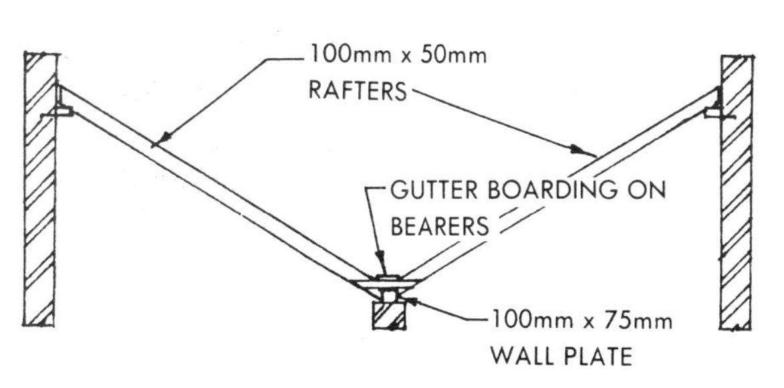

FIG 7 DOUBLE LEAN-TO ROOF

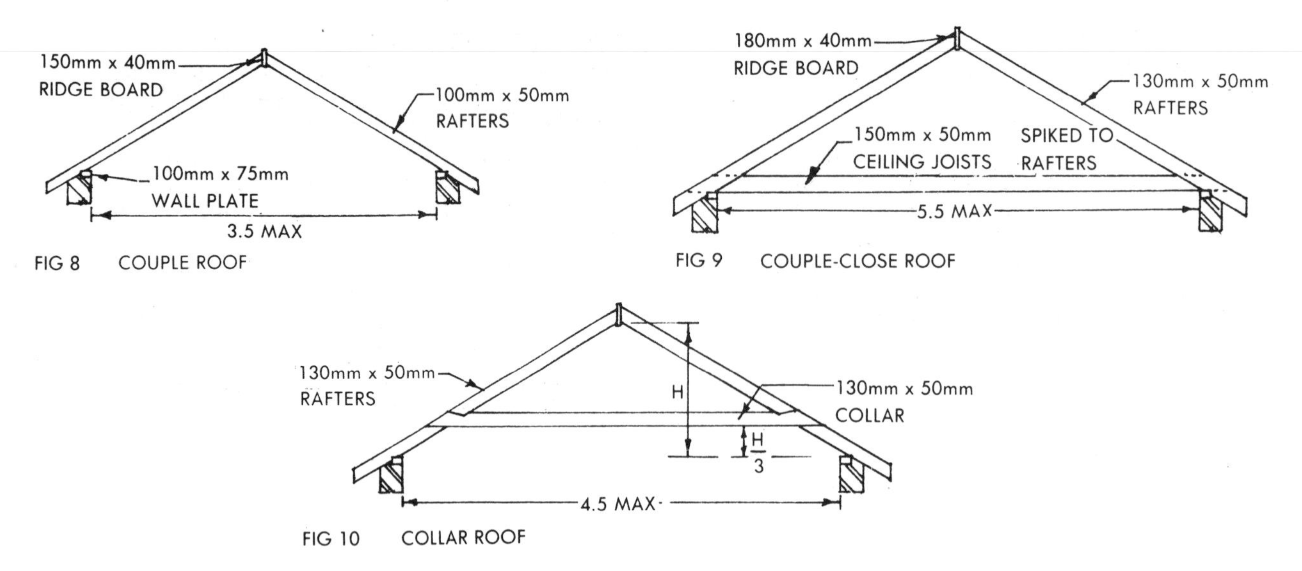

FIG 8 COUPLE ROOF

FIG 9 COUPLE-CLOSE ROOF

FIG 10 COLLAR ROOF

PITCHED ROOFS 2

SINGLE ROOFS

These were illustrated, but further clarification of one or two points is necessary.

Couple roof:- Is a weak roof and has a tendency to splay out at the feet of the rafters. Only suitable for comparatively short spans up to 3-5m, small buildings, sheds etc.

Couple-close roof:- The ceiling joists are nailed to the feet of the rafters and act as ties, preventing the feet from spreading out. May be used for spans up to 5.5m.

Collar roof:- An economical type of roof as the rooms take up part of the roof space, thus reducing the height of the walling. If the collar is to act as an effective tie it must not be placed too high up the roof, and it is usual practive for the underside of the collar not to be higher than ⅓ the height of the roof. The collar may be connected to the rafter by means of a dovetailed halved joint. Fig l, although quite often the collars are simply nailed to the rafters. Figs 2 & 3.

NAILING

It is important that care is taken in spacing and driving nails to ensure that the timber is not split. Fig 2 illustrates an example of a collar nailed to a rafter. If the round wire nails are driven directly into the timber then only two nails should be used, spaced as shown to avoid the possibility of splitting the timber. More nails may be used, thus increasing the strength of the joint, providing the timber is pre-bored. Fig 3. The minimum edge distances and spacing between nails are given.

The diameter of the pre-bored holes should be not greater than four-fifths of the diameter of the nail shank. the example given assumes l00mm nails (4in-No 7 gauge). The holes are only pre-

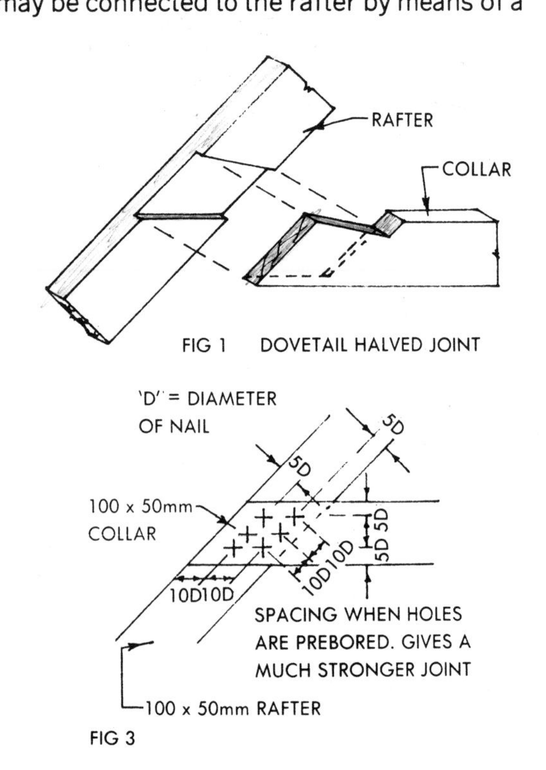

FIG 1 DOVETAIL HALVED JOINT

FIG 3

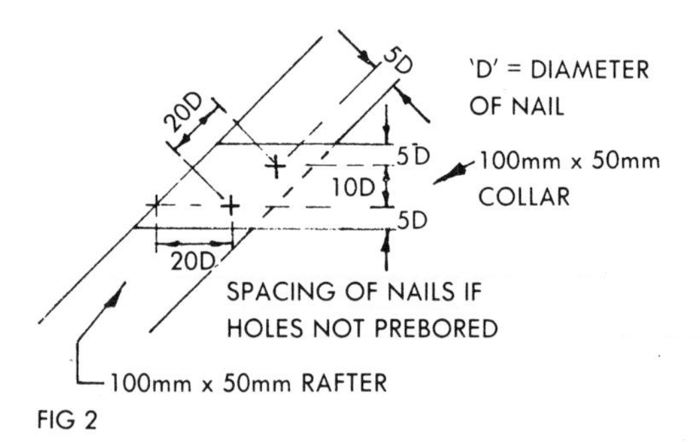

FIG 2

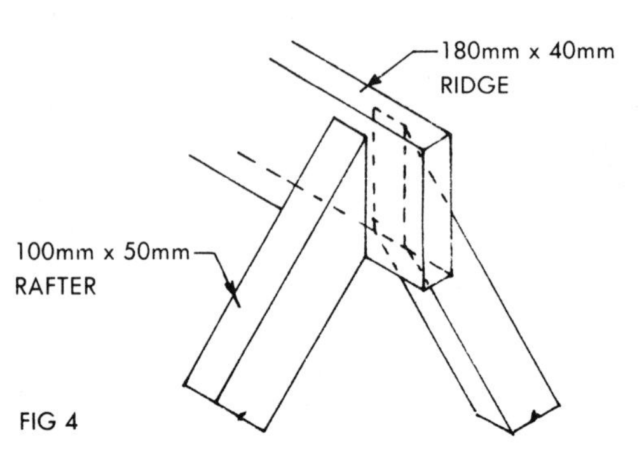

FIG 4

bored in the timber through which the nails pass, the nails being driven into the second without pre-boring as the joint is made.
Ridge Detail Fig 4:- At the ridge the rafters are simply splay cut and nailed to the ridge board. They should oppose each other in pairs as shown.
Rafters:- The size of rafters depends upon the span, the unsupported length, the weight of the roof covering, the spacing apart of the rafters c/c and the timber used. Sizes may vary from 38mm x 75mm (e.g. for a span of 2m, a dead load of 50 kg/m², a pitch between 35° and 40° and rafter spacing 450mm c/c) upwards. Common sizes are 100mm x 50mm. Tables B5 to B20 and B25 to B28 of Approved Document A of the Building Regulations 1985 lay down requirements for roof timbers and contain tables giving sizes for various spans and loadings.

DOUBLE OR PURLIN ROOF
When the roof span exceeds 4.5m single roofs are unsatisfactory as it is necessary to enlarge the timbers to an uneconomical size. Above this span the double or purlin roof may be used and Fig 5 shows an example of this type of roof. Purlins are horizontal timbers fixed under the rafters to provide intermediate support.

Purlins are in turn supported by struts which bear on to a load bearing partition. It is common practice to further stiffen and strengthen the roof by providing collars at every 4th pair of rafters. At party walls or gables the purlins may be supported by corbels, Fig 6.

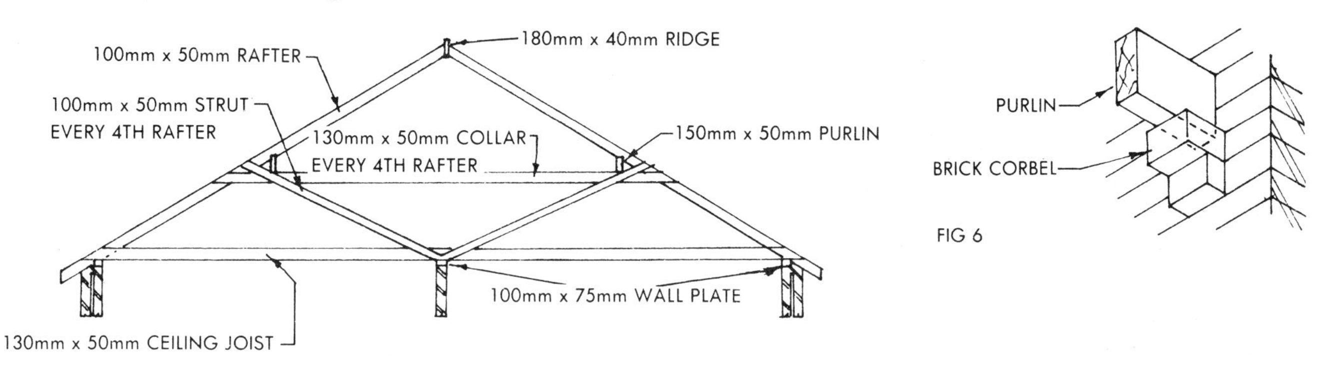

FIG 5 STRUTTED PURLIN ROOF

FIG 6

PITCHED ROOFS 3

PURLIN ROOF DETAILS
The purlin is supported by struts as shown Fig 1 provided at every fourth rafter. The size of the purlin depends upon the weight supported and the spacing of the supports. With struts at intervals of 1.8m a 130mm x 50mm purlin as shown will normally suffice.

Approved Document A of the Building Regulations 1985 contains tables giving sizes of purlins for various spans and loadings.

Over comparatively short spans the ceiling joists will be satisfactory.
In the case of wider spans hangers and binders are introduced (Fig 1) to stiffen and strengthen the joists and avoid having to use timbers of uneconomic size. The hangers are provided at every fourth rafter and connected to the purlin and to the horizontal binders or ceiling plates which are nailed to the ceiling joists.

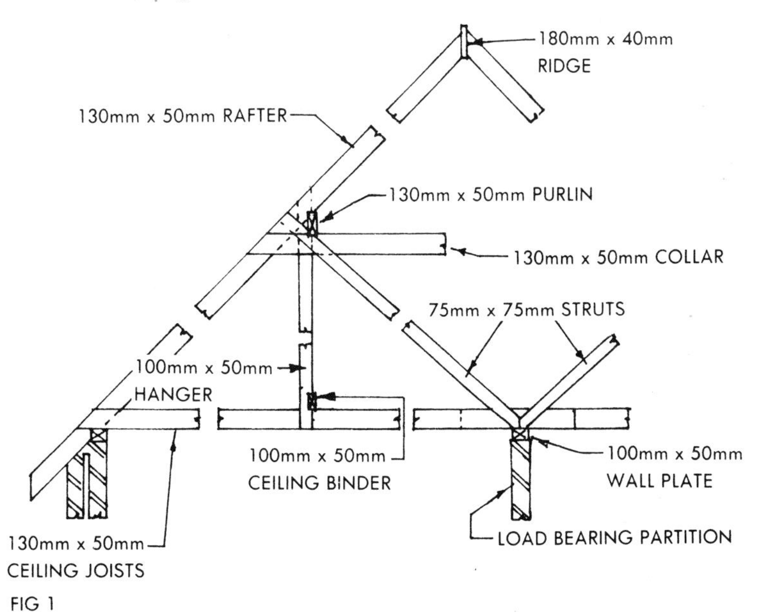

FIG 1

TIMBER TRUSSES

In general a strutted purlin roof may be used for spans up to 7.3m. For wider spans and where load bearing partitions are not available roof trusses may be adopted. The Timber Research and Development Association (TRADA) publish design sheets illustrating a wide range of timber trusses, and Fig 2 illustrates suitable trusses for a roof of 6.9m clear span, with a pitch of 30°. The system is designed to support a roof covering of tiles or slates weighing not more than 73kg/m² on slope plus a superimposed load, including snow of 73kg/m² measured on plan. A ceiling load of 48.8kg/m², including superimposed load has been allowed for. Additional loads e.g. water storage tanks have not been allowed for and these must either be situated over load bearing walls or additional strengthening provided.

N.B. Responsibility for structures made and erected to the design sheets rest with the users and in no way with TRADA. The designs should be used as a whole and in the conditions assumed, variations may entail serious consequences.

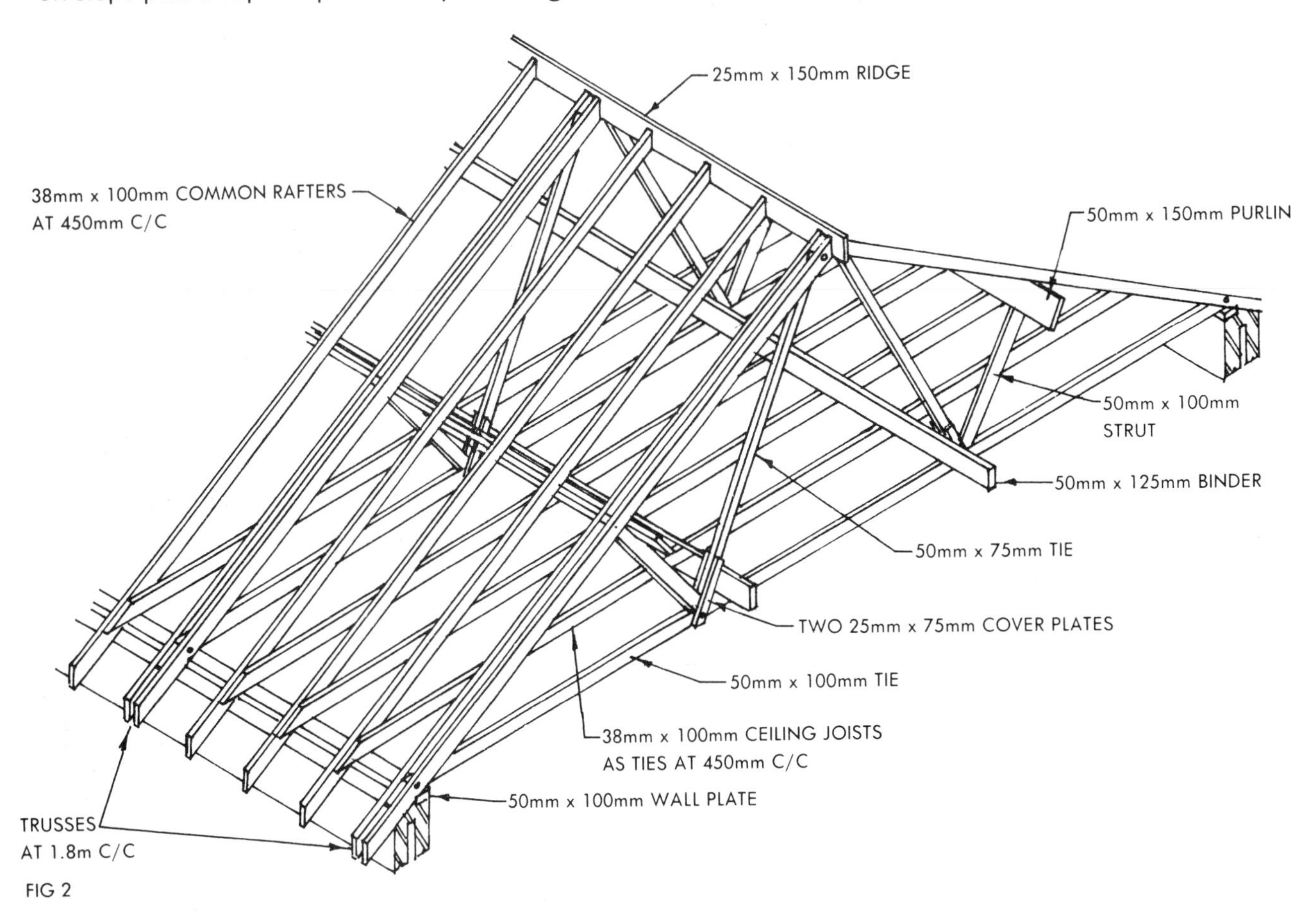

FIG 2

PITCHED ROOFS 4

TIMBER TRUSSES

Fig I shows a 'TRADA' timber truss of 30° pitch for a span of II.Im. The design is for a roof covering of slates or tiles weighing not more than 73kg/m² on slope plus a superimposed load, including snow of 73.2 kg/m² measured on plan. A ceiling load of 48.8 kg/m² including superimposed load has been allowed for. Additional loads, e.g. water storage tanks have not been allowed for. Trusses to be spaced at I.8m c/c with common rafters at 450mm c/c. Timber to be S2-50 grade to CPII2:I97I. The trusses can be made up in sections for transporting and then assembled on site.

Joints are made with bolts plus toothed plate connectors either round or square. Spacing of bolts must be strictly adhered to especially end distances and complete penetration of connectors must be obtained. Bolt holes must have sufficient clearance for the insertion of the bolts but in no case should the hole be greater than I.5mm over the nominal diameter of the bolt.

The strength of a bolted joint is considerably increased by the use of timber connectors. There are a number of types of connector and they are protected against corrosion by being sherardised or galvanised. The bolt holes must be the correct size and connectors holed for bolts of a larger diameter than those in use must not be used. Connectors should not be inseted in wet timber and when tightening the nut care must be taken not to crush the wood.

Toothed plates:- These are round (or square) metal plates with projecting teeth around the edge as shown. Fig 2.

Split rings (Fig 3):-Consist of a steel ring cut at one point to form a split as shown. The split allows the ring to adapt itself to any movement of the timber and thus maintain a tight connection.

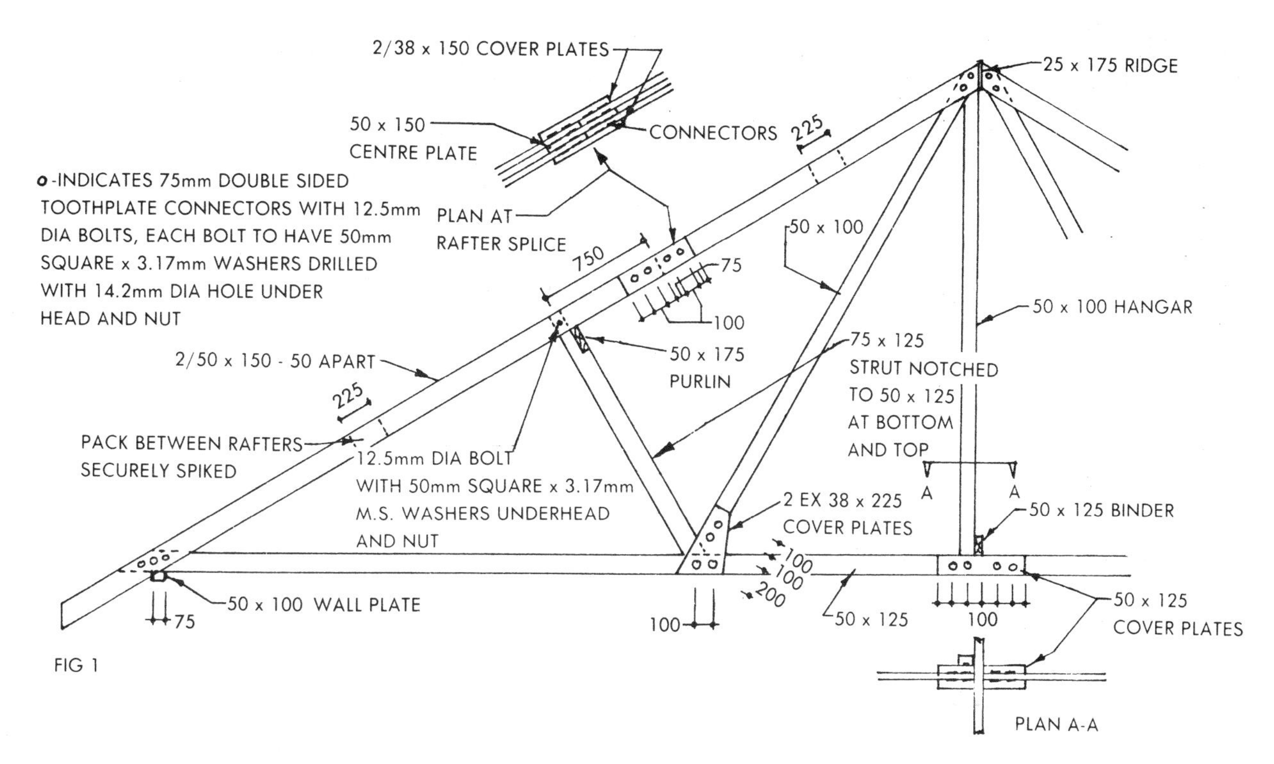

FIG 1

The connectors are fitted into pre-cut circular grooves, and a special tool is available from the manufacturers for cutting the grooves and drilling the bold holes. Split ring connectors have a higher load carrying capacity than toothed plats.

Truss plates:- These are of two types Fig 4, one having projecting teeth and the other having predrilled holes. The plates are made of 16 to 20 gauge galvanised steel plate. The former type should be fixed in place by machine. The latter type fixed by 31.8mm 9 gauge sheradised square-twisted or annular-threaded nails.

The plates must be fitted at the front and back of the joint. Fig. 5

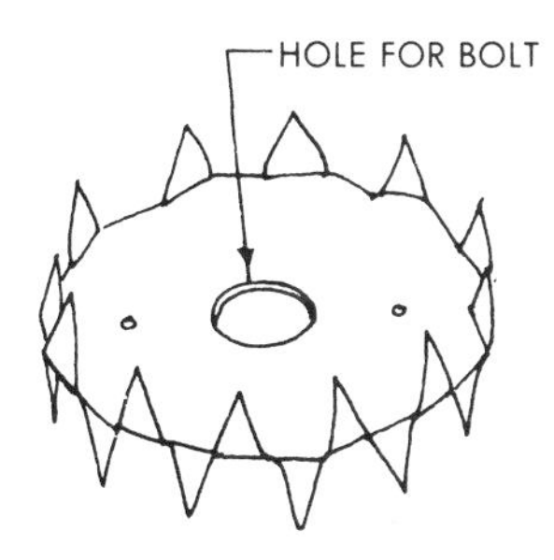

FIG 2 DOUBLE SIDED TOOTHED PLATE CONNECTORS

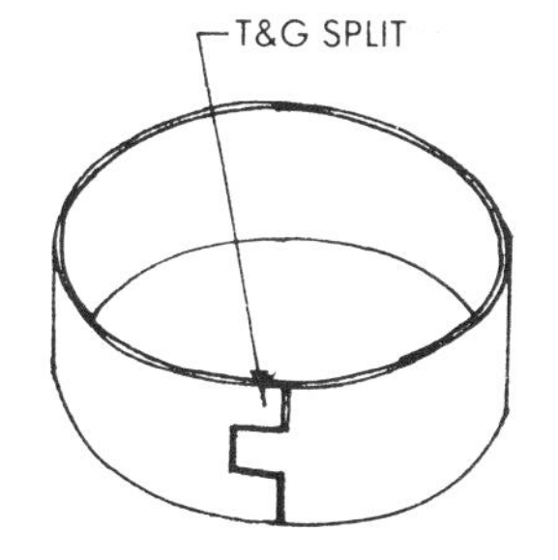

FIG 3 SPLIT RING CONNECTOR

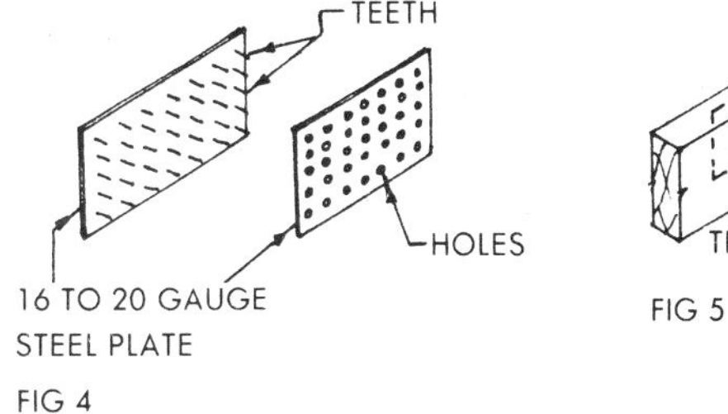

FIG 4

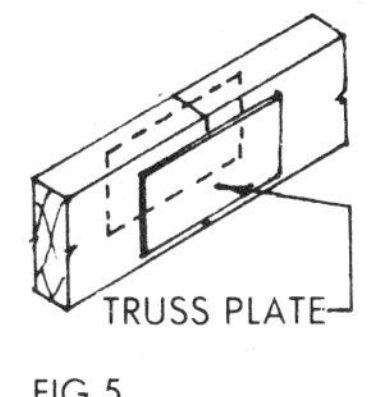

FIG 5

PITCHED ROOFS 5

TRUSSED RAFTERS

'TRADA' trussed rafters have been designed for pitches of 10° (monopitch), 15°, 20° and 25°; and for spans of 5.1m, 6.0m, 6.9m and 8.1m. Fig 1 shows a truss suitable for 6.0m span. The design was intended for a roof covering of a lightweight nature e.g. woodwool slabs and felt weighing not more than 40.8 kg/m^2 on slope, plus a superimposed loading, including snow of 73.2 kg/m^2, measured on plan. A ceiling load of 46.8 kg/m^2 including superimposed load has also been allowed for. The roof covering to be supported directly on the trusses space at 600mm c/c, i.e. no purlins or common rafters. The 20° and 25° pitches are also suitable for tiles or slates weighing not more than 50.3 kg/m^2 on slope. For higher loading the designs may be used at closer spacings; the spacing being reduced by 1mm for each increase of 0.4 kg/m^2, subject to a minimum spacing of 450mm.

The design is based on the use of S2-50 grade timber to C:P 112: 1971.

50mm diameter double sided timber connectors are used at the main connections marked 'O', together with 12.5mm diameter bolts, each bolt to have 50mm square x 3.17mm thick M.S. washers under head and nut.

Nails to be 57mm x 10 gauge driven into pre-bored holes, minimum distance from nail to end of member 31.75mm, to edge of member 15.8mm, spacing between rows of nails 12.5mm, minimum spacing along grain between adjacent nails 31.75mm.

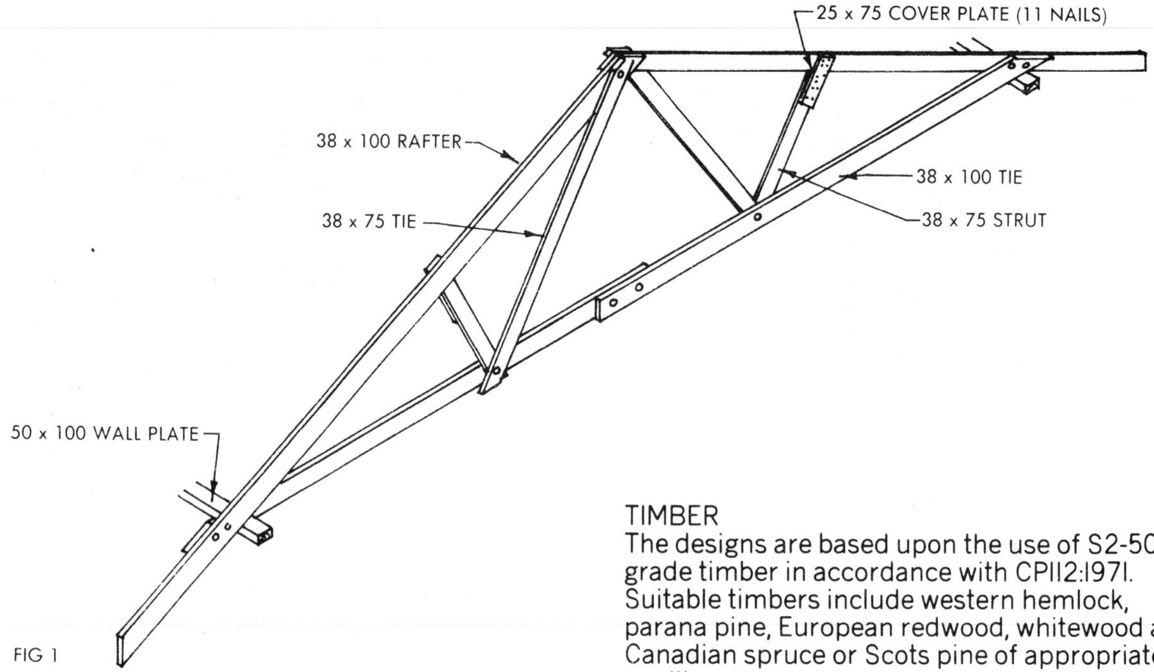

FIG 1

MONOPITCH ROOF
A 'TRADA' trussed rafter suitable for a monopitch roof of 6.0m span and having a pitch of 10° is shown in Fig 2. The system is designed to support a roof covering such as woodwool slabs plus superimposed loads

TIMBER
The designs are based upon the use of S2-50 grade timber in accordance with CP112:1971. Suitable timbers include western hemlock, parana pine, European redwood, whitewood and Canadian spruce or Scots pine of appropriate quality.

N.B. Responsibility for the structures made and erected to the designs rests with the users and in no way with TRADA. The designs should be used as a whole and in conditions assumed, any unauthorised variations may entail serious consequences.

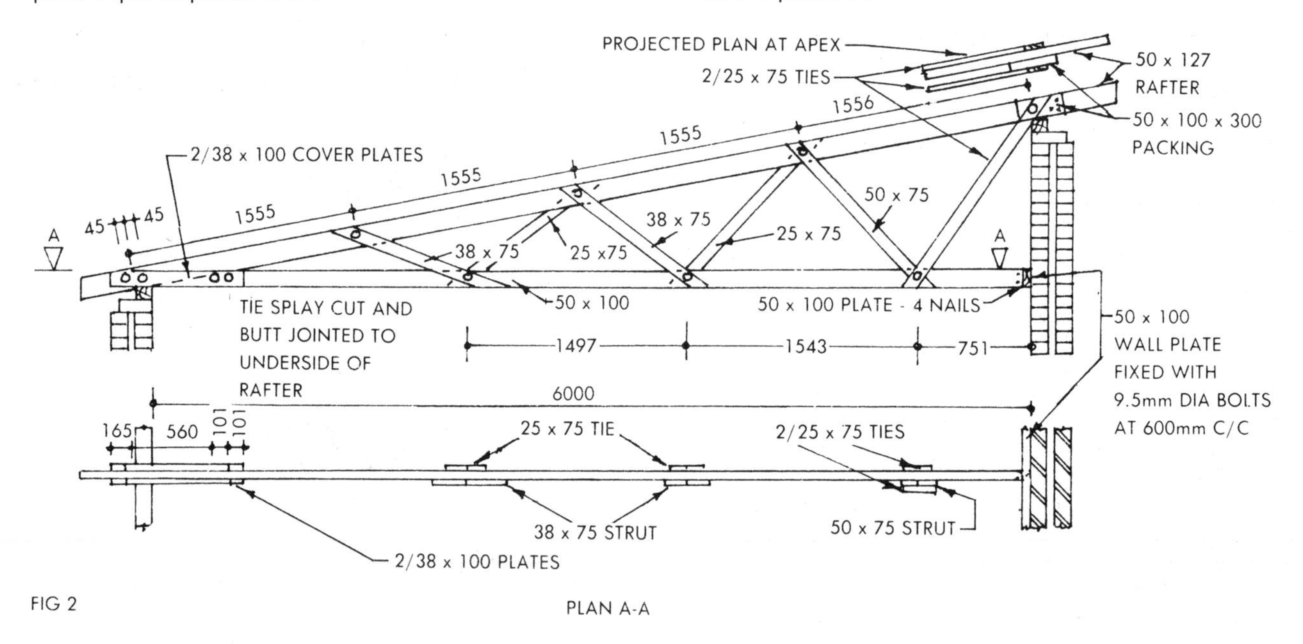

FIG 2

PLAN A-A

PITCHED ROOFS 6

SPROCKETED EAVES
A method of finishing the roof by forming a curve at the eaves. This may be achieved by nailing sprockets to the sides of the rafters Fig l, or by spiking sprockets to the backs of the rafters. Fig 2.

Care must be taken to ensure that the sprockets do not result in too shallow a pitch, as this makes it difficult to sweep the tiles around the sprocket, may lead to water penetrating the roof and cause the tiles to deteriorate fairly rapidly.

HIPPED ROOFS
Where a pitched roof has a hipped end the jack rafters are cut and spiked to the hip rafter Fig 3. As there is a considerable thrust at the foot of the hip, the roof should be strengthened by the provision of an angle tie across the corner. Fig 3 shows one type of tie, which may be dovetail housed to the wall plates or bolted in position.

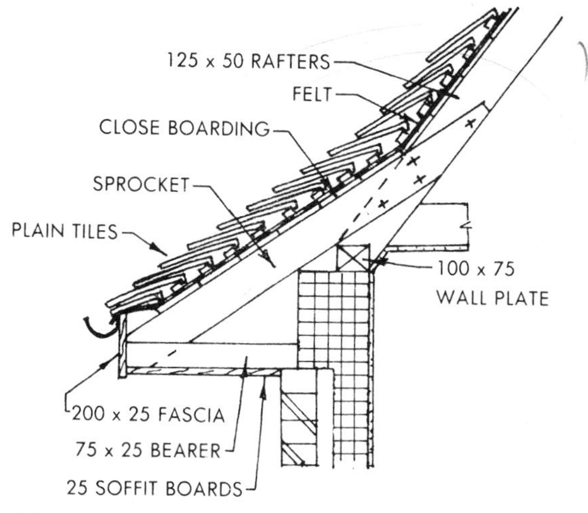

FIG 1

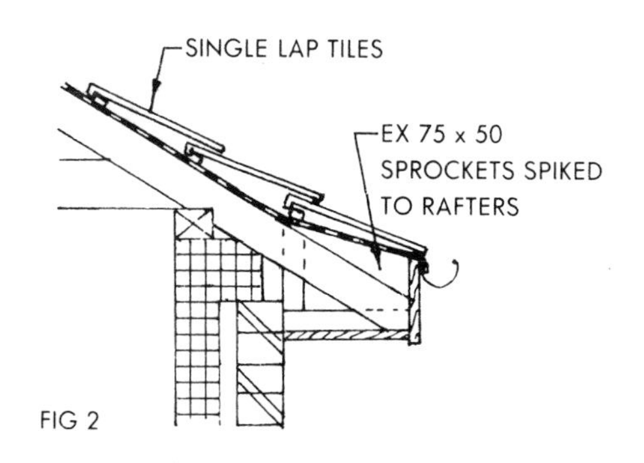

FIG 2

VALLEYS

The pitch of a valley is considerably less than the pitch of the corresponding rafters, and since water from the roof surfaces on either side drains into the valley, it follows that a valley is particularly vulnerable and care must be exercised in the construction. A valley may be finished in a number of ways:-

1. An open valley formed with lead, copper or zinc. Fig 4.
2. A secret valley Fig 5.
3. A mitred valley where slates are close mitred and laid in conjunction with lead soakers
4. A tiled valley, using valley tiles
5. A laced valley
6. A swept valley

Open valley:- The lead is dressed over the boards and tilting fillets as shown, and close copper nailed along the edge. Tile-and-half tiles are used adjacent to the valley gutter, Fig 6, in order to avoid the use of small pieces where the tiles have to be splay cut, thus providing extra security. (No tile should be cut so that it is reduced in width in any part to less than half the width of a tile).

The clear width between the edges of the tiles or slates should be sufficient to allow foot room and 130mm is the minimum distance (Fig 4), although many authorities would consider at least 200mm desirable.

The slates or tiles should overhang the tilting fillets by at least 30mm. If the roof is only felted and battened, and not boarded, it will be necessary to fix boards on either side of the valley intersection for a width of 225mm to receive the lead, copper or zinc.

Secret valley:- The edges of the slates or tiles are close together Fig 5. This type of valley is liable to become blocked, is difficult to clear and is not recommended.

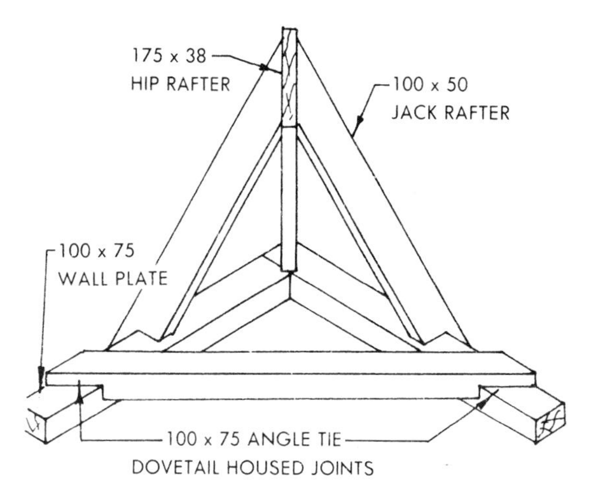

FIG 3 INTERIOR VIEW OF HIP

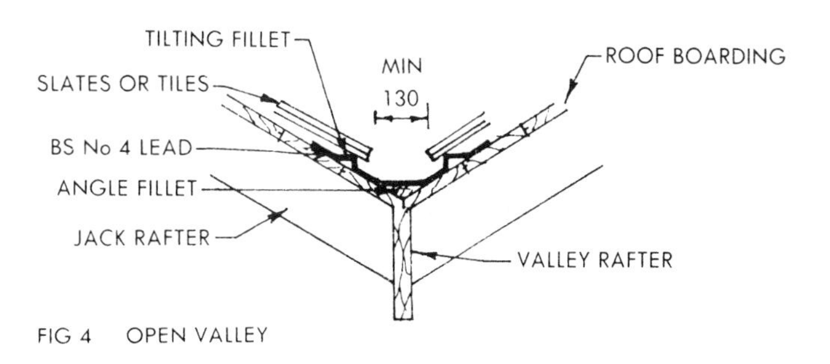

FIG 4 OPEN VALLEY

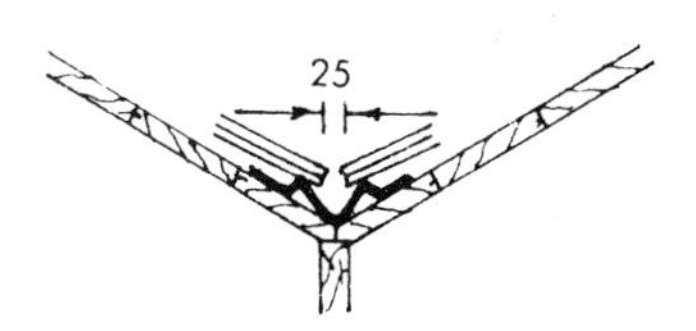

FIG 5 SECRET VALLEY

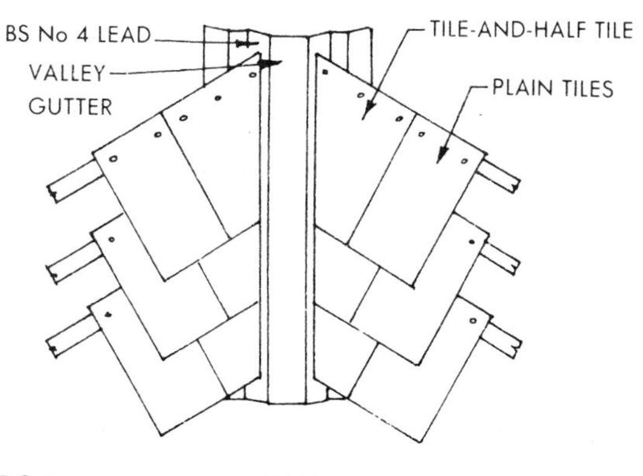

FIG 6 PLAN

PITCHED ROOFS 7

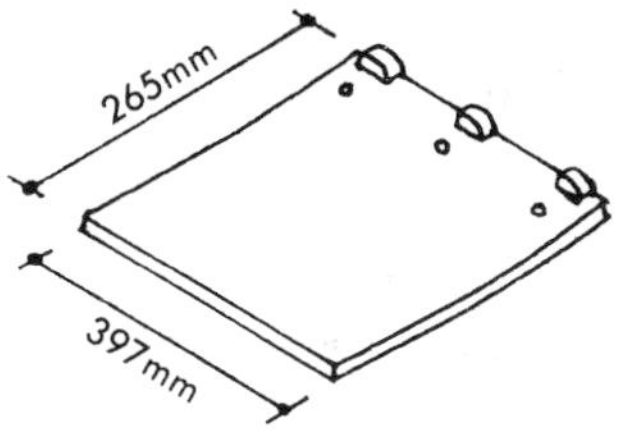

265mm
397mm
TILE AND A HALF TILE

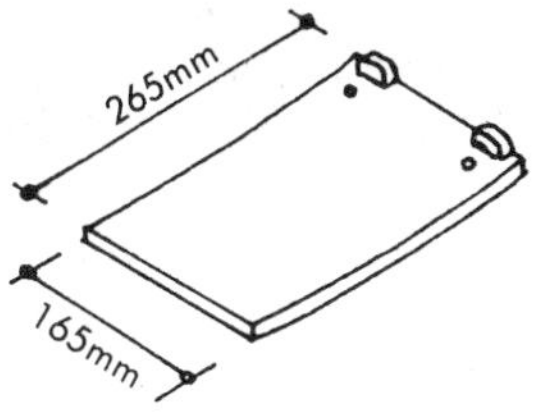

265mm
165mm
PLAIN TILE

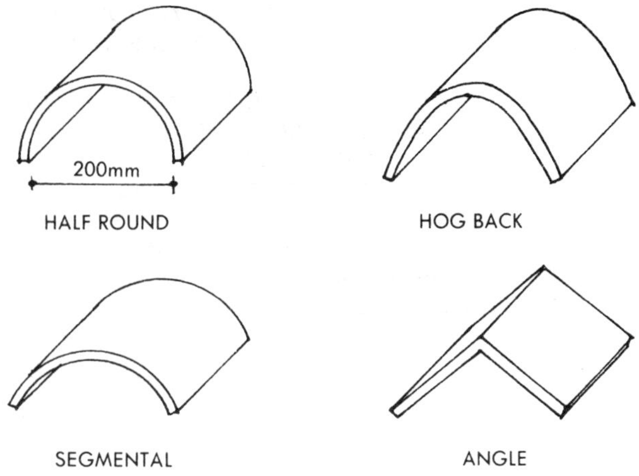

200mm
HALF ROUND

HOG BACK

SEGMENTAL

ANGLE

RIDGE TILES

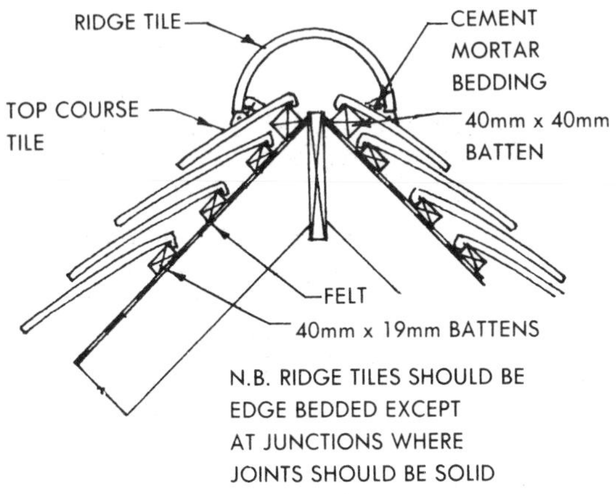

RIDGE TILE
CEMENT MORTAR BEDDING
TOP COURSE TILE
40mm x 40mm BATTEN
FELT
40mm x 19mm BATTENS

N.B. RIDGE TILES SHOULD BE EDGE BEDDED EXCEPT AT JUNCTIONS WHERE JOINTS SHOULD BE SOLID

SECTION THRO' RIDGE

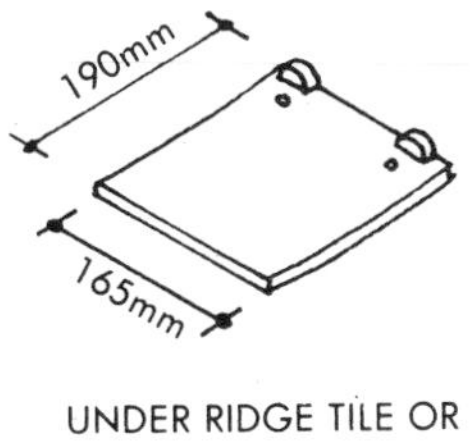

190mm
165mm
UNDER RIDGE TILE OR TOP COURSE TILE

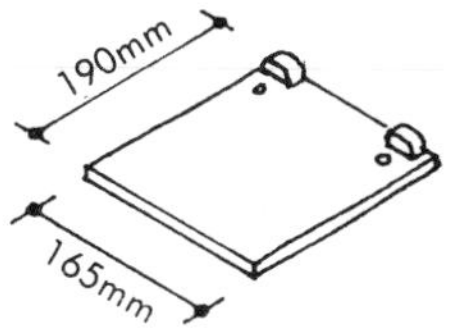

190mm
165mm
UNDER-EAVES TILE

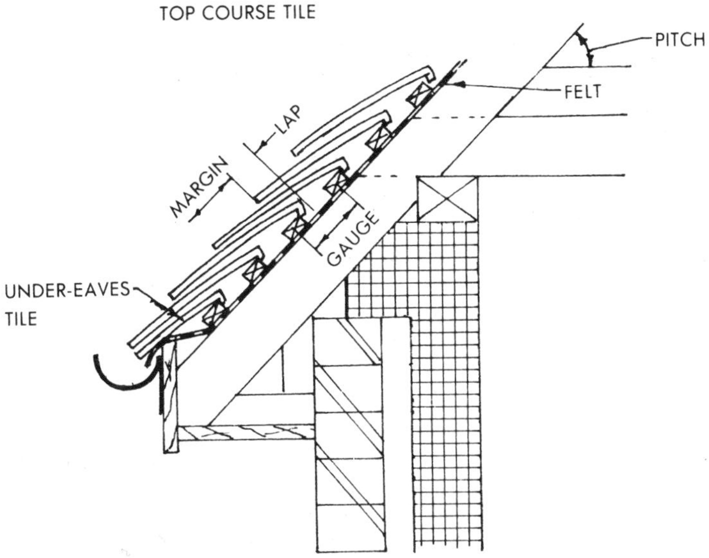

PITCH
FELT
LAP
MARGIN
GAUGE
UNDER-EAVES TILE

SECTION THRO' EAVES

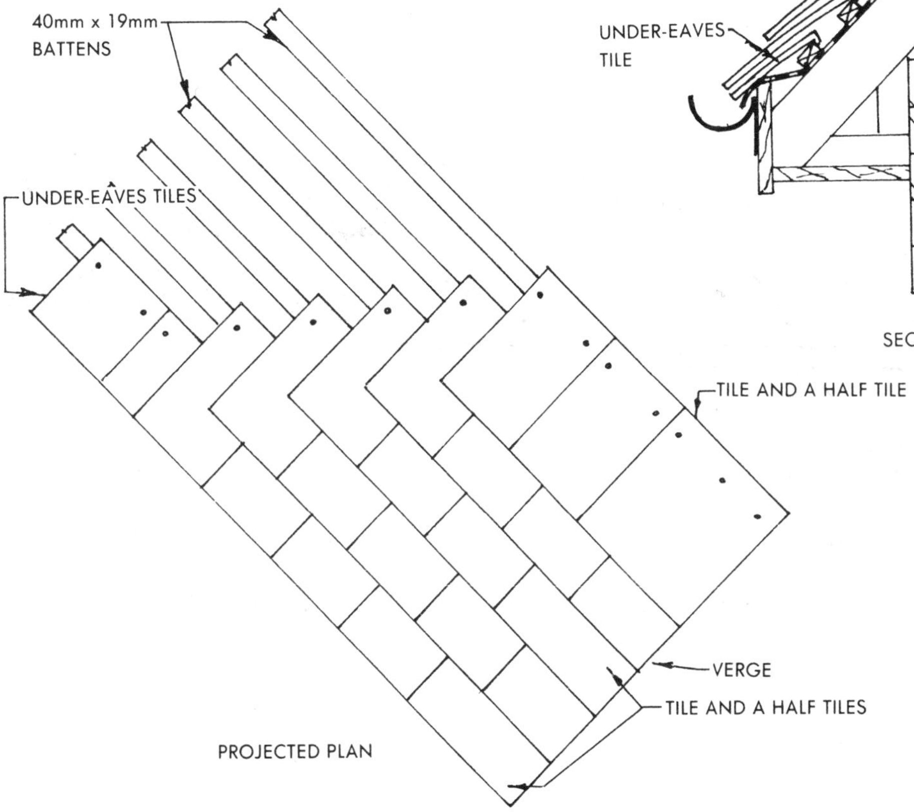

40mm x 19mm BATTENS
UNDER-EAVES TILES
TILE AND A HALF TILE
VERGE
TILE AND A HALF TILES

PROJECTED PLAN

PLAIN TILES

Pitch:- For plain tiles the rafter pitch should not be less than 40°, but the degree of exposure and the durability of the tiles should be considered when deciding on the pitch.

Lap:- Is the amount by which the tails of the tiles in one course overlap the heads of the tiles in the next course but one below. For plain tiles the lap should not be less than 63mm. In exposed positions the lap should be increased to 75mm or more. Under-eaves tiles are required at the eaves as shown in order to maintain the lap.

Gauge:- Is the distance from centre to centre of the battens and is obtained from the formula:-

$$\text{Gauge} = \frac{\text{Length} - \text{lap}}{2}$$

Thus for plain tiles laid to a 65mm lap —

$$\text{Gauge} = \frac{265-65}{2} = 100\text{mm}$$

Margin:- Is the exposed area of each tile on the roof and the length of the margin is the same as the gauge.

Verge:- The finish at a gable end. Tile and a half tiles should be used as shown to maintain the bond. Half tiles should not be used for this purpose.

PITCHED ROOFS 8

VERGES

At the verge the tiles overhang the gable wall 50mm to 75mm, the wider projection being usual for higher buildings. An undercloak as shown Figs 1 & 2 is bedded on the wall and the tile battens carried over this undercloak so as to impart a tilt to the end tiles and ensure that no water from the roof surface runs down the gable.

HIPS

Ridge tiles may be used at the hips Fig 3, but in this case a galvanised or wrought iron hip iron must be fixed at the foot of the hip to give support as shown. When bonnet hip tiles are used, Fig 4, a hip iron is not required. Bonnet hips must be bedded at the tail as shown, the thickness of the bed depending upon the pattern of bonnet hip used. Angular hip tiles Fig 5 are usually close fitting and do not need bedding at the tail, but they should be bedded at the head.

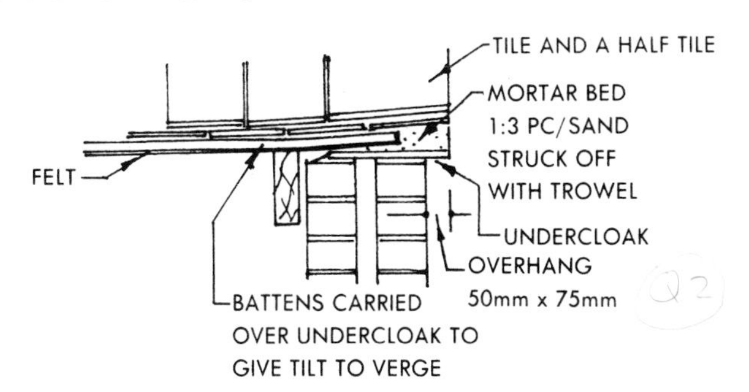

FIG 1 SECTION THRO' VERGE

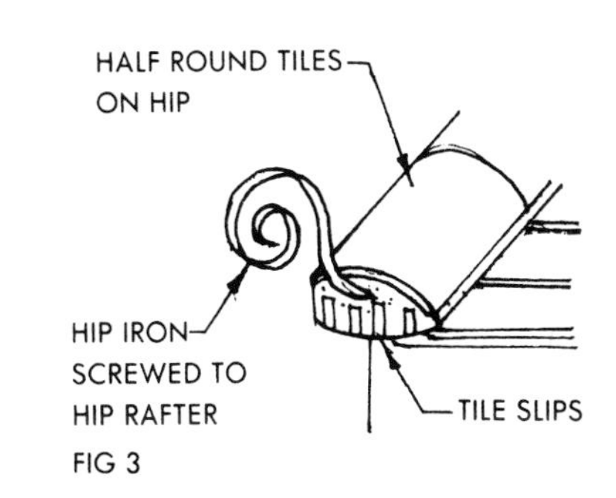

FIG 3

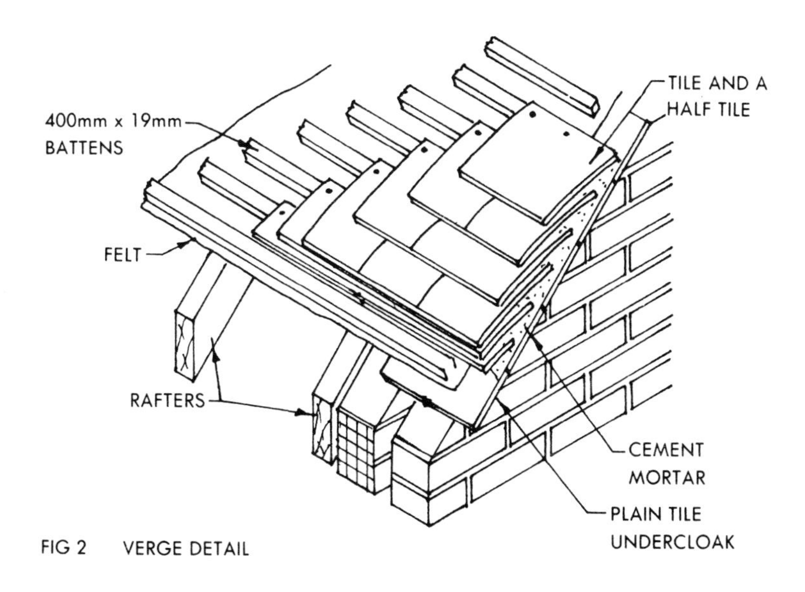

FIG 2 VERGE DETAIL

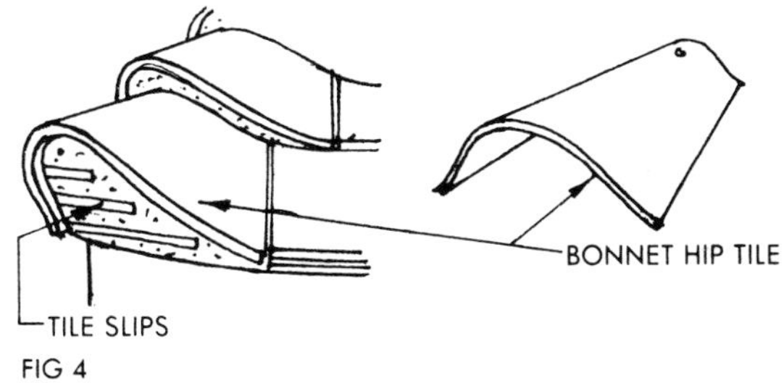

FIG 4

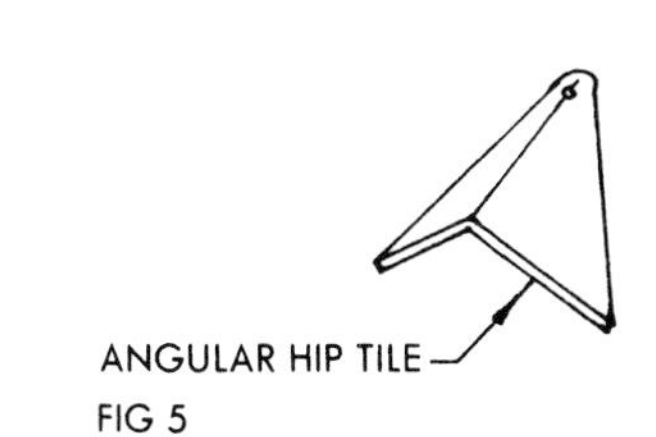

FIG 5

VALLEYS
Alternative methods of dealing with valleys are illustrated below.
Valley Tiles Fig 6:- A very sound type of valley giving a satisfactory appearance. Valley tiles may be angular or rounded form and it is important that they are of the right pitch for the particular roof. Purpose-made valley tiles do not need nailing or bedding.
Swept valley Fig 7:- A valley board must be provided as shown, and a strip of bituminous felt at least 610mm wide should be laid over the valley board and turned into the eaves gutter.
Laced valley Fig 8:- A valley board at least 230mm wide should be provided. No cutting is required, but a high degree of skill is required in forming the valley. N.B. Laced and swept valleys are expensive and need to be constructed by an experienced tiler.

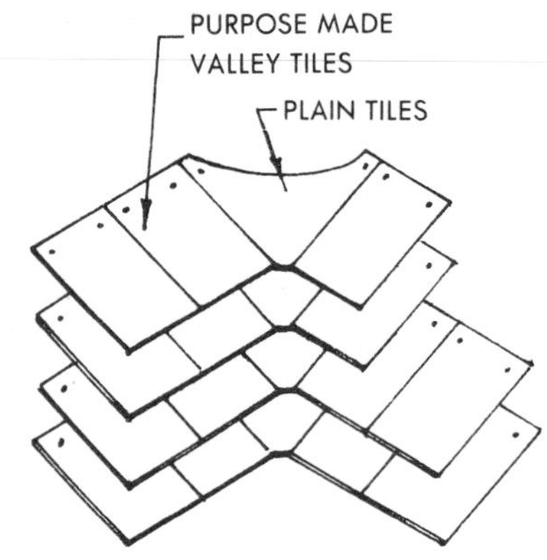

FIG 6 PURPOSE MADE VALLEY

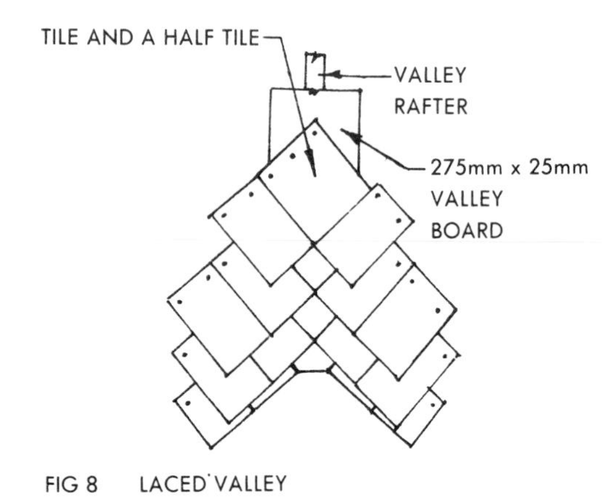

FIG 8 LACED VALLEY

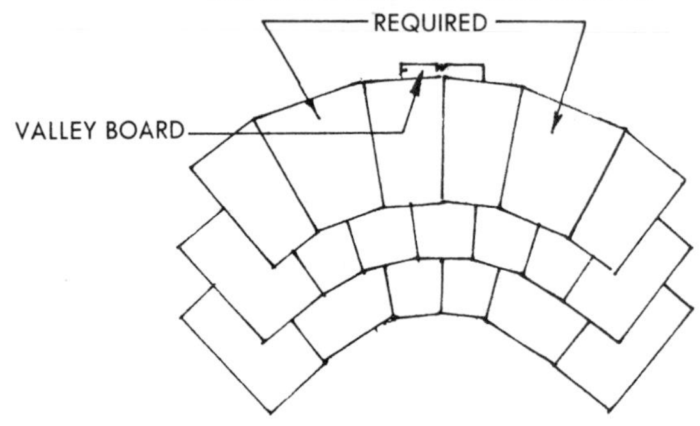

FIG 7 SWEPT VALLEY

PITCHED ROOFS 9

NAILING
Plain tiles should be nailed with two 38mm long nails to each tile, at least every fifth course for pitches up to 40°. For pitches 50° every third course should be nailed with two nails to each tile, for pitches to 55° to 60° every tile should be nailed with one nail to each tile and on pitches steeper than 60° every tile should be nailed with two nails.

Galvanised iron nails have been used for fixing tiles, but their use is not recommended in BS 5534. Nails may be of aluminium, cut copper, copper wire, zinc, or composition nails. Generally 38mm long nails are satisfactory, but for hand made tiles which may be 16mm thick or more, 44mm long nails should be used.

FELT
Sarking felt should be provided under all tiled or slated roofs. The felt improves insulation and ensures that driving rain, snow or leaks from cracked, broken or slipped tiles or slates do not penetrate the roof. Untearable felts on a woven base are available and may be laid direct on the rafters. Thermal insulation may be further improved by the use of aluminium foil faced felts. When felting is laid direct on rafters, end and side laps should be at least 150mm. If laid on boarding side laps of 75mm will suffice. It is good practice to lay a 900mm wide strip of felt over the roof felting at hips and beneath the roof felting at valleys.

ABUTMENTS
Where abutments occur at chimney stacks or parapet walls, soakers and cover flashings should be provided.

The section Fig 2 gives deals of the apron flashing and rear gutter to a stack. Fig 1 illustrates the use of soakers and stepped flashing. The upper horizontal edges of the steps being turned, about 25mm into the bed joints, secured by lead wedges and pointed.

The length of the soakers should not be less than the sum of the gauge and lap, and they are often made 25mm or so longer so as to enable them to be turned over the head of the tile. The width must be sufficient to allow not less than 100mm under the tiles and an upstand of at least 75mm at the abutment.

Tile fillets may be used as a covering to soakers instead of a cover flashing, but in this case the tiles must be bedded in cement mortar into a chase cut into the wall so that the upper edge of the tile fillet is completely covered. The mortar bedding must not be allowed to adhere to the roof surface. Tile fillets should not be simply butted up to the wall, and cement fillets should

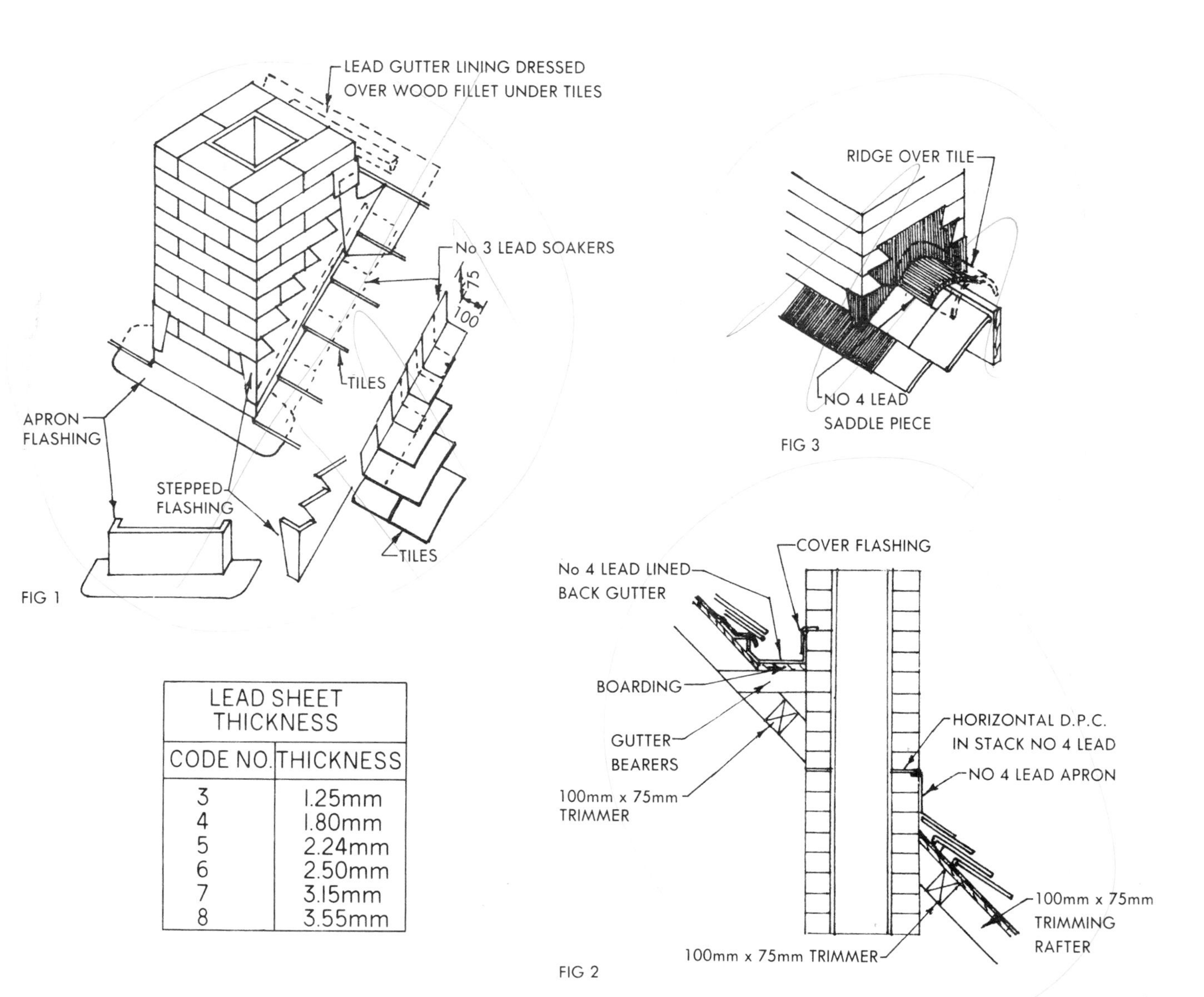

FIG 1

FIG 3

FIG 2

LEAD SHEET THICKNESS	
CODE NO.	THICKNESS
3	1.25mm
4	1.80mm
5	2.24mm
6	2.50mm
7	3.15mm
8	3.55mm

not be used.

The code numbers used for easy identification of the new BS range of thicknesses of lead sheet and strip correspond to the previous weights in lbs per square foot.

Sheet and strip may carry colour markings for easy identification of thickness, as follows:- Code No 3 green, No 4 blue, No 5 red, No 6 black, No 7 white, No 8 orange. An intermediate thickness is indicated by using yellow with one of the colours to show a size thicker. E.g. blue and yellow indicates a size between Code No 4 and Code No 5.

PITCHED ROOFS 10

COUNTER BATTENS

Where the interior of a roof is to be utilised the expense of boarding and counter battening may be justified. This method provides a clean, neat, well insulated roof. If a roof is boarded, felted and battened there is a danger that any water penetrating the tiling (e.g. through a broken tile) would be trapped behind the battens and encourage rot to develop. Counter battens Fig 1 lift the tile battens clear of the felt and allow any water to drain clear.

Counter battens should not be less then 40mm x 6mm.

SINGLE-LAP TILING

There is a wide range of single-lap tiles, common pantiles Fig 2, Italian tiles Fig 3, Spanish tiles Fig 4, single and double Roman tiles Fig 5, interlocking tiles Fig 6 etc.

Single lap tiles are laid with both headlap and sidelap, the dimensions of which are usually fixed by the design of the tile, but the headlap should not be less than 75mm and in conditions of severe exposure may be increased to 100mm or more.

The rafter pitch should not normally be less than 35° for clay tiles or 30° for concrete tiles, but some patterns and qualities of tiles may be less, e.g. the Marley 'Wessex' interlocking tile may be laid on roofs with a rafter pitch as low as 15° concrete tiles complying with BS 473, or 550 are not susceptible to frost.

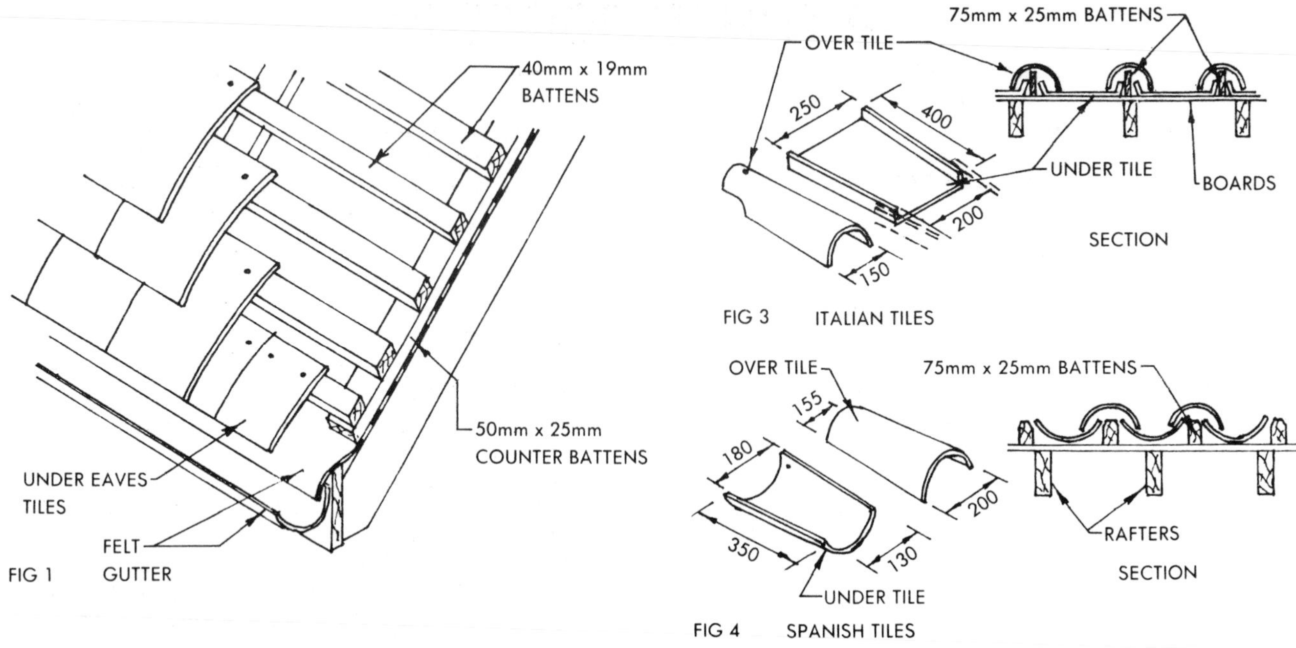

FIG 1 GUTTER

FIG 3 ITALIAN TILES

FIG 4 SPANISH TILES

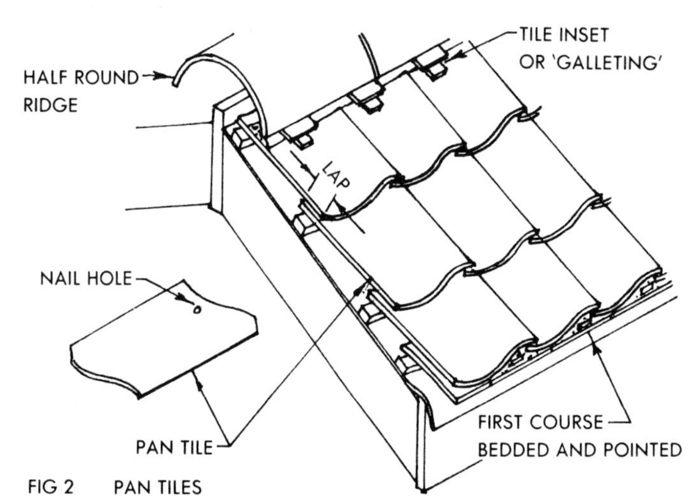

FIG 2 PAN TILES

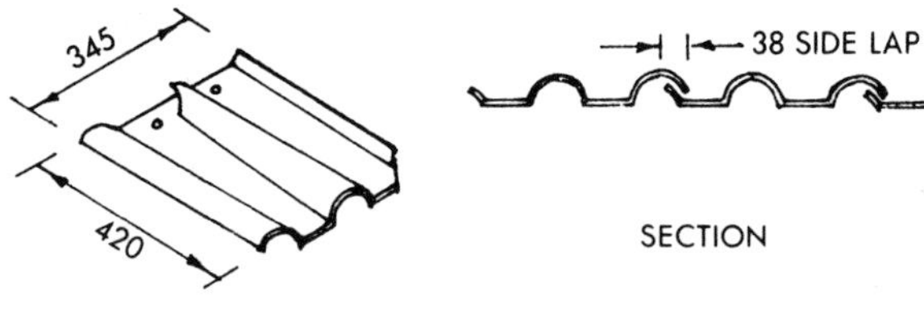

FIG 5 DOUBLE ROMAN TILES

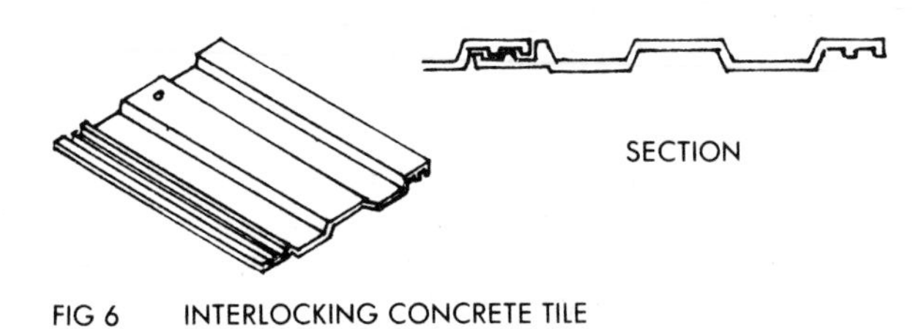

FIG 6 INTERLOCKING CONCRETE TILE

MONORIDGE

For mono-pitched roofs a special ridge tile, the Marley monoridge is available Fig 7. These can be used with virtually any tiling and on roofs of any pitch from 15° to 40°. Each monoridge is tied to the structure by means of a galvanised wire tie, cast in as shown.

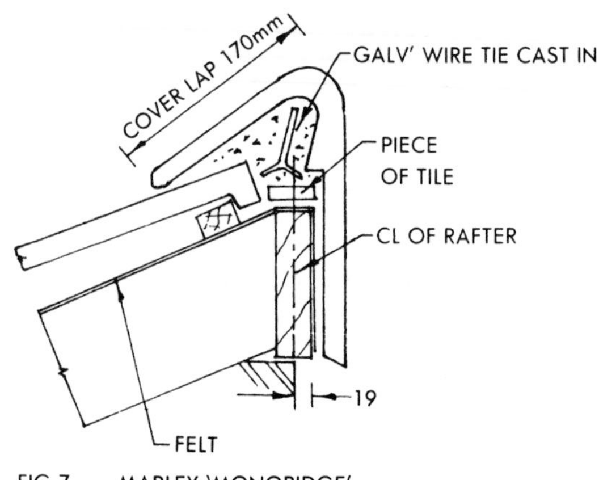

FIG 7 MARLEY 'MONORIDGE'

PITCHED ROOFS II

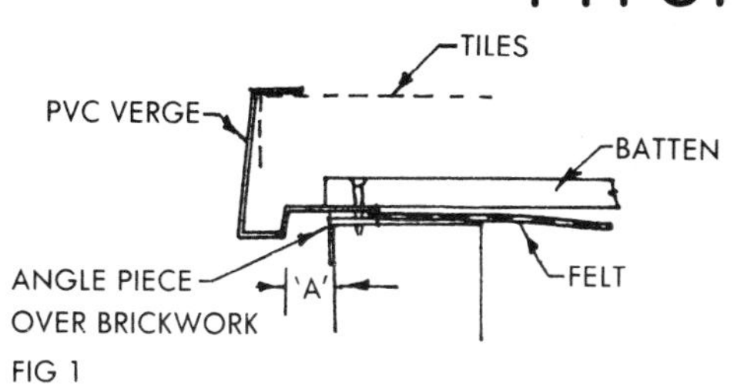

FIG 1

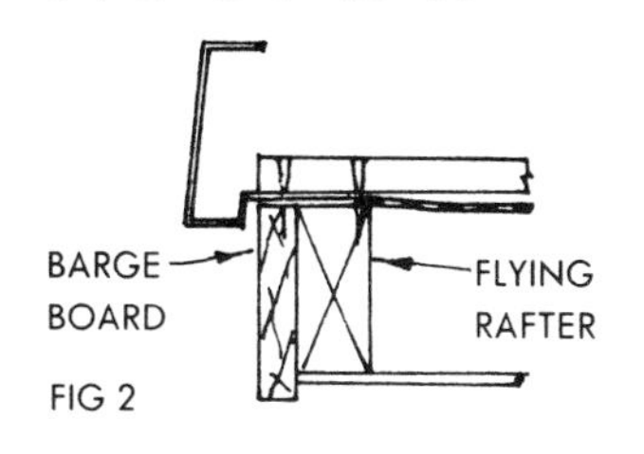

FIG 2

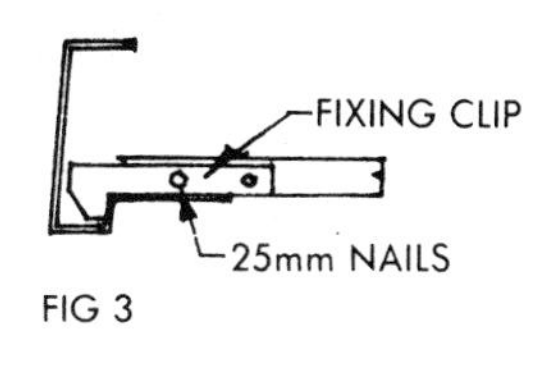

FIG 3

DRY VERGE SYSTEMS

These consist of extruded PVC sections which are laid directly onto bargeboard or brickwork of the gable, and eliminate the use of undercloaks and bedding mortar.

Fig I shows the Marley dry verge fixed to a brick gable, the overhang at 'A' being adjustable. Fig 2 shows the verge fixed over a barge board. Fixing clips Fig 3 are used at the top and bottom of each length. Fig 4 shows the Redland dry verge.

SKYLIGHT

A skylight may be provided to light a landing and staircase or an attic. Typical construction is shown in the elevation Fig 5 and the sections Figs 6 & 7. The bottom rail is thinner than the top rail and stiles and is grooved as shown to allow condensation to drain away. The light may be fixed or hinged.

DORMER WINDOW

Fig 8 illustrates typical construction for a dormer window. The opening in the roof is formed by framing roof timbers in a similar manner to openings in a timber floor.
Firring pieces are provided over the roof joists as shown to support the roof boarding at the required slope.

In the example shown lead sheet is used for covering the roof and cladding the cheeks of the dormer. The thickness of lead recommended is BS Code No 4 to BS 1178, but Code No 5 or Code No 6 may be preferred.

Weights of Lead Sheet		
BS Code No	Kg/m^2	lb/ft^2
4	20.41	4.19
5	20.54	5.21
6	28.36	5.82

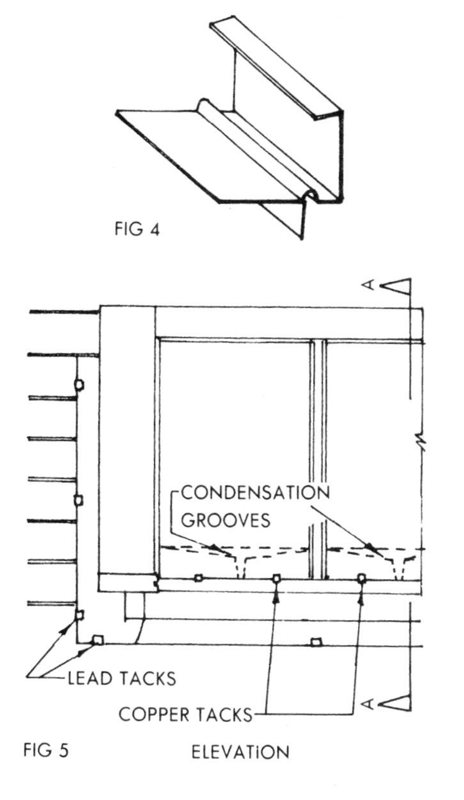

FIG 4

FIG 5 ELEVATION

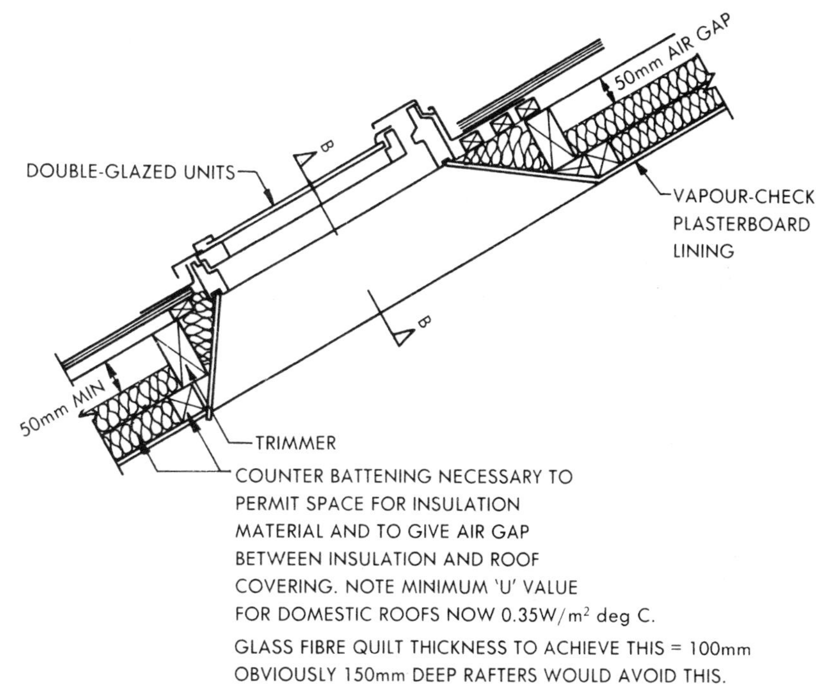

FIG 6 TYPICAL SECTION THROUGH ROOFLIGHT. VARIOUS FLASHING UNITS AVAILABLE TO SUIT DIFFERENT ROOFING MATERIALS

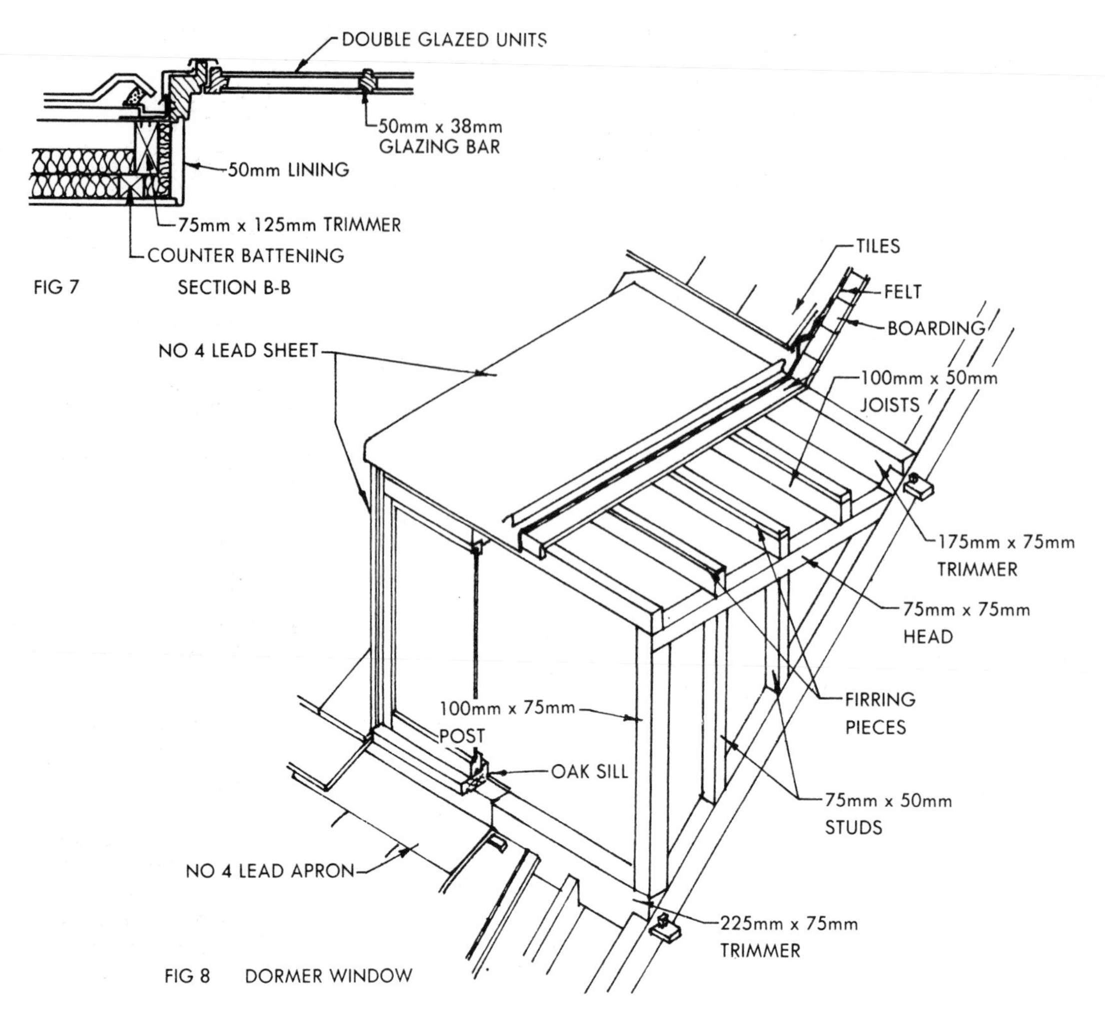

FIG 7 SECTION B-B

FIG 8 DORMER WINDOW

PITCHED ROOFS 12

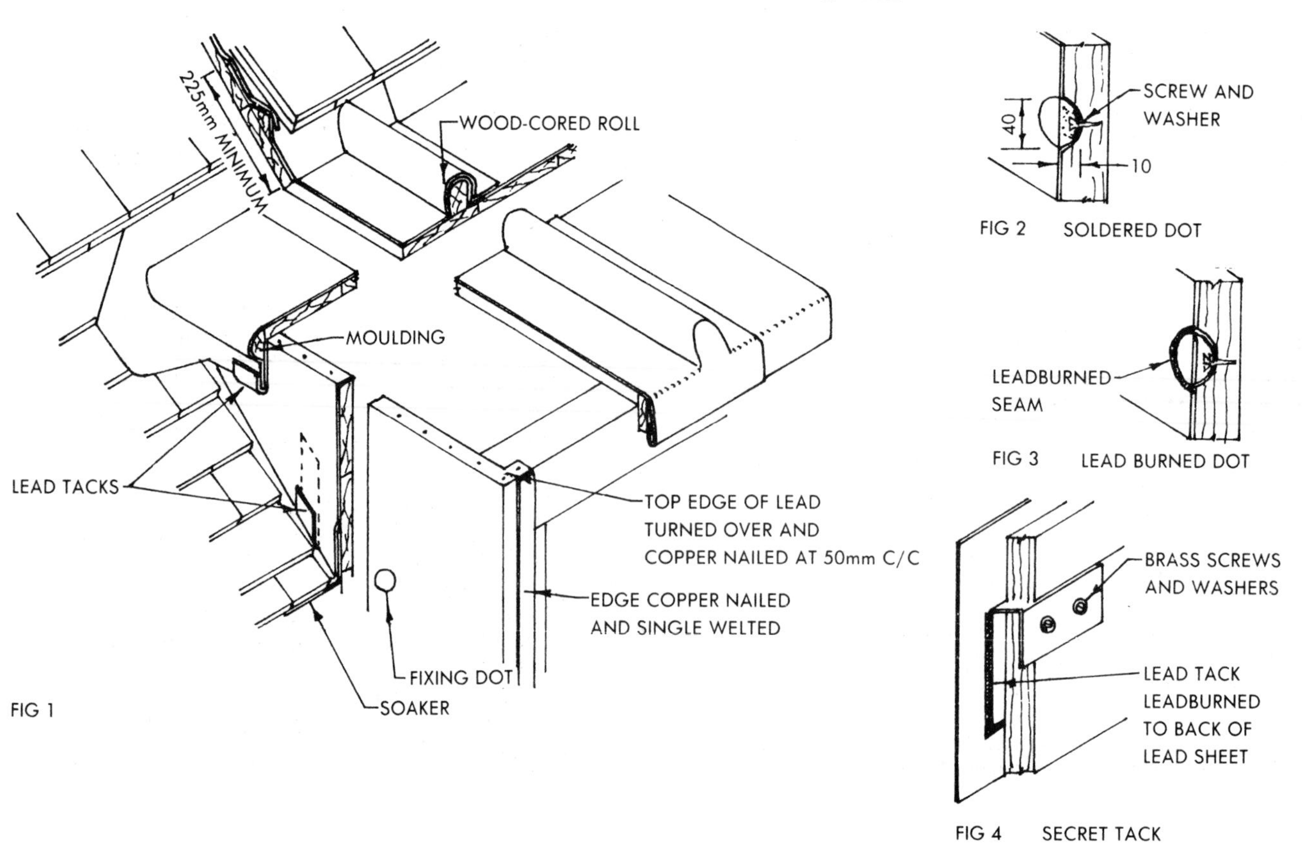

FIG 1

FIG 2 SOLDERED DOT

FIG 3 LEAD BURNED DOT

FIG 4 SECRET TACK

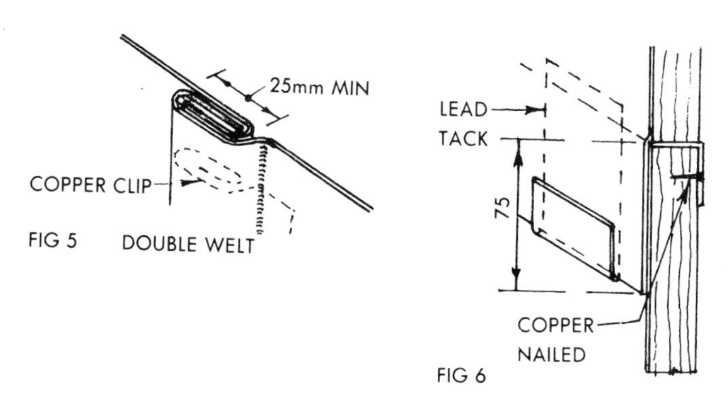

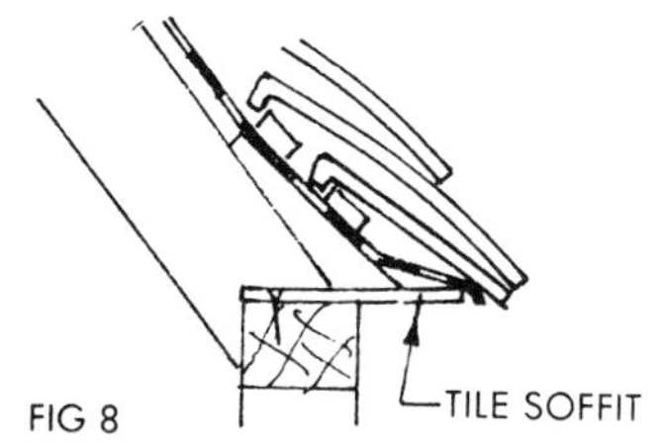

DORMER WINDOW

The minimum recommended thickness of lead flashings and coverings is BS Code No 4 lead, but No 5 lead or No 6 lead may be specified for roofs and cheeks of dormers.

The cheeks may be cladded with a single sheet of lead, but it is recommended that the maximum size of sheet should not exceed 1.2m² in area or 600mm in width in order to allow for thermal movement.

The lead cheeks are turned over at the top Fig 1 and secured by copper nails at 50mm c/c. The front edge is turned round the corner post as shown, close copper nailed and then the edge turned over the nail heads in a single welt. On larger dormers more than one sheet of lead will be required for each cheek and in this case the vertical joints may be made with a double welt, Fig 5, incorporating copper clips at intervals and the horizontal joints made with a 75mm lap, the top edge of the lower sheet being turned through the boarding and secured by copper nails at the back. Fig 6.

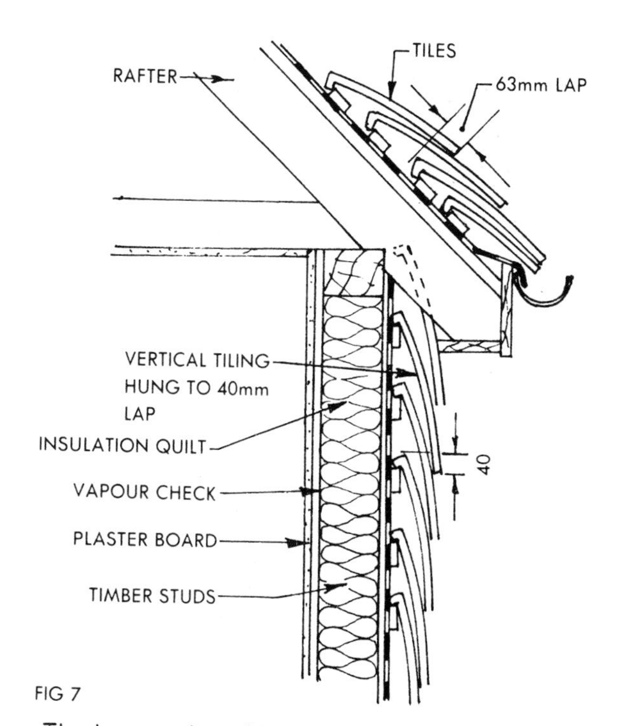

FIG 7

The lower edge of the lead covering to the cheek serves as a cover flashing to the abutment flashing, and should be retained with lead tacks, as shown Fig 1 at 750mm intervals.

Intermediate fixings to support lead cheeks should be dispersed at 500mm intervals where BS Code No 4 lead is used and at 600mm intervals for No 5 lead. These fixings may be a round head brass screw and washer, a soldered dot Fig 2, a lead burned dot Fig 3, or by means of a secret tack. Fig 4.

The lead covering to the dormer roof is constructed as for ordinary lead flats.

TILED DORMER

A detail of the eaves of a dormer with a tiled roof and cheeks is shown in Fig 7. The dormer may have a gutter as shown or alternatively the gutter may be dispenced with and a tiled soffit used as shown in the eaves detail Fig 8.

PITCHED ROOFS 13

SLATING

Slates should comply with the requirements of BS 680 'Roofing Slates'. Slates can normally be obtained from 200mm to 660mm long, and from 150mm to 350mm wide. It is possible to obtain slates up to 910mm (36in) long.

Slates used to be named according to size e.g. 'Ladies' 16″ x 8″ (406mm x 203mm), 'Countess' 20″ x 10″ (508mm x 254mm), 'Duchess' 24″ x 12″ (610mm x 305mm) but it is common practice to refer to them by size, e.g. 20″ x 10″ (508mm x 254mm).

Slates are laid in a similar manner to plain tiles and the same terms are used, but as slates have no nibs, each slate must be secured by two nails. Fig 1.

Lap:- Is the amount by which the tails of slates in one course overlap the heads of slates in the next course but one below. The amount of lap must be the recommended lap for various pitches. In conditions of severe exposure the laps given should be increased.

Margin:- Is the exposed area of each slat, and the length of the margin is the same as the gauge. Fig 1.

Nailing:- Slates may be centre nailed Fig 1, or head nailed Fig 2. In the latter case there is a tendency for the slates to lift in high winds and this may result in breaking at the nail holes and slipped slates. For this reason it is usual to headnail only the small sizes of slates. The advantage claimed for head nailing is that there are two thicknesses of slate covering the nails.

The nails should be at least 30mm from the edge of slate and at least 25mm from the head of the slate. Figs 1 & 2.

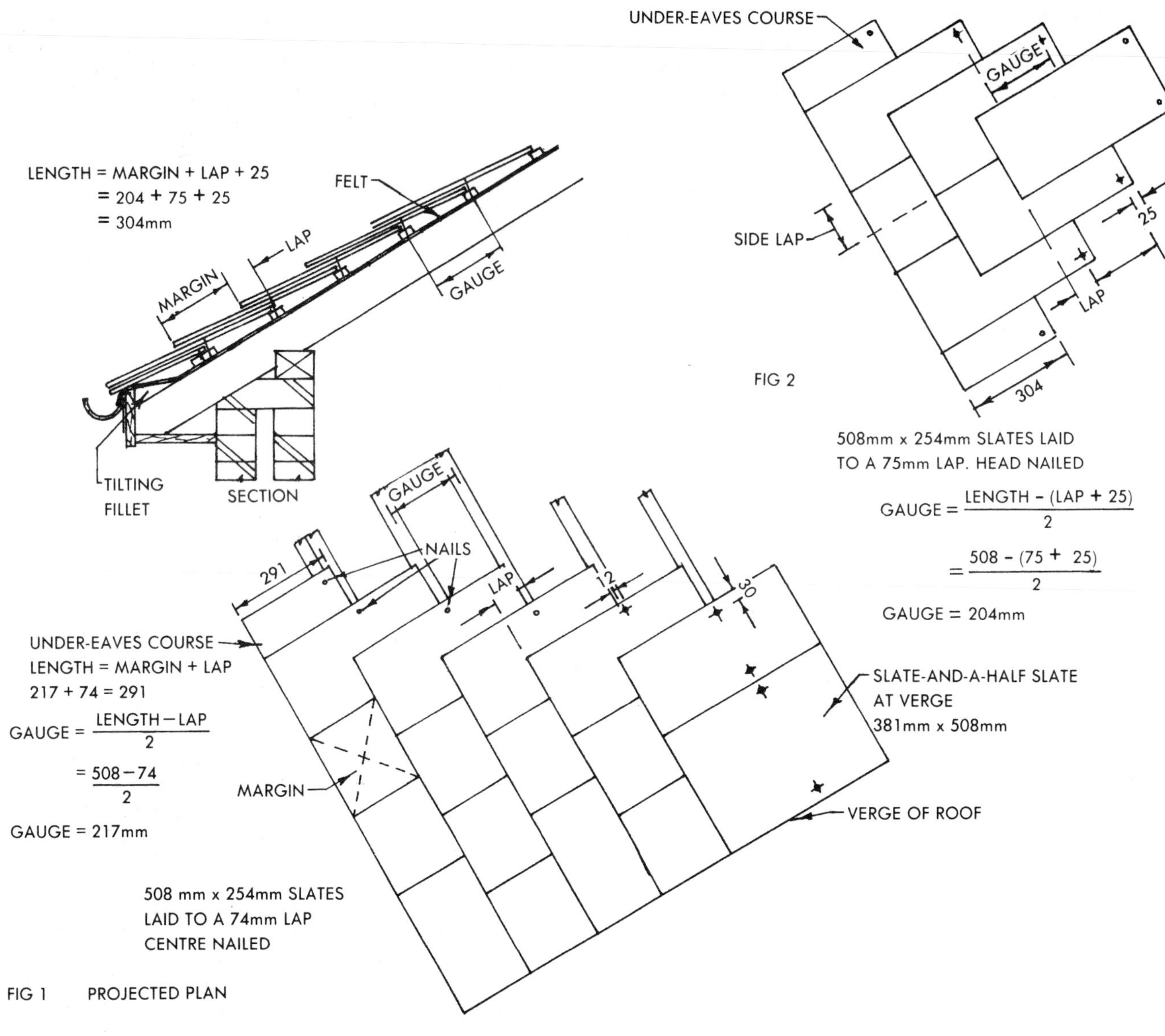

FIG 1 PROJECTED PLAN

Gauge:- Is the distance apart of the nail holes in one slate to those in the next slate Figs I & 2. For centre nailed slates the gauge is obtained from the formula:-

$$\text{Gauge} = \frac{\text{Length} - \text{lap}}{2}$$

Fig I. For hand nailed slates the formula for obtaining the gauge is:-

$$\text{Gauge} = \frac{\text{Length} - (\text{Lap} + 25\text{mm})}{2}$$

Margin:- Is the exposed area of each slate, and the length of the margin is the same as the gauge. Fig I.

PITCH	MINIMUM LAP(mm)
45°	65
40°	65
35°	75
30°	75
25°	90
20°	115

PITCHED ROOFS 14

RIDGE DETAILS

Ridges on slate roofs may be finished with a half-round, hogs-back or angle ridge in a similar manner to a tiled roof as shown in Fig I.
An alternative method is the lead ridge shown in Fig 2. The old fashioned 'slate-roll and wing' ridge is little used today.

HIP DETAILS

The slates at the hip are close-cut and mitred as shown Fig 3. Extra wide slates are required to allow for cutting and slate-and-a-half slates are commonly used. Half-round, hogs-back or purpose made angle tiles may be used to finish the hip, and these are bedded and pointed in a similar manner to a tiled hip. A hip hook or hip iron should be provided for security.

An alternative method which provides a neat finish, is to use close-cut and mitred slates and lead soakers Fig 3.

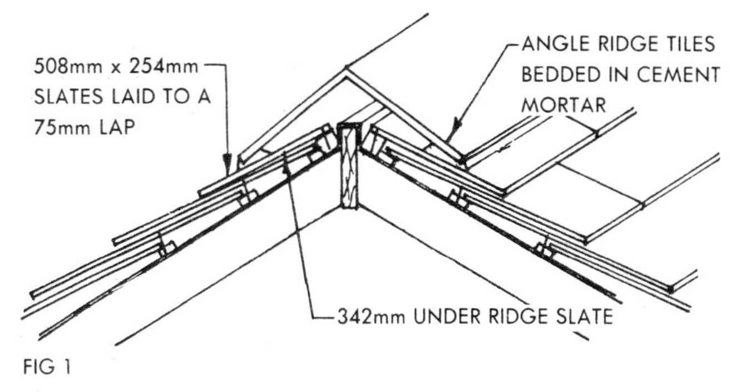

FIG 1

LEAD SLATE
Where pipes, e.g. a soil pipe penetrate a slated or tiled roof the roof may be made weathertight by using a lead slate as shown Fig 4. The lead slate is made from BS Code No 4 lead, and is made to course in with the slates or tiles. The upstand is made into a short length of pipe with the seam lead burned, the lower end being splayed to the angle of pitch and lead burned to the base sheet. When the lead slate is in position the upstand is dressed closely to the surface of the pipe, providing an effective weathertight seal. If the roof is simply felted and battened then boarding should be provided immediately below the lead slate as shown, to give support. (Alternatively, copper, nuralite or zincon may be used).

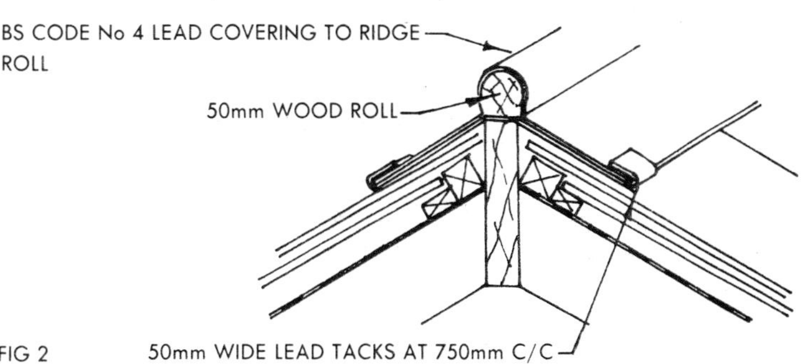

FIG 2

NAILING
Nails for slating may be yellow metal, copper, aluminium alloy or zinc. For thick heavy slates 63mm long nails should be used, but for lighter slates shorter nails may be used. In no case should the nails be less than 32mm long. Each slate should be twice nailed, the holes being driven through the slate from the bed so that the resulting spalling on the back forms a countersinking to take the heads of the nails. It is most important to ensure that the nails will be durable under the conditions to which they will be exposed, and in certain conditions, e.g. in chemical works, processing plants, retort houses etc it may be necessary to resort to the use of lead nails, stainless steel nails or clips etc.

FELTING
The roof should be felted in a similar manner to tiled roofs. A 900mm wide strip of felt should be laid from top to bottom of hips and valleys, over the ordinary felting at the hip and under the ordinary felting at the valley.

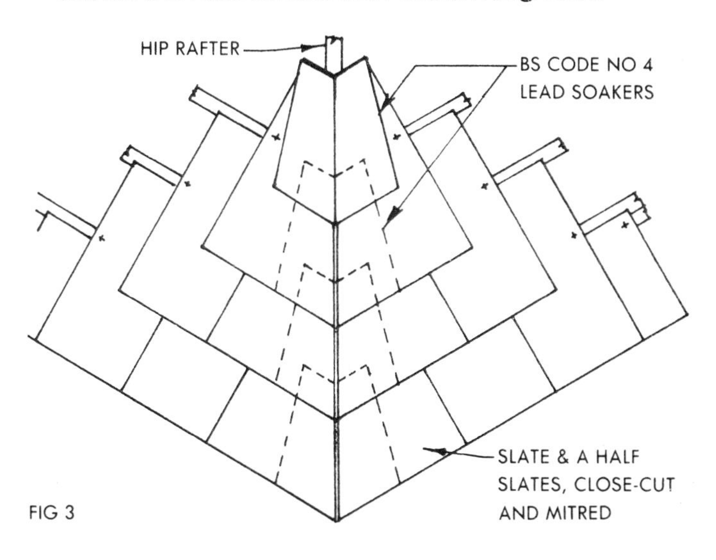

FIG 3

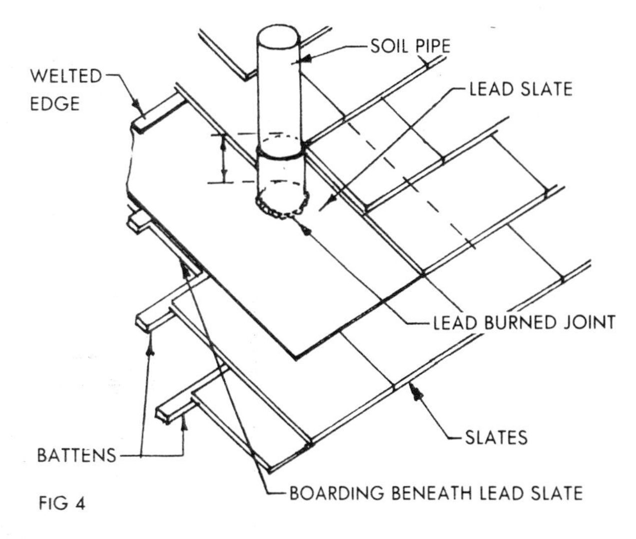

FIG 4

FLAT ROOFS I

Flat roofs may be constructed of timber, concrete or steel decking. In practice the roof surface is not finished level, but given a slight fall to allow water to drain away to some form of gutter. Diagram I of Approved Document F of the Building Regulations indicates the need for roof ventilation to pitched and flat roofs and the need to provide at least 50mm air space above the insulation.

Fall:- The gradient or slope of the roof surface is determined by the type of roof covering used, e.g. a built-up felt roof should have a fall of not less than 1:60, where as a lead covered roof may have a fall as little as 1:120, although 1:80 is probably more common.

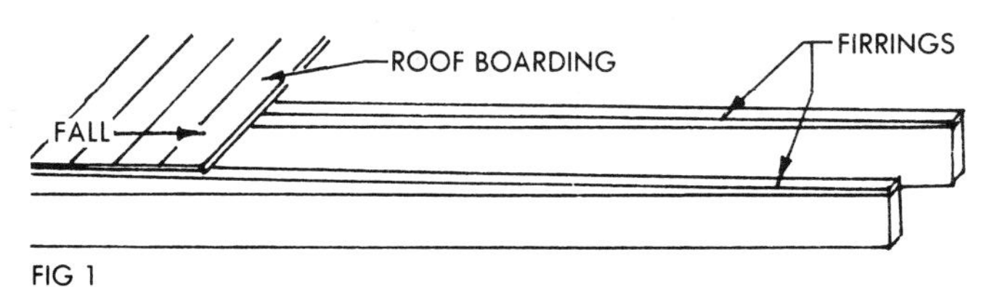

FIG 1

TIMBER FLAT ROOF

A timber flat roof is constructed in a similar manner to an upper floor. The required surface slope is achieved by using tapered firring pieces, either nailed along the joists as Fig I or across the joists as Fig 2. Plain edged rough boarding is often used for cheapness, but tongued and grooved boarding, blockboard etc may also be used. If the roof covering is of lead sheet then the roof boarding should preferably be laid diagonally, but if this is not considered practical the boarding should be laid with the grain running in the direction of the fall. In this case firrings will be fixed as shown Fig 3. All nail-heads should be punched down below the surface of the boarding, sharp corners trimmed off and high spots planed down. It is recommended that boards should be below 15% moisture content when laid, and ventilation of the roof space is essential. Timbers may be treated with a preservative, but for built-up felt roofs creosote must not be used as it is injurious to bitumen.

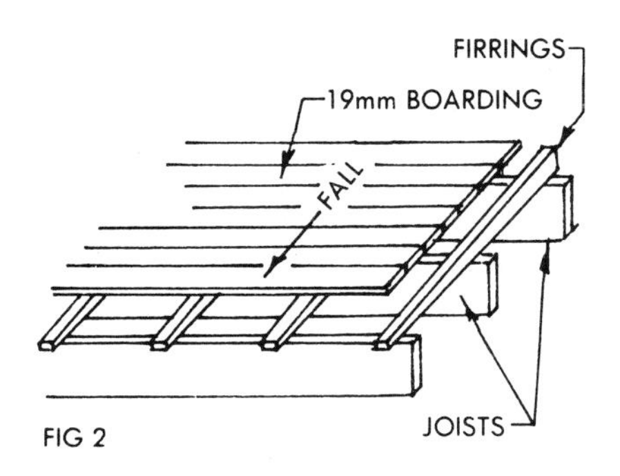

FIG 2

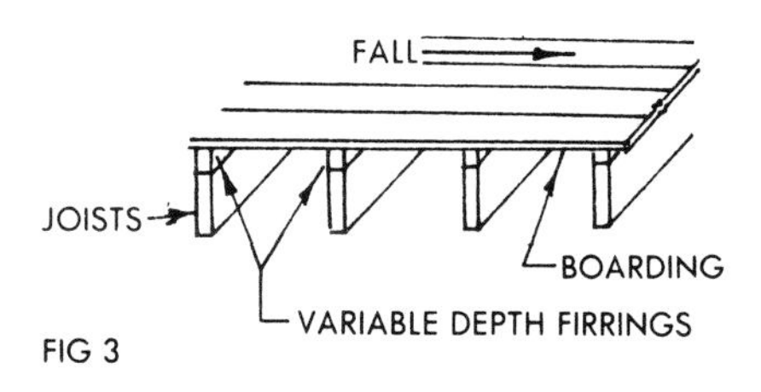

FIG 3

BUILT-UP FELT ROOFS

Built-up felt roofing usually consists of three layers of felt laid breaking joint and bonded together with a bitumen adhesive, generally poured hot on boarded roofs (T&G boards, plain boards or plywood). The first layer of underfelt should be random nailed with 19mm galvanised clout nails at 150mm c/c, each length of felt overlapping the adjacent sheet by 50mm. Figures 4, 5 and 6 show various methods of constructing a flat roof while meeting the requirements of the Building Regulations 1985.

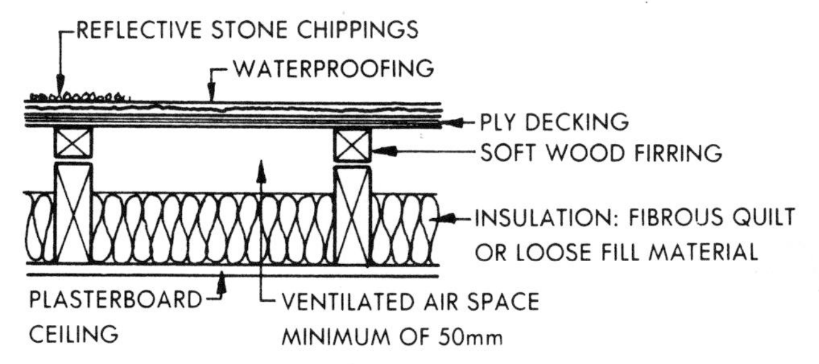

RATE OF MOISTURE VAPOUR CAN BE REDUCED BY INTRODUCTION OF VAPOUR CHECK AT CEILING POSITION: i.e. FOIL-BACKED PLASTERBOARD

FIG 4 COLD ROOF

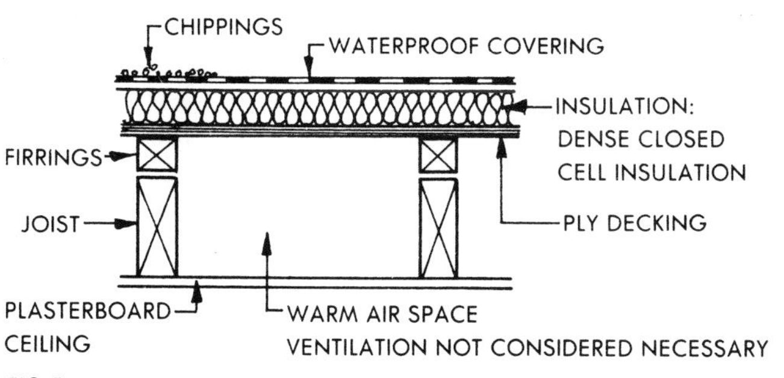

FIG 5 WARM ROOF

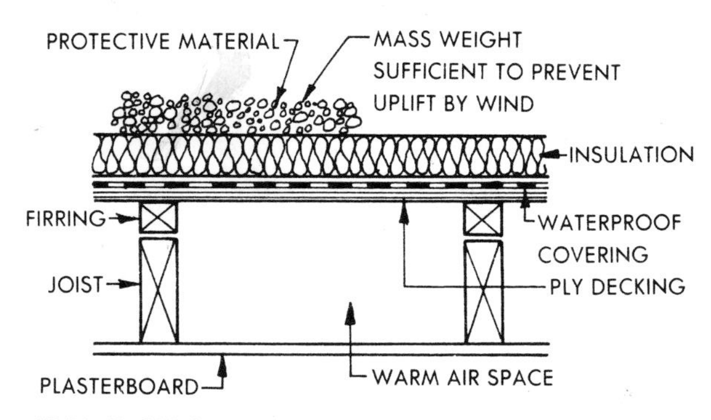

ADVANTAGES OF THIS FORM OF CONSTRUCTION ARE THAT THE MEMBRANE IS PROTECTED FROM EXTREMES OF CLIMATE AND FROM MECHANICAL DAMAGE BY SUBSEQUENT TRAFFIC

FIG 6 INVERTED ROOF (OR PROTECTED MEMBRANE ROOF)

On flaxboard or compressed strawboard, the joints between the boards should be taped with suitable tape and the first layer, either random nailed with aluminium serrated nails or bedded with approved mastic. The second layer and the cap sheet are fixed with adhesive, care being taken to ensure that the joints in successive sheets are staggered. The roofs can be finished with 12mm stone chippings (limestone, granite, gravel, calcined flints) applied over an adhesive coating. This protects the cap sheet, provides additional fire resistance and increases solar reflection.

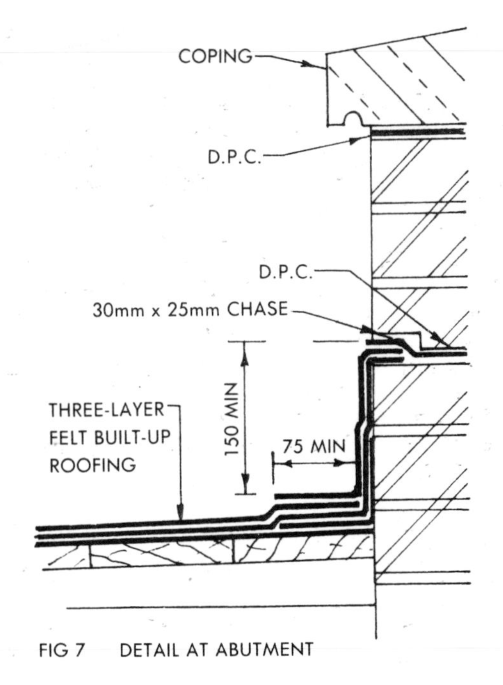

FIG 7 DETAIL AT ABUTMENT

FLAT ROOFS 2

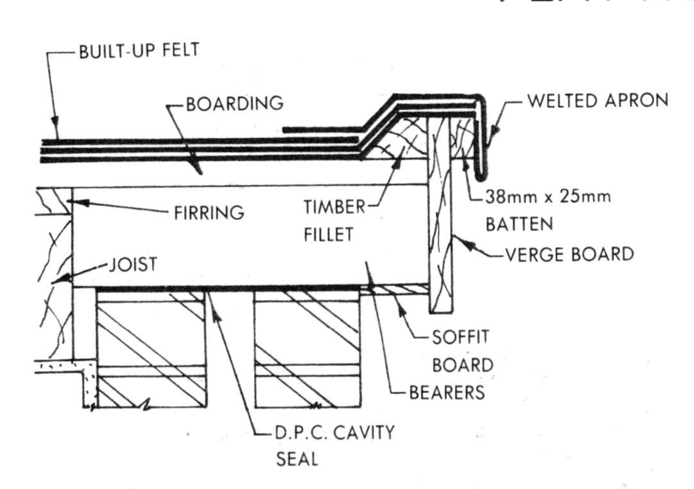

FIG 1 VERGE DETAIL

BUILT-UP FELT ROOFS

Fig No 1 on the left shows a typical construction at a verge. It is necessary to raise the verge 30mm or so above the level of the roof surface to stop rain running over the edge. The arrangement of the welted apron is similar to that at the eaves of the roof and at these positions the felt should be secured with 19mm galvanised clout nails at 50mm c/c.

CONCRETE ROOFS

Reinforced concrete roofs are constructed in a similar manner to R.C. floors and may be of solid, hollow pot or self-centering type. It is common practice to finish the surface of the structural concrete level and to provide the required fall by means of a suitable screed.

ROOF SCREEDS

Sand/cement screeds may be used as a topping over structural concrete to provide the required falls, but as concrete roofs are poor insulators, lightweight screeds, incorporating lightweight aggregate, or aerated concrete, may be used.

Moisture trapped in roof screeds and in the structural concrete may take a very long time to dry out (months, in some cases). This moisture may cause subsequent faults to develop, rotting of timber, insulating material etc. Ventilated screeds may be provided by forming shallow channels, bridged by cover strips, in the screed which connect to roof ventilators. Alternatively if the base layer of felt is partially bonded by forming a grid of unbonded channels (Fig 2), air and vapour pressure can circulate underneath the felt and escape through the vents, which are spaced at not more than 6.0m intervals in each direction. A number of patent ventilators are now available and Fig 3 shows the 'Anderson breather vent'. Manufactured by D. Anderson & Son Ltd.

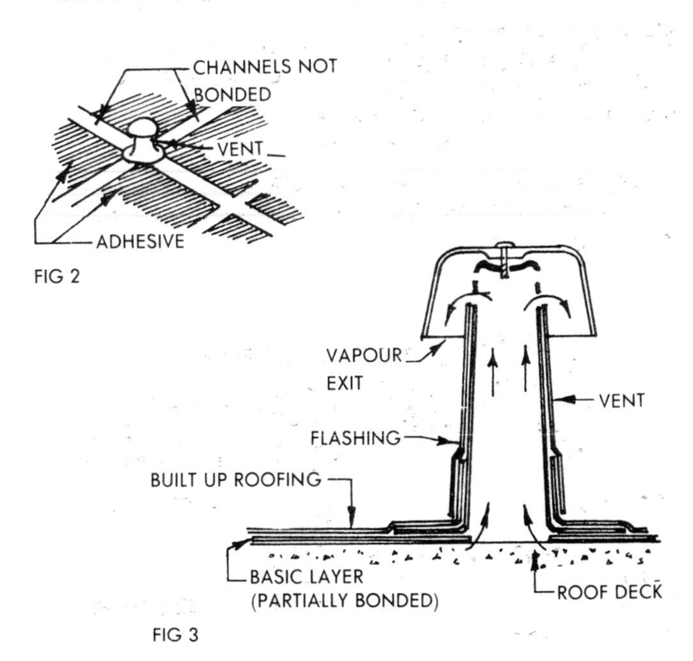

FIG 3

VAPOUR BARRIERS
Warm air inside a building usually contains more water vapour at a higher pressure than the outside air. Water vapour may pass through the structural roof and condense on the underside of the waterproof roof covering. This may result in rotting of insulating material, blistering and damage of the roofing felt and staining on ceilings. If insulating material becomes saturated it ceases to provide effective thermal insulation. To prevent this situation arising a vapour barrier should be provided. Suitable materials include:- bitumen felt laid in hot bitumen, aluminium foil, polythene sheeting, bitumen laminated fibre insulating board and external quality plywood. Gloss oil paint on plaster and zinc oil paint on wood may also provide an effective vapour barrier.

In practice it is difficult to provide a fully effective vapour barrier, because of leakages at wall and floor junctions, through joints, and where ceilings are pierced by pipes, electric points etc. Boards incorporating a vapour barrier e.g. foil backed plaster board, should have joints sealed with tape or a sealing compound. In the case of a timber roof since vapour may penetrate into the roof space it should be ventilated.

An effective construction for a concrete roof is shown Fig 4. The ceiling battens should be treated with preservative. An alternative method is to use dry insulation instead of a lightweight screed, Fig 5., preferably using a rot proof material e.g. mineral/glass wool.

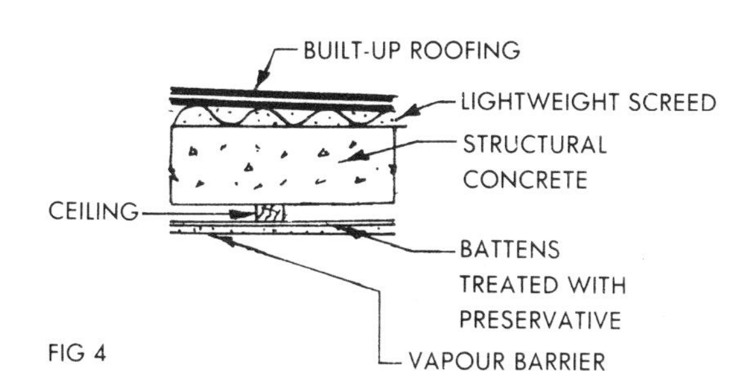

FIG 4

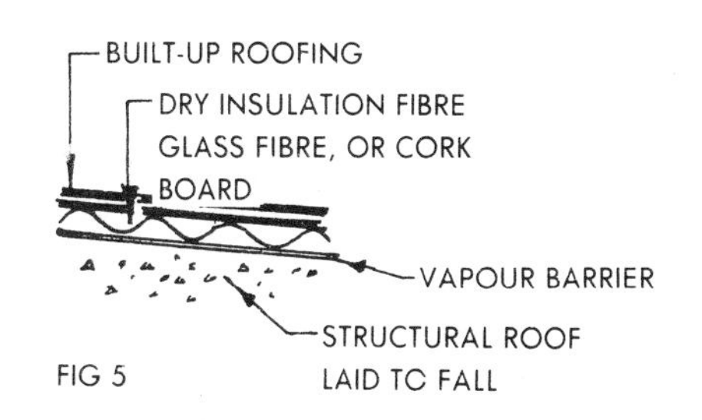

FIG 5

FLAT ROOFS 3

ASPHALT ROOFING
Mastic asphalt is manufactured from natural rock asphalt or limestone aggregate and asphaltic cement. The asphalt is supplied in solid blocks which are heated on site to a plastic condition and then spread over the surface. On flat roofs two coats of asphalt each 10mm thick are used, the joints in the second coat are staggered and should overlap joints in the base coat by at least 150mm. The minimum fall on an asphalt flat is 1:80.

The asphalt is laid over sheathing felt to BS 747: 1977 'Specification for roofing felts', this isolates the asphalt from the structure and allows for differential movement. Screeds should be as dry as possible and ventilators as previously described may be used. Heat transmission through the roof may be reduced by providing a suitable reflective surface, e.g. white spar chippings, light coloured tiles or a suitable paint.

Abutments should be finished with a 13mm skirting in two coats, at least 150mm high Fig 1.

The edge of a roof may be finished with an asphalt apron Fig 2. An alternative method is to use an extruded alloy roof edging, such as 'alutrim' Fig 2.

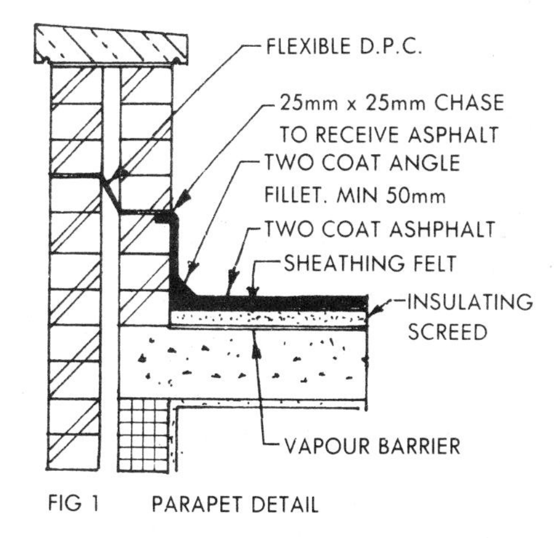

FIG 1 PARAPET DETAIL

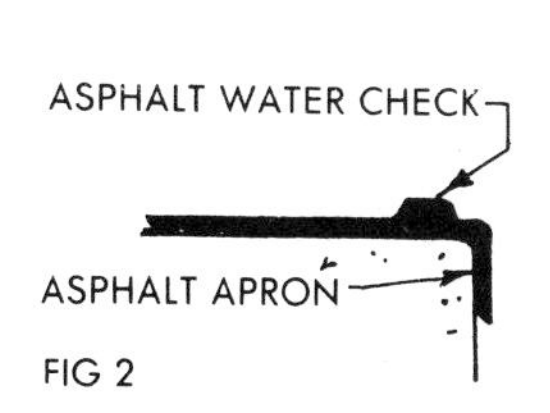

FIG 2

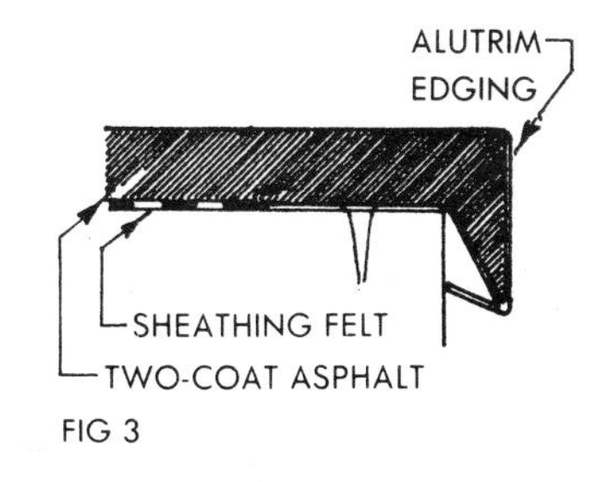

FIG 3

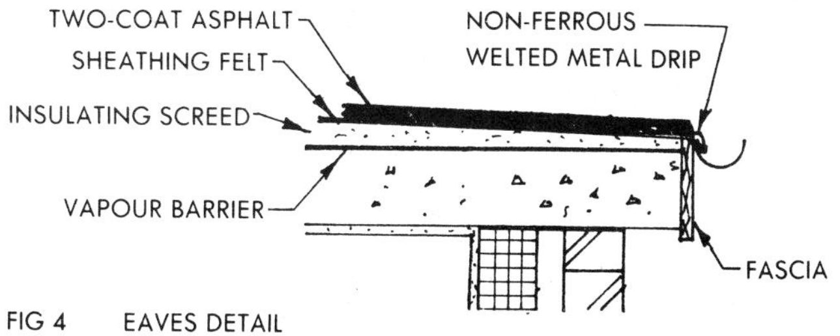

FIG 4 EAVES DETAIL

LEAD FLATS

Roof boards should preferably be laid diagonally, but as this involves wasteful cutting they may be laid with the grain running in the direction of the fall. All nail heads must be punched down below the surface of the boards and sharp corners trimmed off.

The surface of concrete roofs should have a smooth finish and be laid to appropriate falls. It is desirable to allow a minimum fall of 1 in 120, although this is often increased and 1 in 80 is sometimes preferred.

Underlay:- An underlay must be used to allow the lead to move freely and to minimise surface irregularities. Brown No 1 inodorous felt to BS 747 is recommended and should be laid with all joints close butted, preferably across the fall of the roof.

Bay sizes:- Since the coefficient of expansion of lead is comparatively high the size of lead sheets should be limited and pieces not greater than 2m² should be used. The maximum distance between joints across the fall of the roof, which are constructed as drips must not exceed 2400mm, and between joints running with the fall of the roof, constructed as rolls, 800mm. A recommended spacing is to make the distance between drips 2100mm and place roles 675mm apart.

Wood-cored rolls:- The adjoining edges of the lead sheets are dressed over a shaped wooden core, the underlap being nailed with flat-headed copper nails at 150mm c/c, the overlap extending at least 30mm on to the surface of the adjoining bay to form a splash lap. Fig 5.

Drips:- These are normally 60mm deep as Fig 6, but a minimum of 40mm is permissible if an anti-capillary groove is provided as Fig 7, and the lead of the undercloak is dressed into the groove as shown.

The top edge of the drip is rebated to receive the end of the undercloak as shown, which is fixed with flat headed copper nails.

Thickness of lead:- For small areas with no traffic BS Code No 5 lead may be used, for larger areas or with traffic BS Code Nos 6 or 7 should be used.

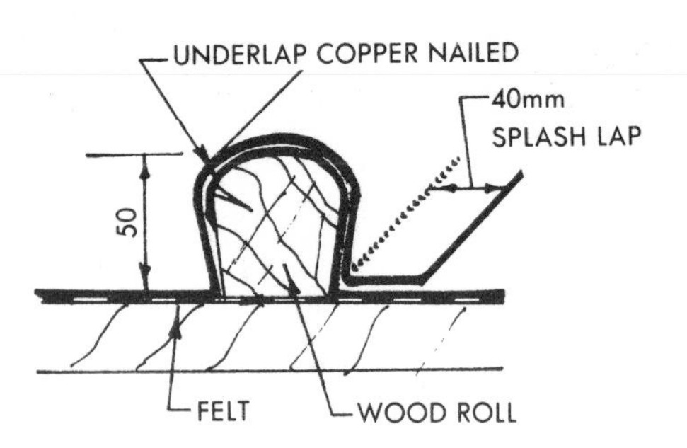

FIG 5 SOLID ROLL

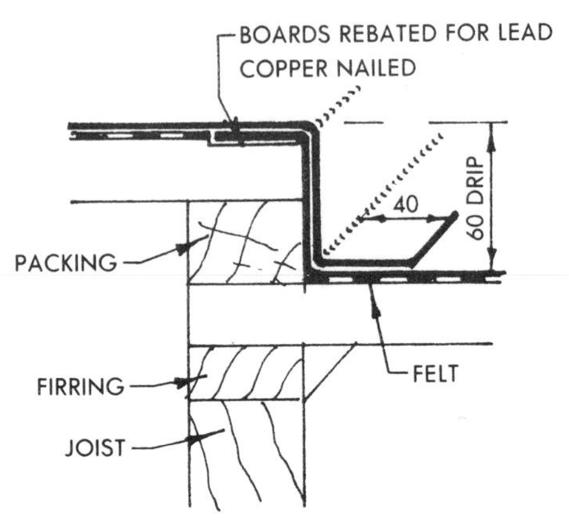

FIG 6 DETAIL OF 60mm DRIP

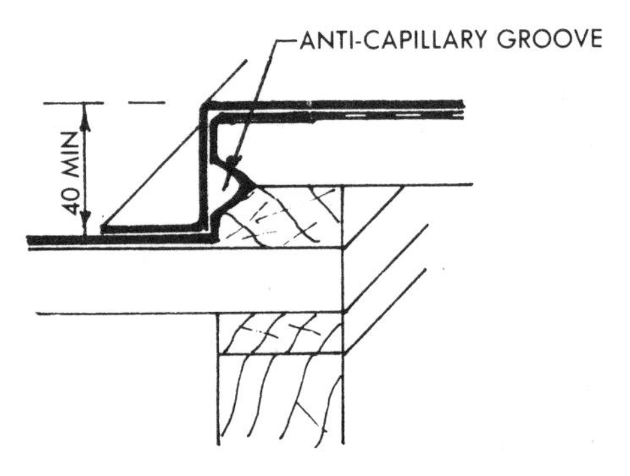

FIG 7 DETAIL OF 40mm DRIP

FLAT ROOFS 4

LEAD FLATS

The length of one piece of apron flashing should not exceed 2100mm and joints in length should be lapped 150mm. The lower edge of the apron flashing is secured against wind lift by lead tacks at 750mm intervals. Fig 3. At internal and external angles the upstand is bossed , or cut and lead-burned to shape.

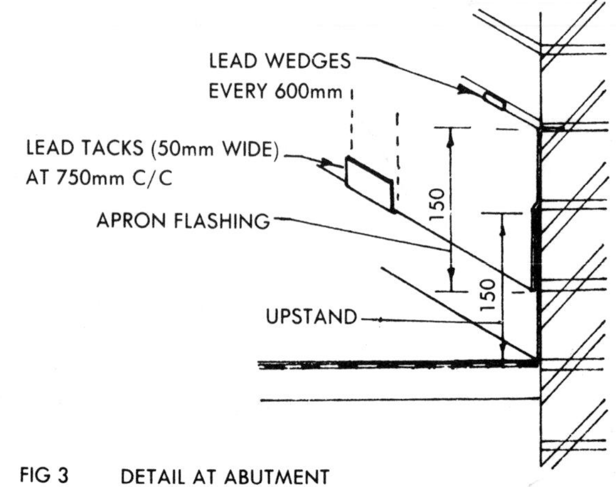

FIG 3 DETAIL AT ABUTMENT

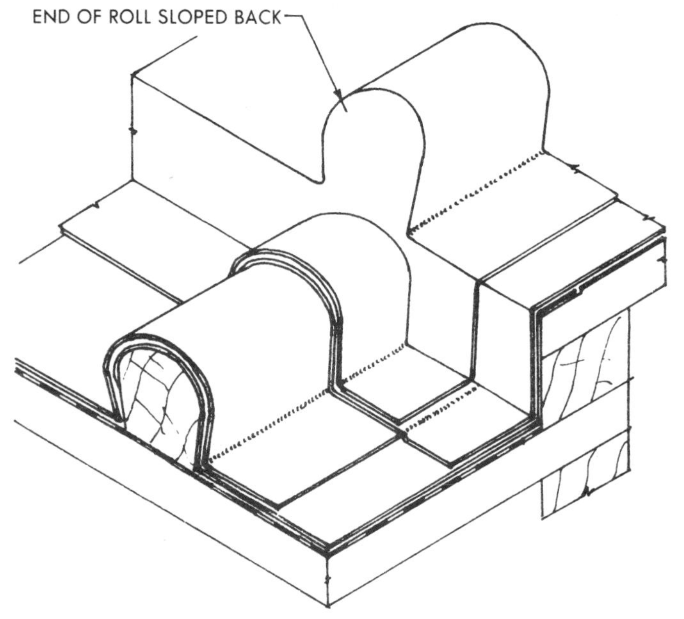

FIG 1 JUNCTION OF ROLL AND DRIP

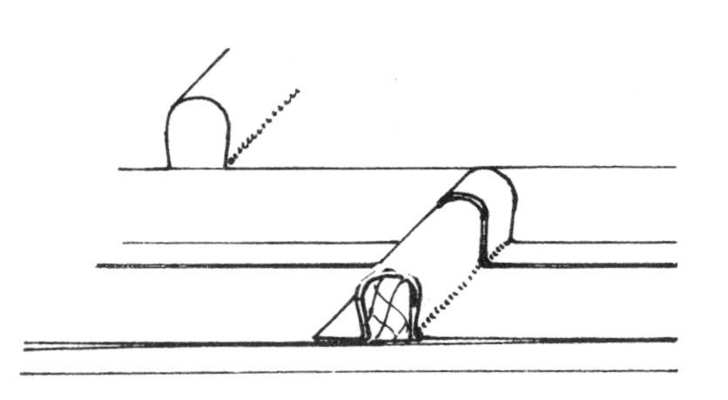

FIG 2 STAGGERED ROLLS

It is recommended that where possible a staggered layout of rolls be adopted , (Fig 2) as this simplifies the construction at the junction of rolls and drip (Fig I).

Where the roof is contained by parapet walls the lead flat will fall to a parapet gutter (Fig 4). the gutter may discharge to a rainwater head on the outer face of the wall (Fig 5) or to a catchpit or cesspool connecting to a pipe situated internally. Fig 6 shows a section through a typical cesspool. In case of an ordinary eaves the lead is dressed down into the eaves gutter.

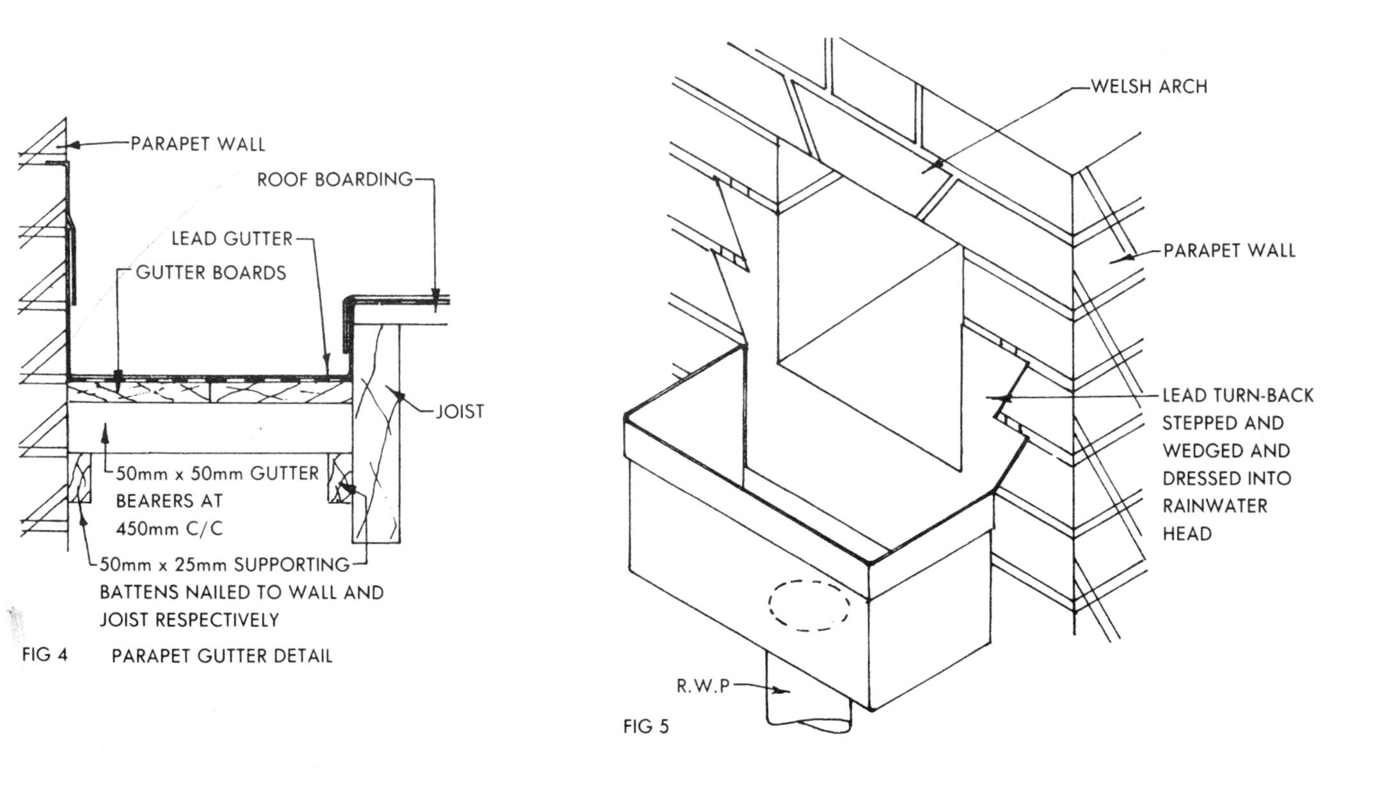

FIG 4 PARAPET GUTTER DETAIL

FIG 5

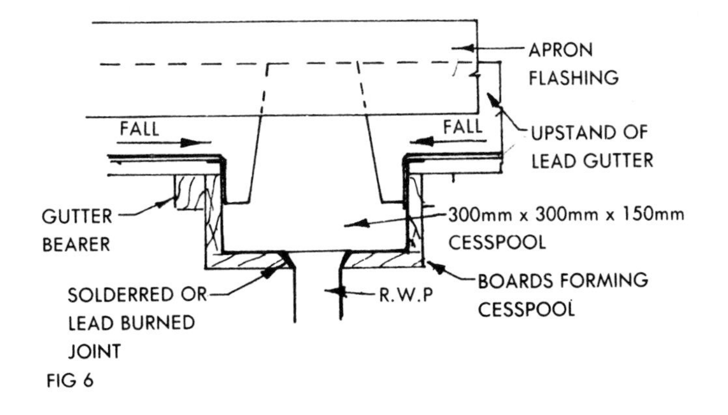

FIG 6

FLAT ROOFS 5

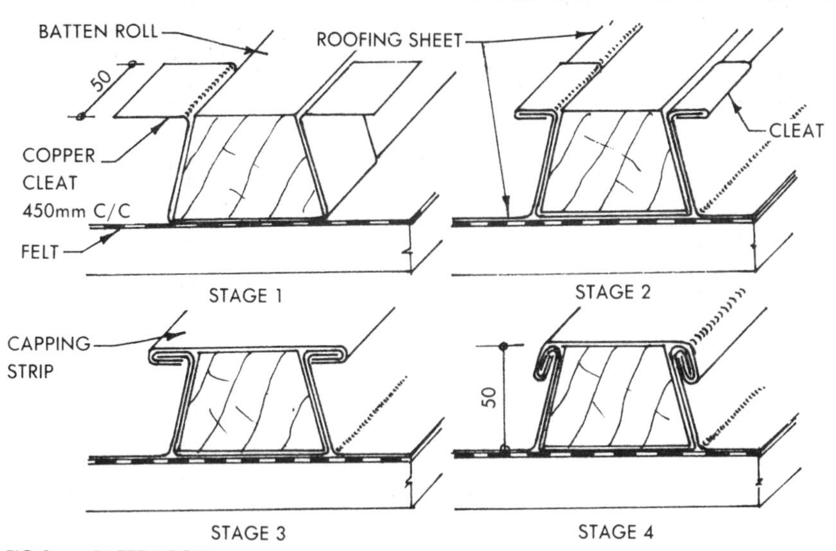

FIG 1 BATTEN ROLL

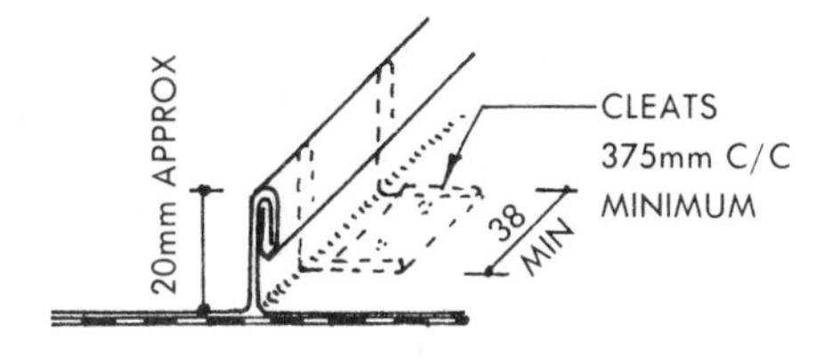

FIG 3 STANDING SEAM

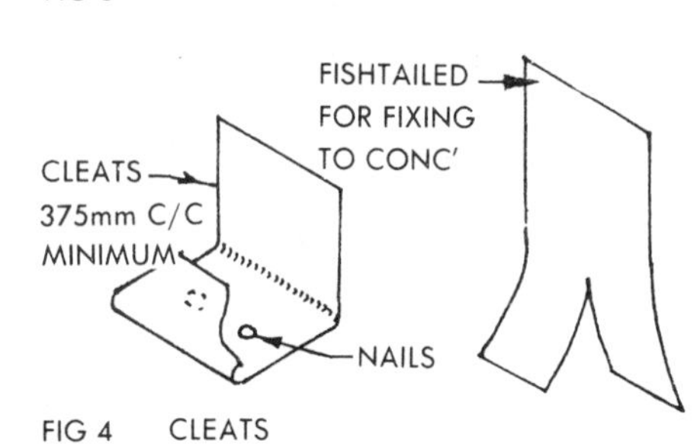

FIG 4 CLEATS

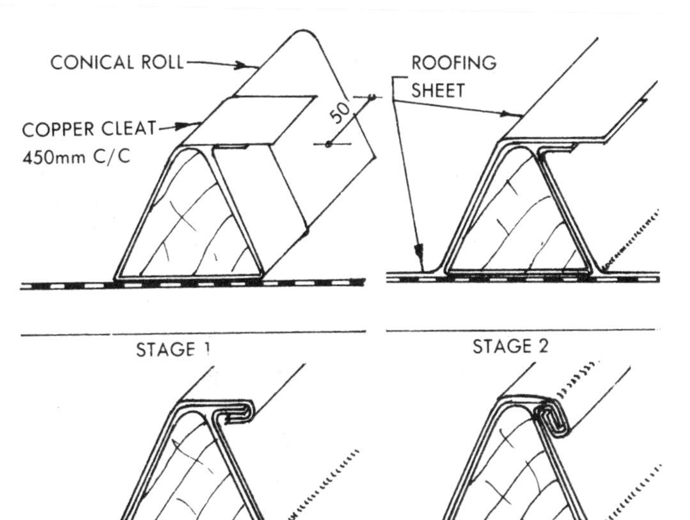

FIG 2 CONICAL ROLL

COPPER ROOFING

For flat roofs preparation is similar to that for a lead covered roof, firrings being used to provide the falls.

Copper sheet and strip to BS 2870 should be used.

On exposure to atmosphere an oxide film or patina is formed on the surface of copper. This patina usually bright green in colour, not only protects the copper but is generally considered to enhance the appearance.

Sheet copper roofs are constructed with sheets jointed along the edges and secured by copper clips inserted in the folds of the joints. Figs I, 2, 3 and 4.

Joints in the direction of the fall are either formed into a standing seam, Fig 3 or dressed to wood rolls. Batten rolls may be used, Fig I, or conical rolls Fig 2. Spacing of rolls is given below.

Thickness of copper mm	Bay width standing seams mm	Bay width wood rolls mm
0.45	525	500
0.60	525	500
0.70	675	650

The minimum fall for any copper roof is I in 60. Drips should be used on roofs of 5° pitch or less and should be spaced at not more than 3m c/c and be 65mm deep Figs 7 & 8. Since the length of each sheet of copper should not exceed I.8m joints are required between drips and are formed by means of flattened welted seams. For pitches greater than 45° a single lock welt (Fig 5) will suffice, but for pitches below this pitch a double lock cross welt (Fig 6) is essential.

An underlay of approved felt is necessary to provide a satisfactory base for the copper sheet. The felt also provides a degree of sound and thermal insulation.

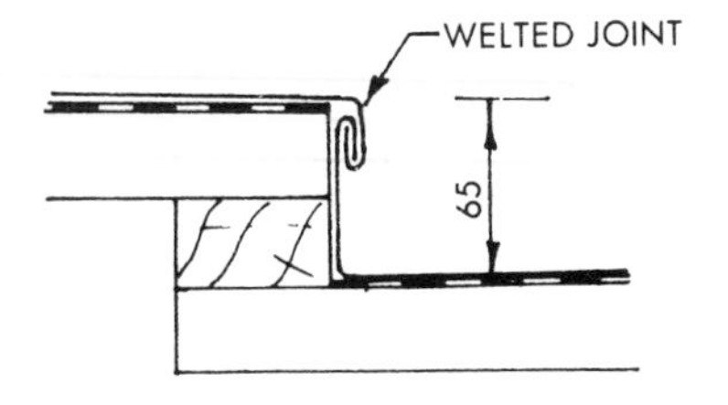

FIG 7 SECTION THRO' DRIP

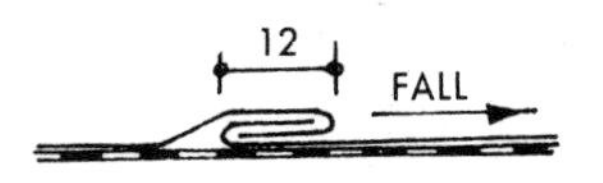

FIG 5 SINGLE LOCK WELT

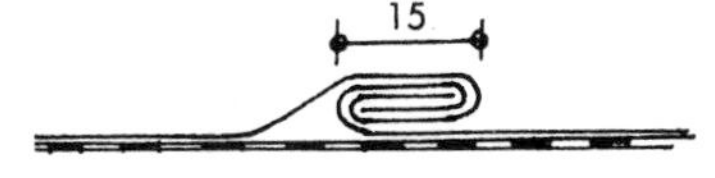

FIG 6 DOUBLE LOCK CROSS WELT

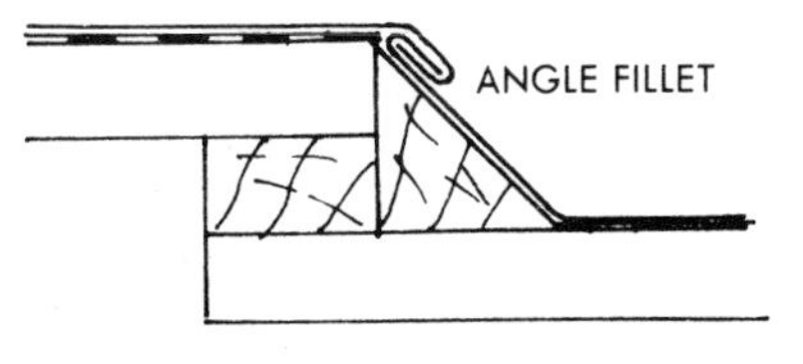

FIG 8 SECTION THRO' SPLAYED DRIP

FLAT ROOFS 6

COPPER ROOFING
The layout of rolls, drips and cross welts is shown in Fig I. Spacing of rolls is as given in the table. Distance between drips should not exceed 3.0m. Double lock cross welts should be staggered in adjacent bays to avoid the difficulty of welting too many thicknesses of copper. The roles may also be staggered to facilitate finishing at the ends of the rolls.

Where the roof pitch is 20° or less, double lock cross welts should be sealed by painting the edges of the copper with boiled linseed oil before baking the joint.

The choice of system (standing seam, batten roll or cornical roll) depends on the architectural treatment required, but where roofs are subjected to foot traffic the wood roll system should be used. If traffic is concentrated on one particular area duckboards at least 375mm wide should be provided to prevent damage to the metal roofing.

Cleats:- For securing the main roof covering cleats should not be less than 50mm wide. On boards the cleats are fixed with two copper or brass nails or screws, for standing seams.
For fixing to concrete the cleats may be fishtailed

With batten or conical wood rolls the cleats are taken underneath the rolls and fastened there before turning up the sides Fig 4.

At ridges, verges, eaves etc the cleat may consist of a continuous fixing strip or lining plate running the full length of the weathering.

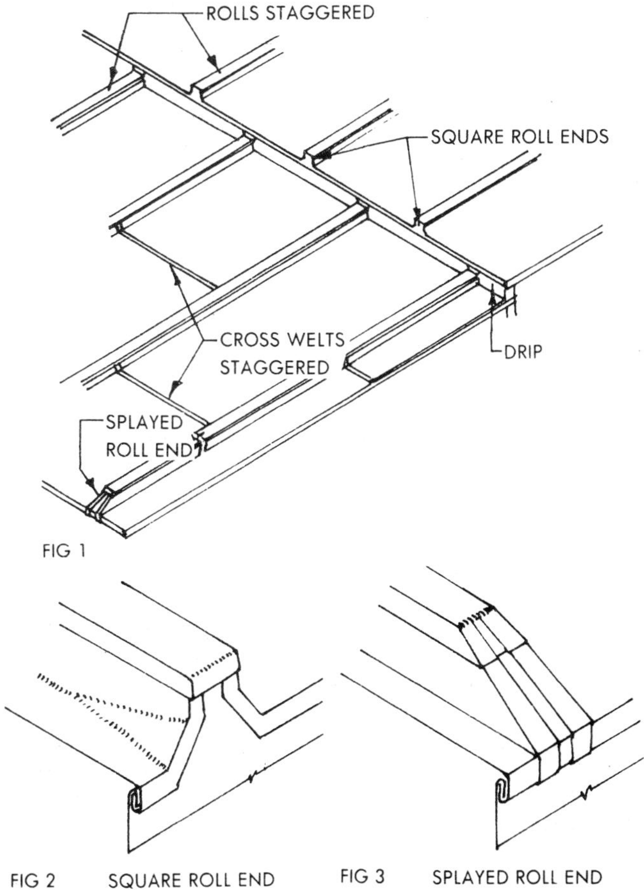

FIG 1

FIG 2 SQUARE ROLL END

FIG 3 SPLAYED ROLL END

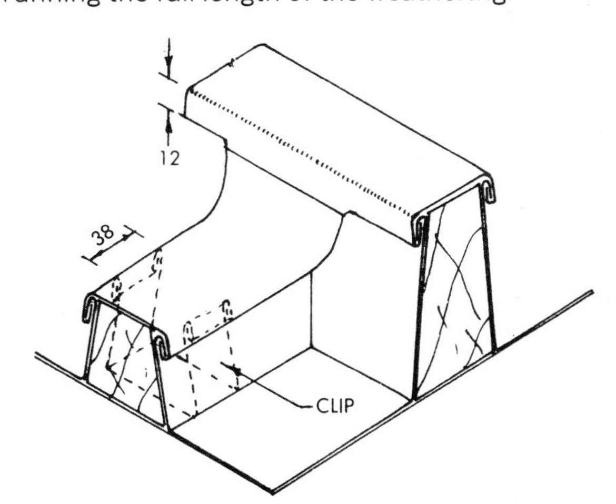

FIG 4 DETAIL AT RIDGE (RIDGE 38mm HIGHER THAN INTERSECTION WITH TOP OF ROLL ABUTTING)

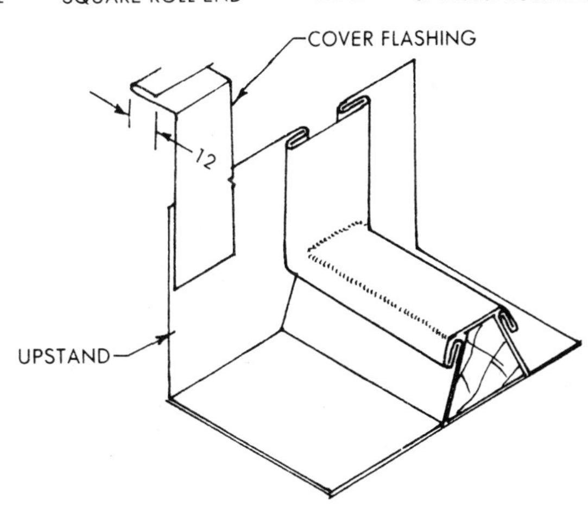

FIG 5 ABUTMENT DETAIL

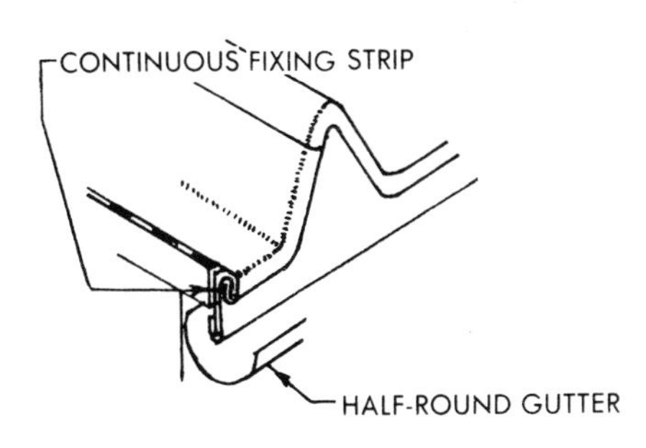

FIG 6 CONICAL ROLL WITH SQUARE END

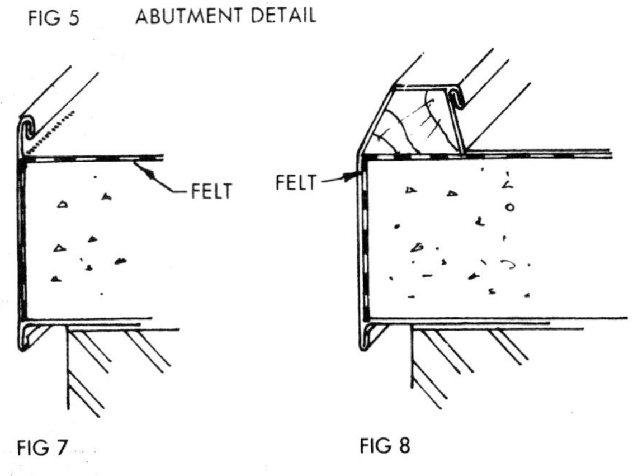

FIG 7

FIG 8

VERGE DETAILS

FLAT ROOFS 7

ZINC ROOFING

Standard zinc sheets are 914mm (3ft) x 2.133m (7ft) or 2.438m (8ft) and should comply with BS 849 'Plain sheet zinc roofing'. The minimum thickness should be English zinc gauge No 14 (21 S.W.G. approx).

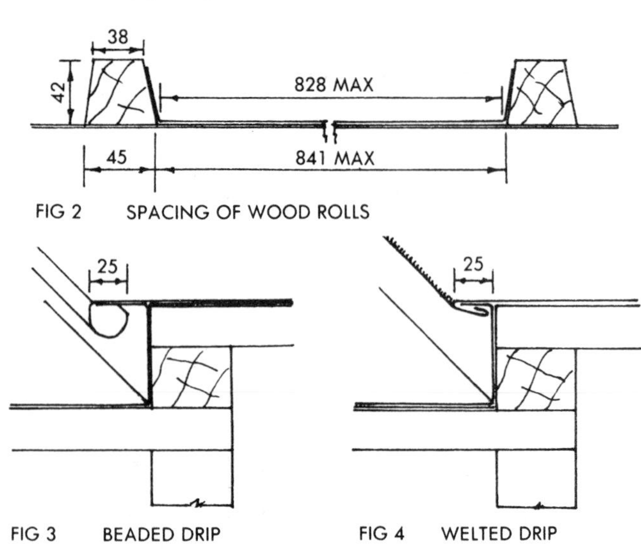

FIG 2 SPACING OF WOOD ROLLS

FIG 3 BEADED DRIP

FIG 4 WELTED DRIP

Thermal movement:- Zinc has a relatively high coefficient of expansion and care in spacing of rolls and drips is necessary.
Rolls:- Shaped wood rolls (Fig 1) are used to provide joints running with the fall of the roof. These rolls should be spaced at a maximum distance of 841mm apart (Fig 2), and the roof sheets should be 828mm wide as shown. Thus allows for a 38mm turn-up and provides 6.3mm space for movement along each side of the bay. The sheets are secured by sheet zinc clips. Fig 1.
Drips:- Should be spaced at 152mm less than the length of the roof sheets. i.e. 2.285m when the sheets are 2.438m long. Drips should be not less than 50.8mm deep and may be beaded (Fig 3) or welted (Fig 4).
Capping:- The capping is a standardised machine-formed component made to suit the standard size roll. It is usually supplied in lengths of 2.438m but should not be fixed in lengths of more than 1.219m. Fig 5. The capping is secured by holding down clips nailed to the battens. Fig 6. A verge detail incorporating a drop apron is shown Fig 7.

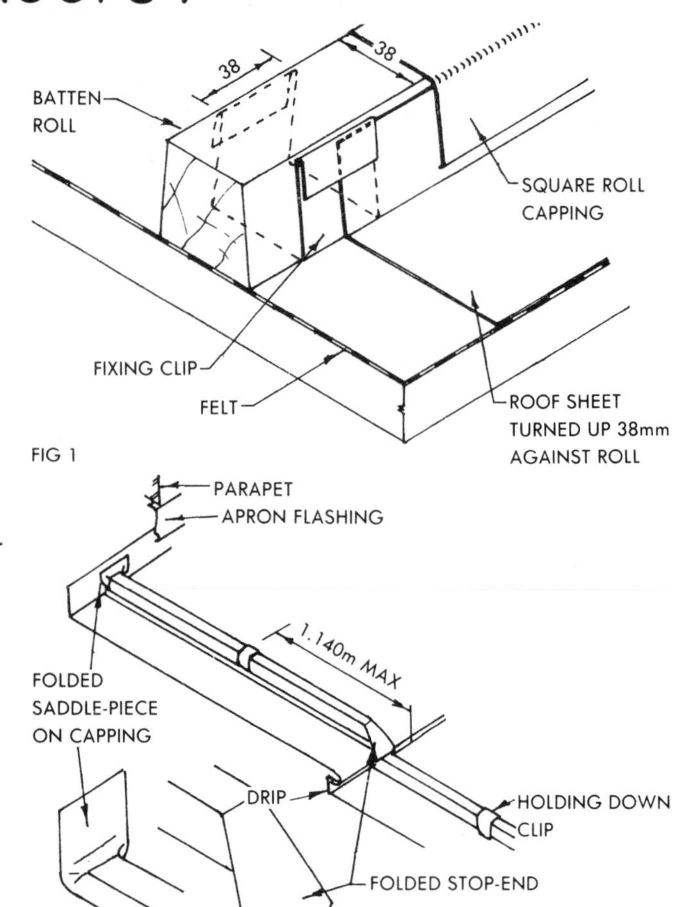

FIG 1

FIG 5 ROLL DETAILS

Boarding:- Most well seasoned softwoods do not affect zinc, but it must not be laid in direct contact with or receive drainage from, Western red cedar, oak or sweet chestnut.
Felt:- Suitable underlays are open textured felts such as defined under Type 4A BS 747.
N.B. Zinc is susceptible to electro-chemical corrosion in contact with copper or if receiving drainage from copper.

ALUMINIUM ROOFING

Sheets up to 3.66m x 1.22m are available but common sizes used are 457mm x 1.8m and 610mm x 1.8. It is inadvisable to use sheets larger than 1.8m x 1.2m as trouble may ensue from wind suction. Jointing and fixing is similar to copper using batten rolls. Extruded aluminium alloy rolls as Fig 8 may also be used.

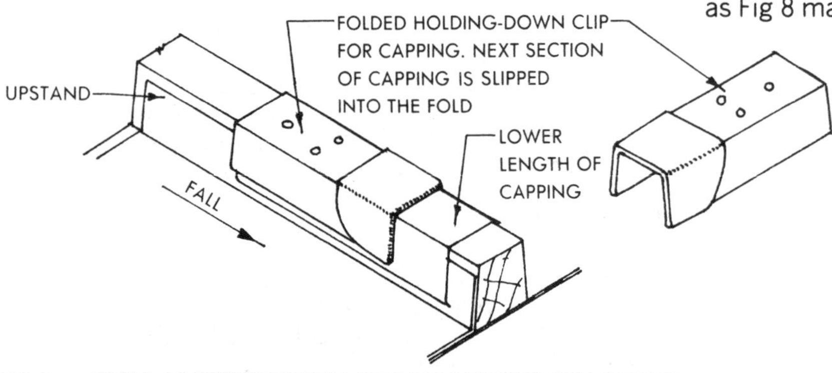

FIG 6 DETAIL OF HOLDING-DOWN CLIPS FOR SECURING ROLL CAPPING

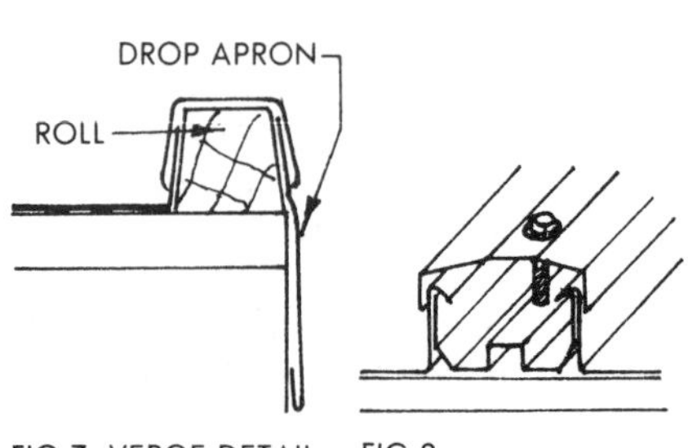

FIG 7 VERGE DETAIL FIG 8

FIREPLACES & FLUES I

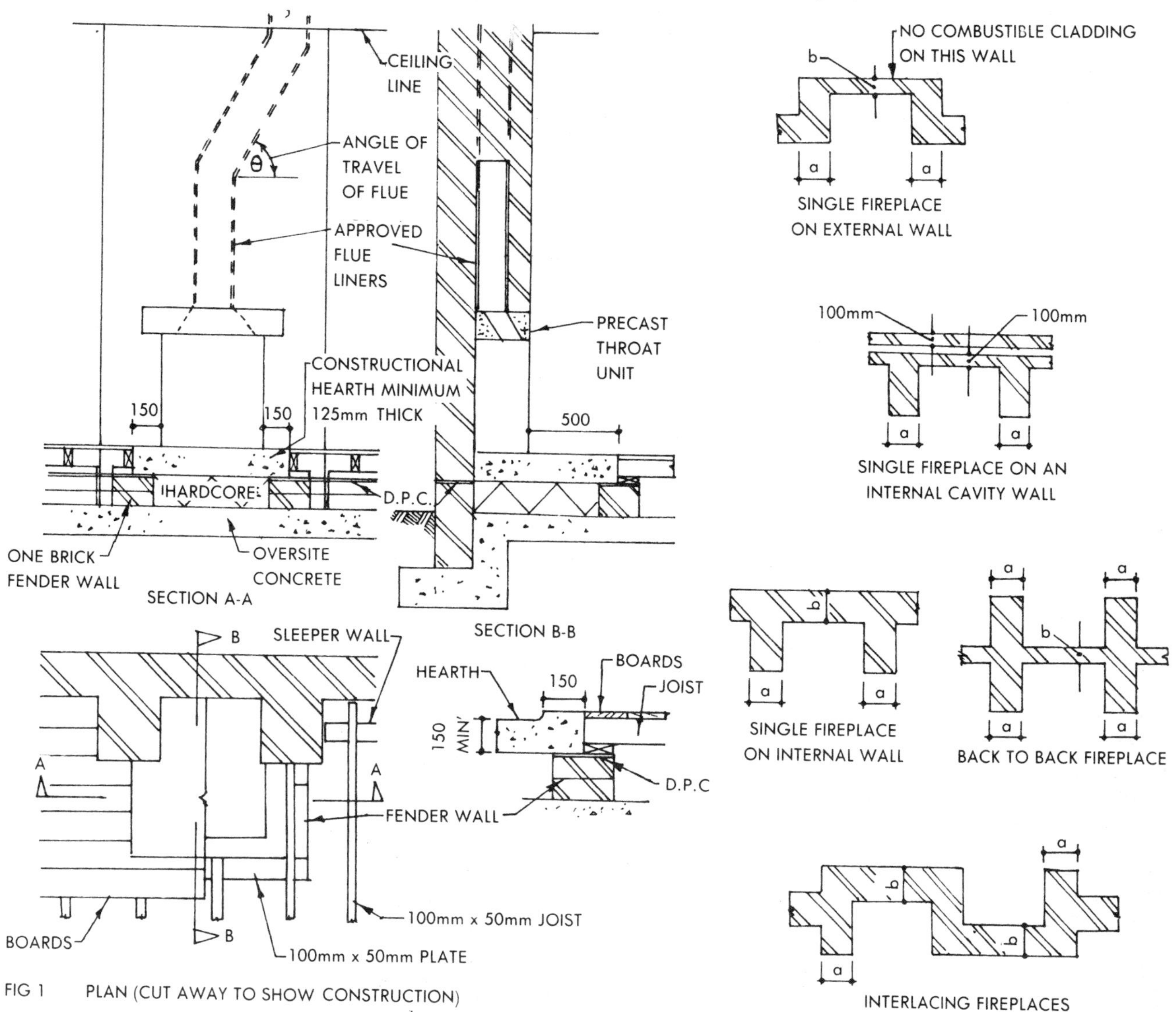

FIG 1 PLAN (CUT AWAY TO SHOW CONSTRUCTION)

REQUIREMENTS OF THE BUILDING REGULATIONS 1985

Approved Document J part C illustrates how Regulation J3 can be met. Diagrams 4, 5 and 6 of the Approved Document should be referred to.

The jamb on each side of the fireplace recess 'a' to be not less than 200mm thick (L3). If the recess is on an internal wall the back 'b' must be not less than 200mm thick, if on a cavity wall each leaf must be not less than 100mm thick and any such thickness must extend the full height of the recess.

If the recess is on an external wall and no combustible cladding is carried across the back of the recess, the back 'b' may be a solid wall less than 200mm thick but not less than 100mm. N.B. Although the back of the recess, or a flue on an external wall may be 100mm, if reasonable insulation is to be achieved a minimum thickness of 200mm is recommended.

Constructional hearths:- Shall be not less than 125mm thick, extend within the recess to the back and jambs, project not less than 500mm in front of the jambs and extend not less than 150mm on each side of the fireplace opening. Any part of the exposed surface of the hearth which is not more than 150mm horizontally from the adjoining floor, must be not lower than the floor level. Thus in the case of a sunken hearth the edges would be as Fig I.

If the hearth is constructed otherwise than in conjunction with a fireplace recess, it shall be of such dimensions as to contain a square having sides measuring not less than 840mm (J part C 1.25)

Chimneys:- Any chimney serving a Class I appliance★ shall be lined with approved liners having rebated or socketed joints, or be constructed of concrete flue blocks made of, or having inside walls of kiln-burnt aggregate and high alumina cement, and so made that no joints between blocks other than bed joints adjoin any flue. N.B. Under certain conditions a flexible flue liner may be used. (L6).

Flues:- Any flue serving a Type I appliance shall be of the following sizes (Table 2 Approved Document J, the Building Regulations 1985); square-185mm; circular-225mm diameter with an absolute minimum of 200mm.

Flue travel:- As far as possible bends in a flue should be kept to a minimum. Where bends are necessary, the angle of travel 'Θ' should be as steep as possible, preferably at least 60° and in no case less than 45°. It is good practice to take the flue as high as possible above the opening before commencing the bend, as shown.

★A Class I appliance is one using solid fuel or oil, having an output rating not exceeding 45kW.

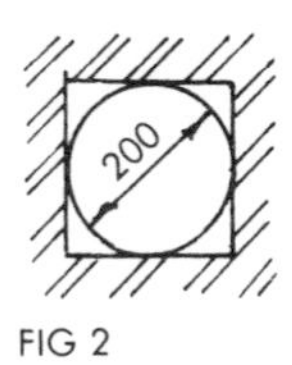

FIG 2

FIREPLACES & FLUES 2

100mm MIN

150

150

SECTION B-B

40mm MINIMUM

225mm x 50mm TRIMMED JOISTS

B

B

500

225mm x 75mm TRIMMER JOIST

BRIDGING JOIST

BOARDS

TRIMMING JOIST

PLAN (CUT AWAY TO SHOW CONSTRUCTION)

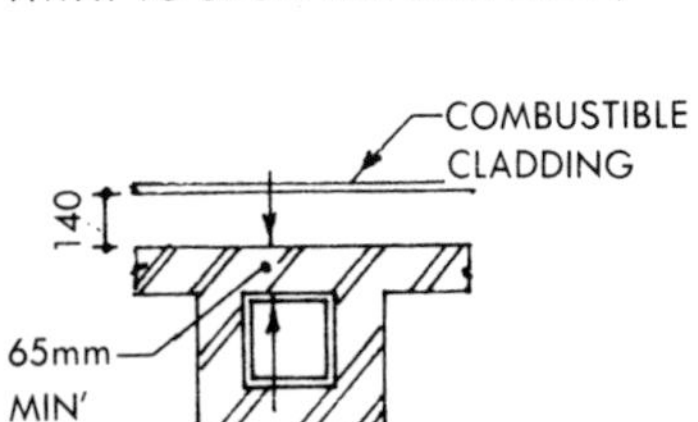

REQUIREMENTS OF THE BUILDING REGULATIONS APPROVED DOCUMENT J PART C

No combustible material, other than timber fillets supporting the edges of a hearth where it adjoins a floor, shall be placed under a constructional hearth serving a Class I appliance within a distance of 250mm measured vertically, from the upper surface of the hearth, unless such material is separated from the underside of the hearth by an air space of not less than 500mm.

Where the thickness of solid non-combustible material surround a flue in a chimney serving a Class I appliance is less than 200mm, no combustible material, other than a floorboard, skirting board, dado rail, picture rail, mantle-shelf or architrave, shall be so placed as to be nearer than 40mm to the outer surface of the chimney (Approved Document J Part C).

If chimney forms part of an external wall and is constructed of approved blocks and there is a distance of not less than 140mm between the flue and any combustible cladding adjoining the outer surface of that part of the chimney which separates the flue from the external air, such part may be less than 100mm thick but not less than 65mm thick.

If a chimney serving a Class I appliance is built of bricks or blocks, any flue in the chimney shall be surrounded and separated from any other flue by solid material not less than 100mm thick excluding the thickness of any flue lining. (Diagram 8(a) Approved Document J Part C).

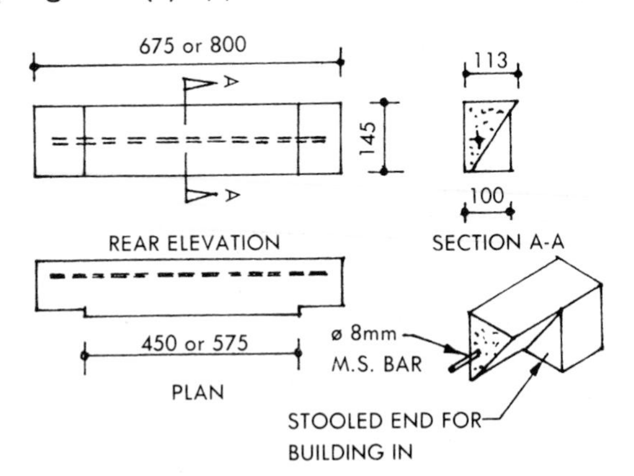

STANDARD LINTEL B.S. 1251: 1970
Length to suit size of opening
675mm for 450mm opening (330mm fire)
800mm for 375mm opening (400mm or 450mm fire)

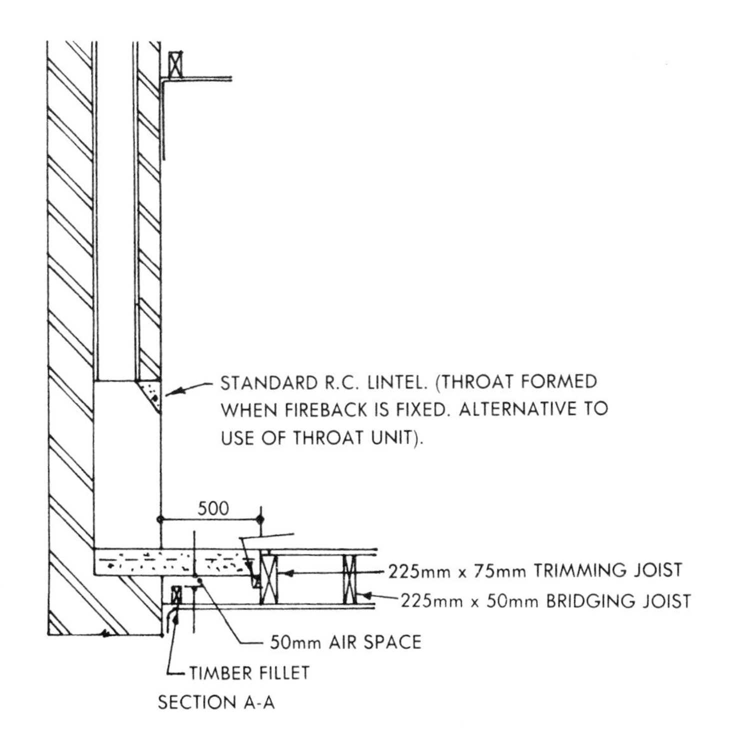

SECTION A-A

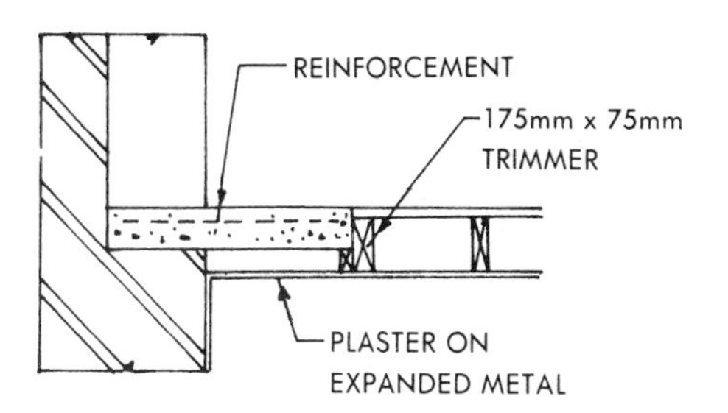

ALTERNATIVE CEILING WHERE SPACE DOES NOT PERMIT USE OF TIMBER FILLET

If flue wall is less than 200mm no combustible material may be placed in the wall nearer than 200mm to a flue. It is presumed that in the example shown the trimmer joist is supported by a corbel or joist hanger and is not built into the wall.

FIREPLACES & FLUES 3

HEIGHTS OF STACKS AND POSITIONS OF FLUE OUTLETS

1. The outlet of any flue serving a Class I appliance shall be so situated that the top of the chimney (exclusive of any chimney pot) is not less than 1m above the highest point of contact between the chimney and the roof (J 1.9).
2. Where the pitch of the roof on both sides of the ridge is not less than 10° and the chimney passes through the roof at or within 600mm of the ridge, the top of the chimney may be less than 1m but not less than 600mm above the ridge.
3. 1m above the top of any part of a window or skylight capable of being opened, or any ventilator or similar, which is situated in any roof or external wall of a building and is not more than 2.3m, measured horizontally, from the top of the chimney. (J 1.9).
4. Part D of Approved Document A of the Building Regulations 1985 refers to the proportions for masonry chimneys above the roof surface and the height to width relationship is given in Diagram D1 of this section. Where a chimney is not adequately supported by ties or securely restrained in any way, its height if measured from the highest point of intersection with the roof surface, gutter, etc., should not exceed 4.5W where W is the least horizontal dimension of the chimney measured at the same point of interseciton and H is measured to the top of the chimney pot or other flue terminal.

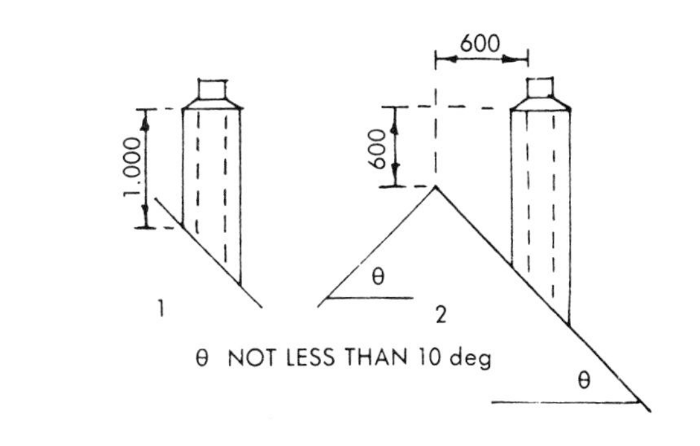

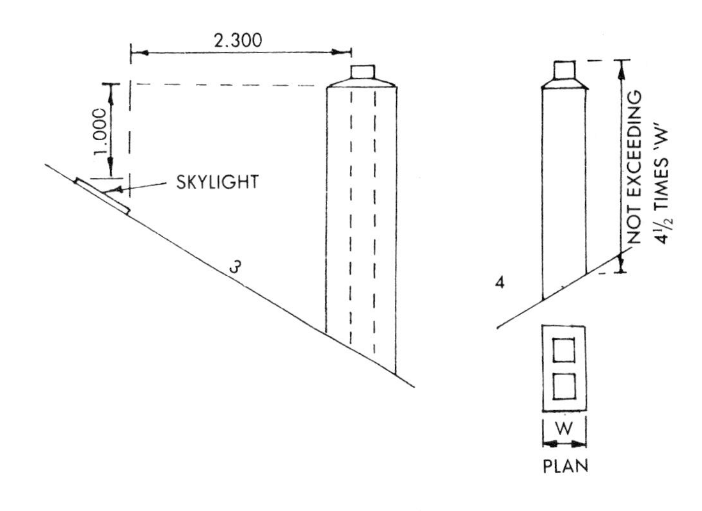

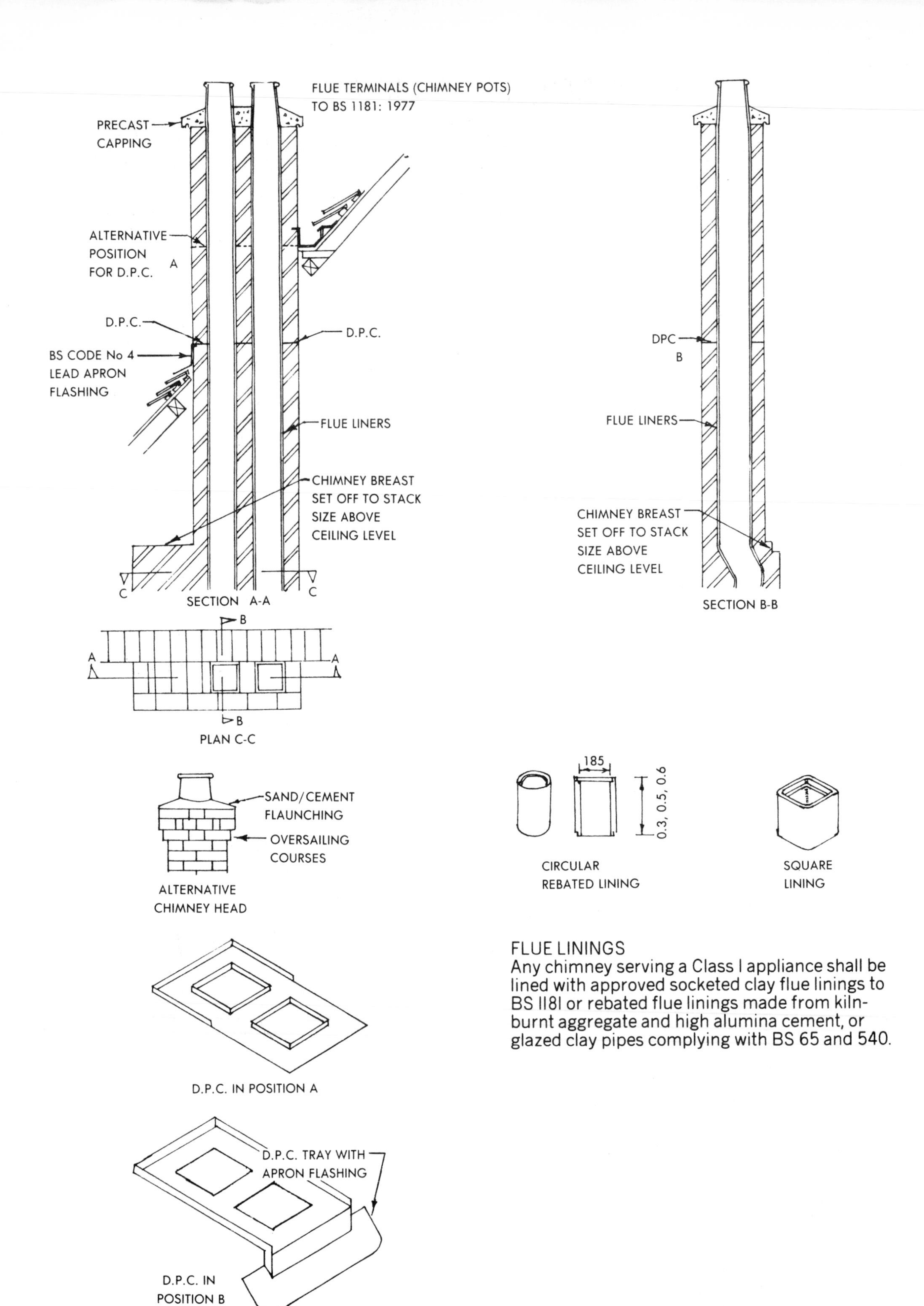

FLUE LININGS

Any chimney serving a Class I appliance shall be lined with approved socketed clay flue linings to BS 1181 or rebated flue linings made from kiln-burnt aggregate and high alumina cement, or glazed clay pipes complying with BS 65 and 540.

FIREPLACES & FLUES 4

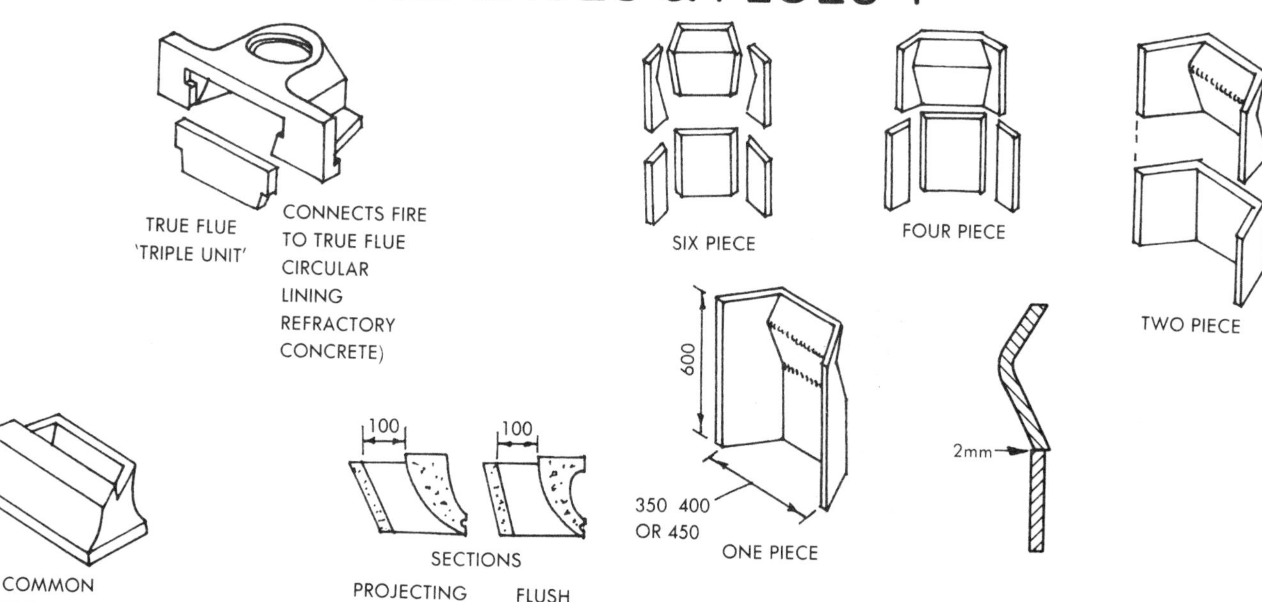

THROAT UNITS

Preformed concrete or fireclay throat units are available and provide an efficient method of making the connection from fire to flue. A wide range is available to suit flush or projecting surrounds, back boilers etc.

FIREBACKS

One-piece firebacks are available as shown but it is preferable to use two, four or six piece firebacks as there is less risk of cracking. Asbestos tape may be used to ensure a durable joint between sections. Setting the top parts back 2mm or so protects the lower edge against flame.

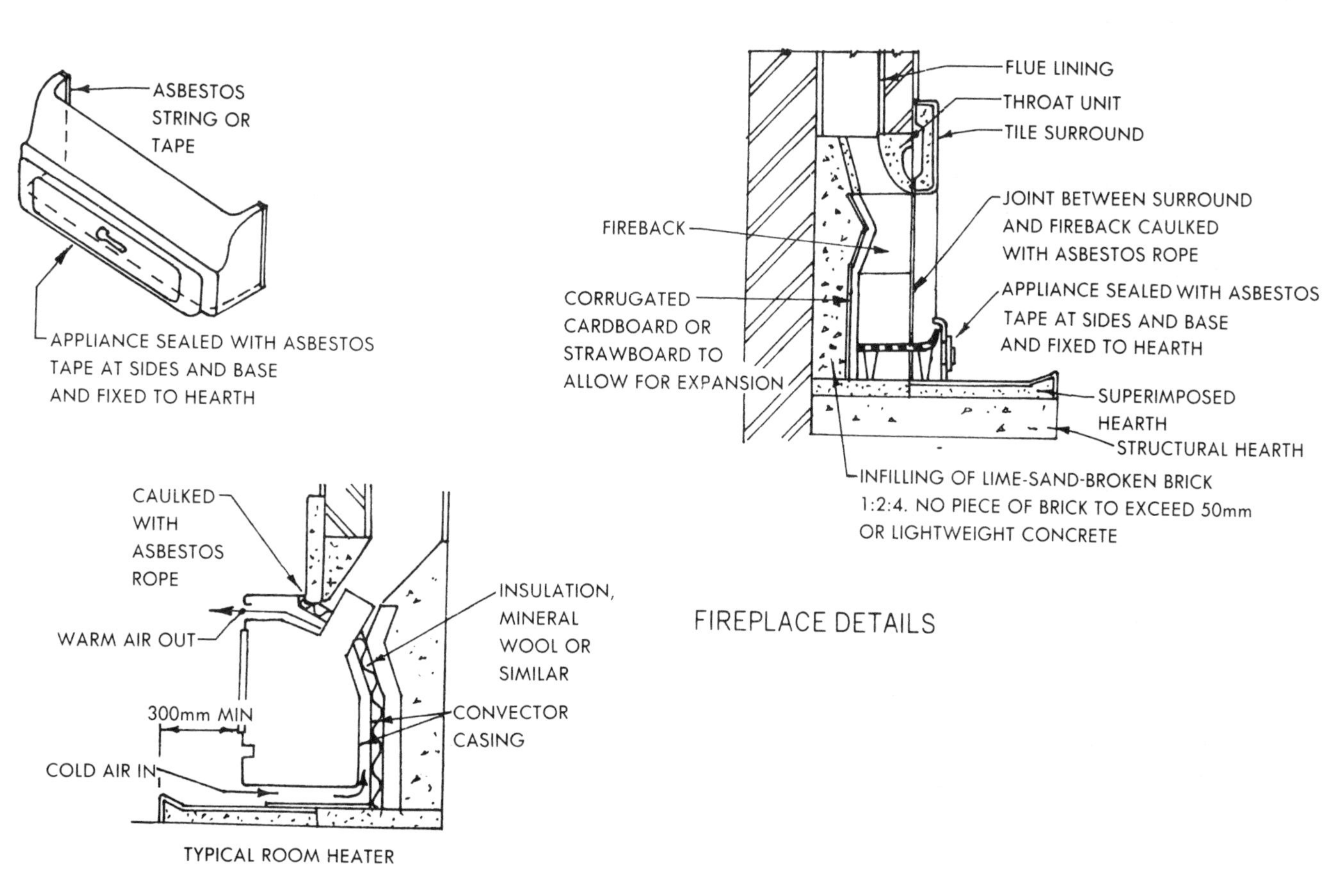

FIREPLACE DETAILS

TYPICAL ROOM HEATER

STAND-IN UNIT WITH CHIMNEY SEAL, PROVIDES SPACE HEATING BY CONVECTION AND RADIATION

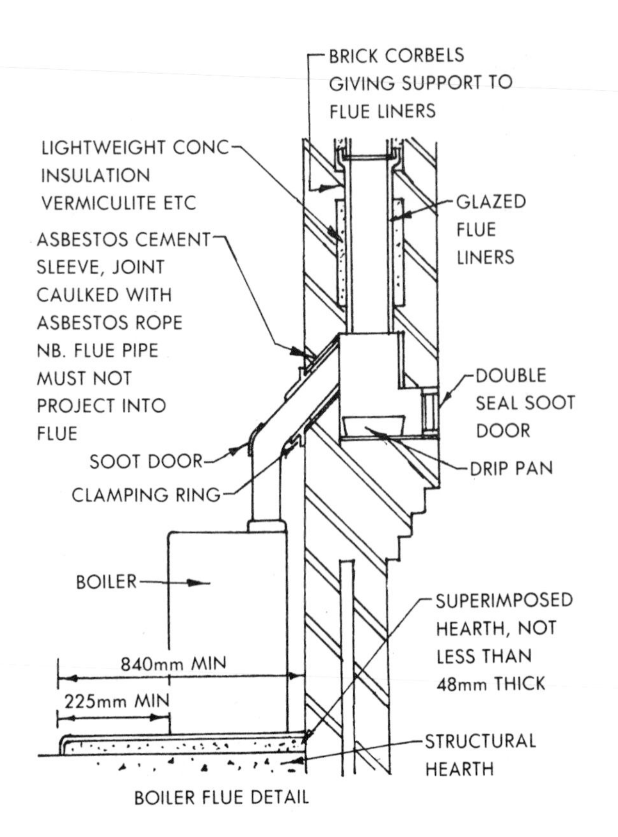

BOILER FLUE DETAIL

HEARTHS
The requirements for constructional hearths in conjunction with a fireplace recess were outlined. If the hearth is constructed otherwise than in conjunction with a fireplace recess, (e.g. as in the case of the hearth for the appliance shown opposite) it shall be of such dimensions as to contain a square having sides measuring not less than 840mm. (Reg' J3)

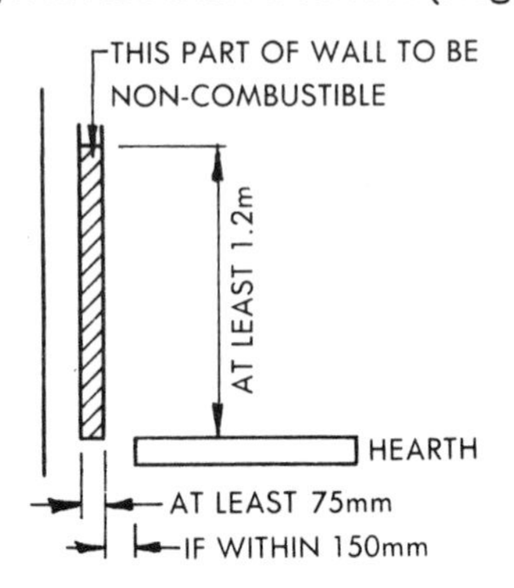

WALLS
The requirements for the construction are laid out in Approved Document J, Part C, item 1.28 of the Building Regulations 1985. Walls which do not form part of a fireplace recess but are within 150mm of the hearth should be of solid non-combustible material at least 75mm thick for a height of at least 1.2m above the top surface of the hearth.

FIREPLACES & FLUES 5

FLUES FOR GAS FIRES
Approved Document J2 of the Building Regulations 1985 deals with appliances. It describes provisions for a) cooking appliances, b) balanced-flued appliances, c) decorated log or other solid fuel fire effect gas appliance (decorative appliances) and d) other individual, natural draught, and open-flued appliances.

Part A deals with the provisions for introducing air to the appliances, and part B covers the provisions for discharging the products of combustion.

SIZE OF FLUE
Flues should have the following sizes: a) no dimension should be less than 63mm, or b) flues for decorative appliances should have no dimension across the axis less than 175mm, or c) flues for gas fires should have a cross-sectional area of at least 12000mm^2, or d) any other appliances should have a flue with a cross-sectional area at least equal to the size of the outlet from the appliance and if the flue is rectangular, the greater dimension should not be more than five times the lesser.
Materials :- Any chimney serving a gas appliance not being an appliance ventilation duct, shall be either:-

a) lined with any one of the following (i) acid resistant tiles embedded in and pointed with, high alumina cement mortar; or (ii) pipes which comply with the requirements of the Building Regulations or (iii) glazed rebated or socketed clay flue linings complying with BS 1181, jointed and pointed with high alumina cement mortar.
b) constructed of dense concrete blocks made of, or having inside walls made of, high alumina cement, and in either case jointed and pointed with high alumina cement mortar.

N.B

1. In certain cases a chimney for a gas appliance may be constructed of clay bricks or dense concrete blocks and unlined, providing, that the flue serves only one appliance and complies with the requirements of the Building Regulations in respect of type of appliance and length of flue.
2. Under certain conditions a flexible flue liner may be used.

Two types of True Flue gas flue blocks are shown above. The type X 150 has an aspect ratio 1:1:4 while the Type X 115 has an aspect ratio of 1:3:6. Where possible the aspect ratio should be near unity, but this is not always feasible and the slimmer block provides a useful alternative.

A wide range of blocks are available and the systems are flexible and adaptable. Flues may terminate in brick clad stacks as shown, or alternatively a flue terminal may be used.

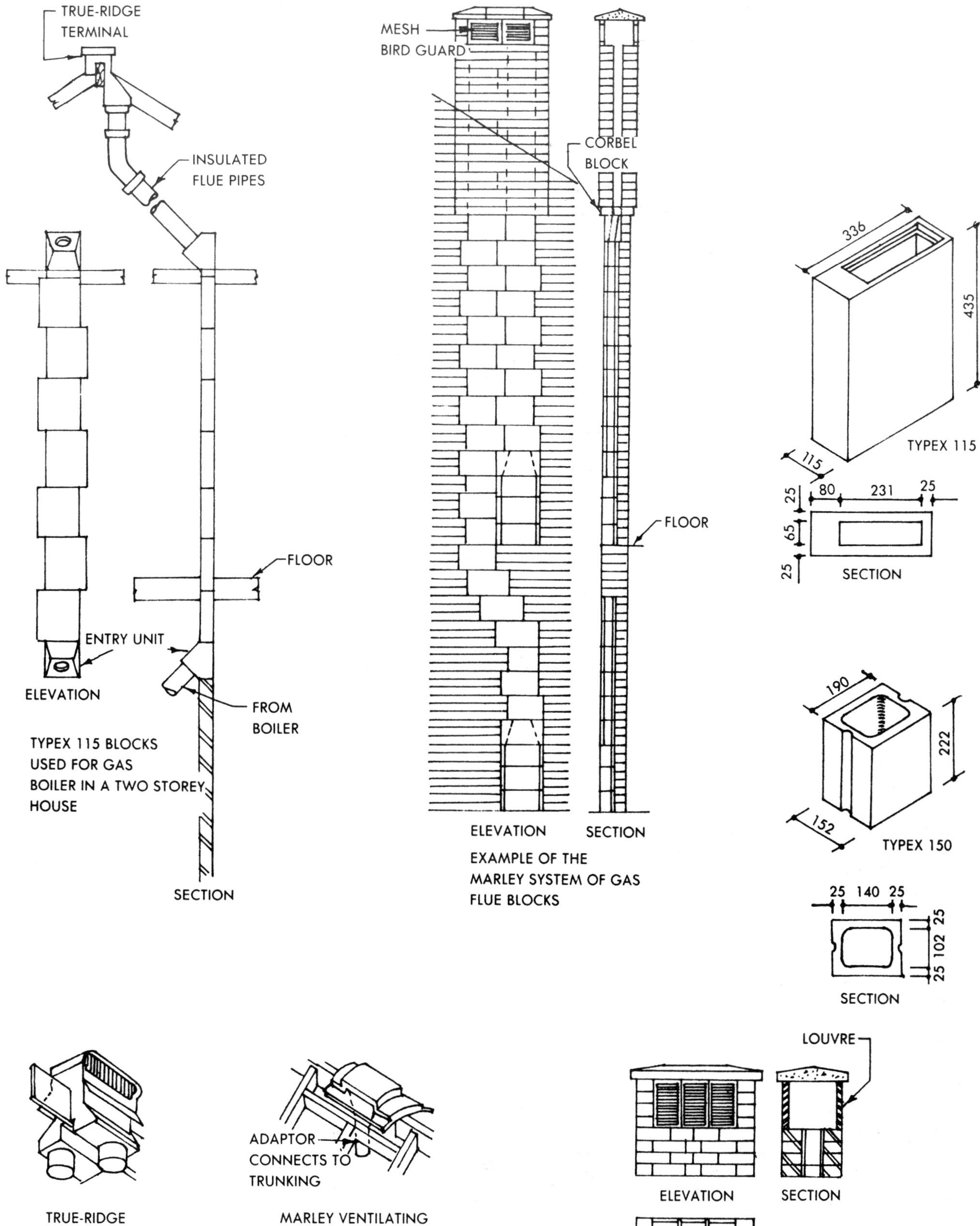

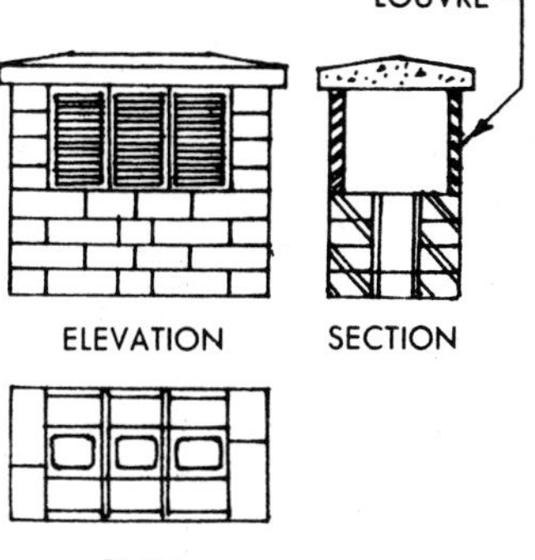

TERMINALS
The outlet of a gas flue should be fitted with a terminal, which allows free discharge, minimises down draught, prevents entry of any matter which might restrict the flue, and allows a current of air to pass freely across at all times.

Flue terminals should be built in to the stack to give an embedment of not less than 150mm, the flaunching, or one quarter the length of the terminal whichever is the greater.

CONCRETE BLOCKS

BLOCK	Length x height		Thickness work size
	Co-ordinating size*	Work size‡	
TYPE 'A' Block density† Not less than 1500 kg/m³	mm 400 x 100 400 x 200	mm 390 x 90 390 x 190	mm 75, 90, 100 140 and 190
	450 x 225	440 x 215	75, 90, 100, 140, 190 and 215
TYPE 'B' Block density less than 1500 kg/m³	400 x 100 400 x 200	390 x 90 390 x 190	75, 90, 100 140, and 190
	450 x 200 450 x 225 450 x 300 600 x 200 600 x 225	440 x 190 440 x 215 440 x 290 590 x 190 590 x 215	75, 90, 100 140, 190 and 215
TYPE 'C' Block density less than 1500 kg/m³	400 x 200 450 x 200 450 x 225 450 x 300 600 x 200 600 x 225	390 x 190 440 x 190 440 x 215 440 x 290 590 x 190 590 x 215	60 and 75

BLOCK WALLS
Concrete blocks are manufactured from a binder (cement or lime or both) and a range of aggregates. BS 6073: Part I (Precast concrete blocks) classifies blocks under three types as shown in the table

†*Block density*:- The density calculated by dividing the weight of a block by the overall volume including holes and cavities.

**Co-ordinating size*:- A size of the space, bounded by co-ordinating planes, allocated to a component, including the allowance for joints and tolerances.

‡*Work size*:- A size of a building component for its manufacture to which the actual size should conform within specified permissible deviations.

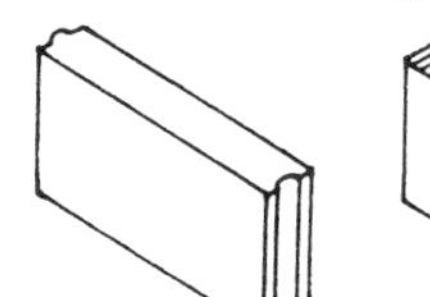

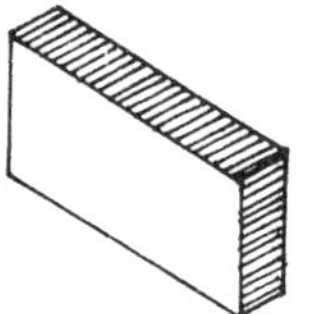

DENSE & LIGHTWEIGHT AGGREGATE BLOCKS

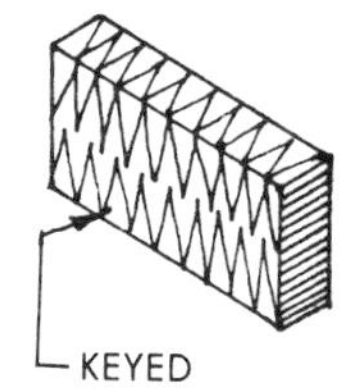

KEYED FACE AERATED BLOCK

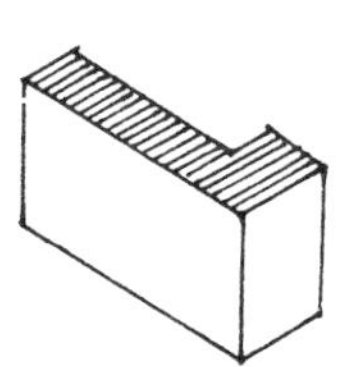

REVEAL BLOCK (CAVITY CLOSER)

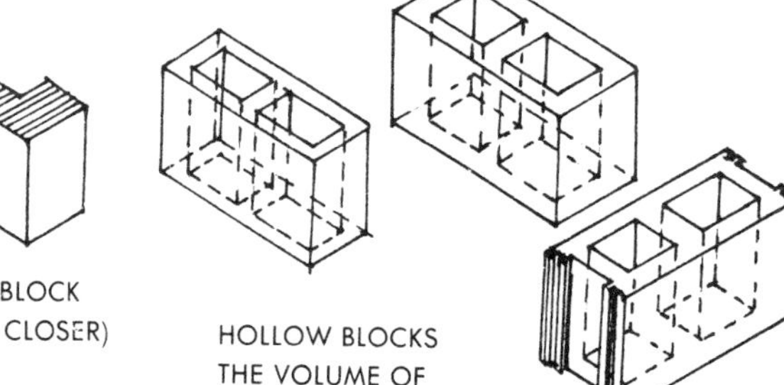

HOLLOW BLOCKS THE VOLUME OF THE VOIDS IN THE BLOCK MUST NOT EXCEED 50% OF THE GROSS VOLUME (NOT AVAILABLE IN AERATED CONCRETE)

SOLID BLOCKS CAN HAVE END GROOVES, FINGER HOLES ETC UP TO 25% OF THE GROSS VOLUME OF THE BLOCK

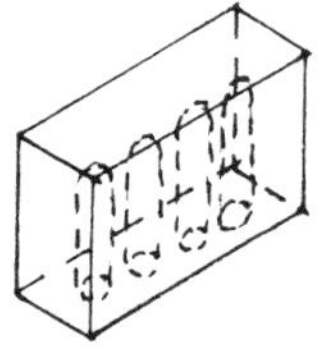

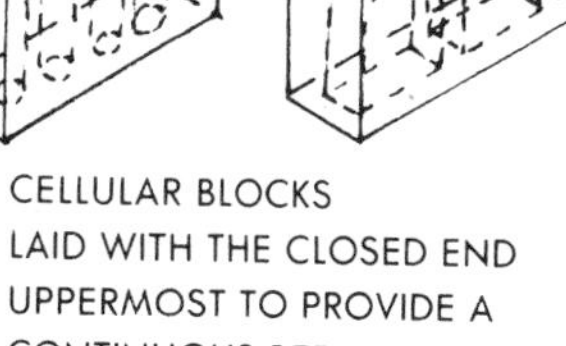

CELLULAR BLOCKS LAID WITH THE CLOSED END UPPERMOST TO PROVIDE A CONTINUOUS BED FOR THE NEXT SPREAD OF MORTAR (NOT AVAILABLE IN AERATED CONCRETE)

Type 'A' Blocks:- For general use in building including use below ground level D.P.C.

Type 'B' Blocks:- For general use in building including use below ground level D.P.C. in internal walls and the inner leaf of external cavity walls. In other positions such as below ground level D.P.C. in the outer leaf of external walls, they should be solid, hollow or cellular, Type 'B' blocks made with dense aggregates complying with BS 882 or BS 1047 only, or Type 'B' blocks with an average compressive strength of not less than 7.0N/mm², other Type 'B' blocks may also be used below ground level in these situations even though they do not satisfy these requirements, if the manufacturer supplies authoritative evidence that the blocks are suitable for the purpose for which they are used.

Type 'C' Blocks :- Primarily for internal non-loading bearing walls.

Solid blocks :- Can have end grooves, finger holes etc up to 25% of the gross volume of the block.

MORTAR

Concrete blocks should be built with mortar that is easily workable, stiffens up quickly as the work proceeds, has good adhesion to the blocks and develops sufficient compressive strength in the wall or partition. As a general principle, the mortar should not be stronger than the material of the blocks and preferably slightly weaker, so that if any shrinkage cracking occurs it will take place in the mortar joints and not in the blocks. A 1:2:9 mix (by volume) of cement, lime, sand (in cold weather 1:1:6 is suitable). Alternatively a 1:5 masonry cement, sand mix may be used, or a 1:7 cement, sand mix with plasticizer. (See also Sheet No 37).

PRECAUTIONS

(i) Concrete blocks shrink on drying out and should be covered in transit and protected on site.
(ii) If damp blocks are used or moisture is absorbed during building, the wall should be allowed to dry before plastering to reduce risks of shrinkage cracks.
(iii) Long lengths should be avoided, walls should have vertical joints every 6m filled with mastic.
(iv) In thin walls use storey height door frames.

PARTITIONS I

HOLLOW CLAY BLOCKS

Made of clay or of diatomaceous earth. The sides of the blocks are grooved as shown to provide a key for plastering. Fig I. Blocks are lightweight, not affected by moisture, provide good thermal insulation and fire resistance. Fixing to these blocks, e.g. shelving, built-in furniture etc. presents some difficulties, but special toggle fixings can be used and fixing blocks are available which can be nailed or screwed. Clay blocks are difficult to cut but half and three-quarter blocks are available to obviate cutting.

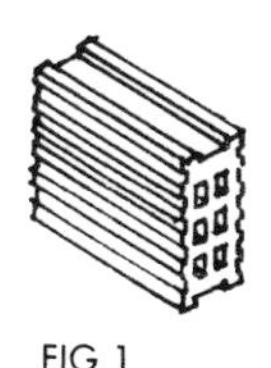
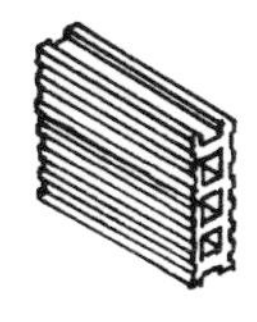

FIG 1

Junctions of partitions and main walls:- Block partitions are usually bonded into indents which are left in the main wall as shown. Fig 2. An alternative method is to build ties of expanded metal into the main wall, leaving them projecting so that they can be built into the joints of the partition wall as shown Fig 3.

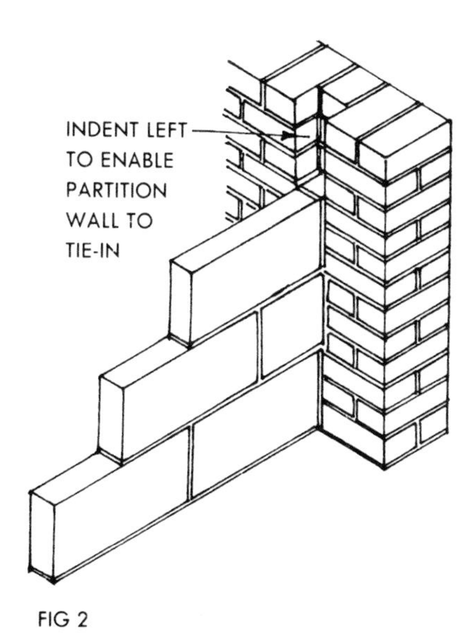

FIG 2

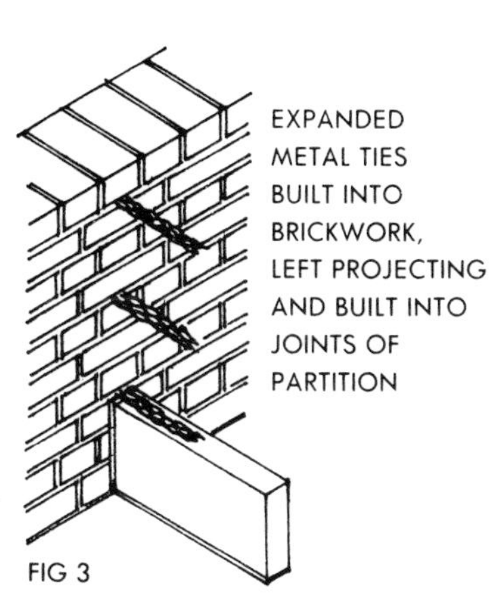

FIG 3

EDGE ISOLATION

Figs 4, 5 & 6 illustrate three methods of isolating a block partition from the main structure by means of a gap or a layer of resiliant material. This is sometimes resorted to in order to reduce the effects of possible structural movement (e.g. deflection of a floor under load), or to reduce sound transmission.

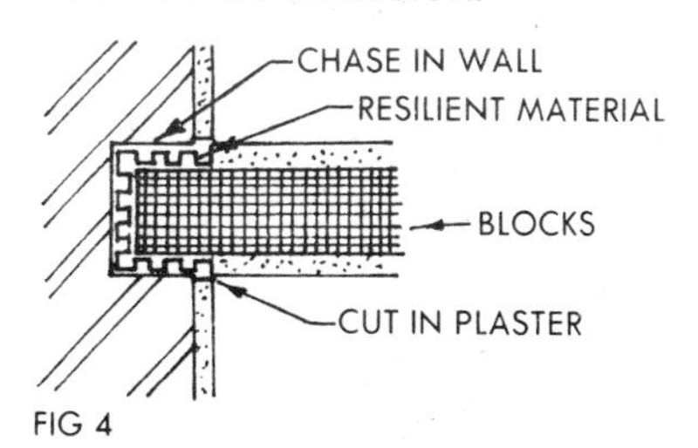

FIG 4

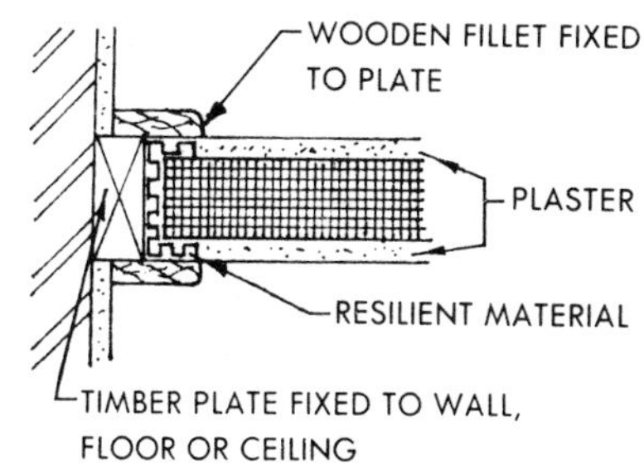

FIG 5

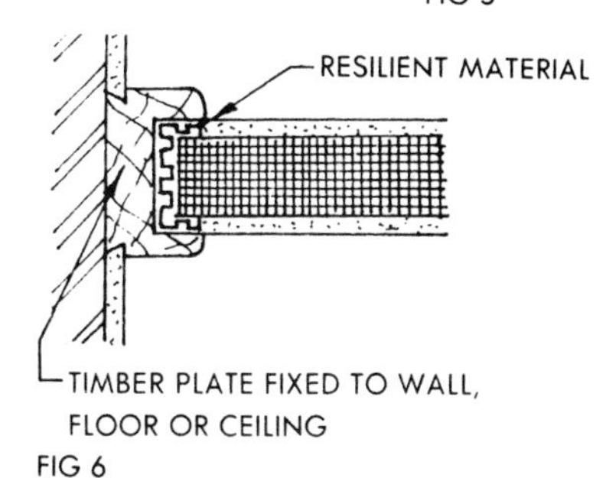

FIG 6

DOOR FRAMES

For a door opening in a block partition where the blocks are less than 75mm thick, it is advisable to use a storey height frame, fixed top and bottom. Fig 7, thus increasing stability.

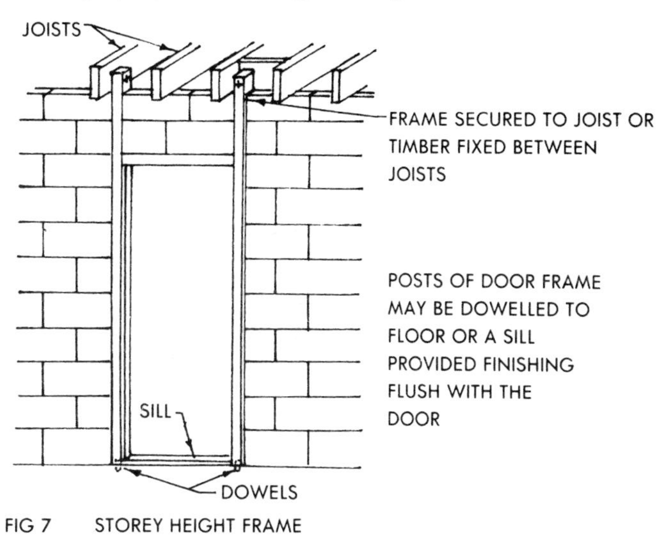

FIG 7 STOREY HEIGHT FRAME

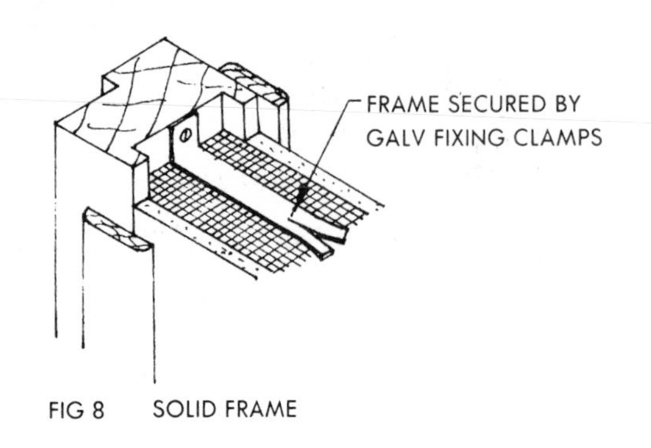

FIG 8 SOLID FRAME

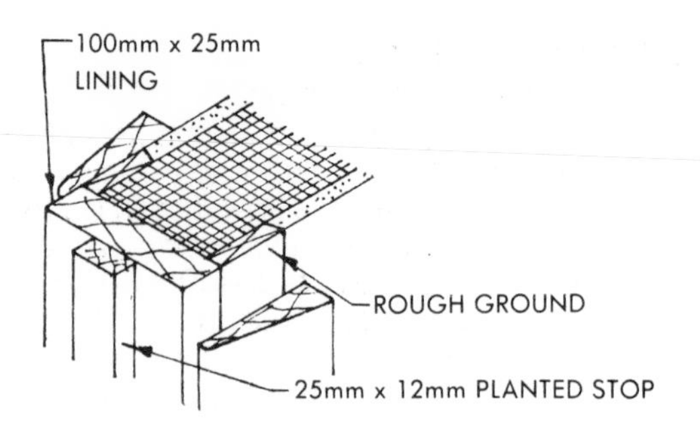

FIG 10 DOOR LININGS AS ALTERNATIVE TO FRAME

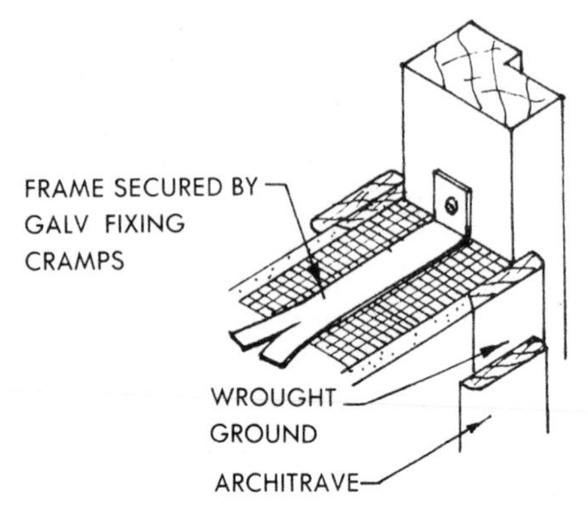

FIG 9 TIMBER FRAME WITH APPLIED GROUNDS

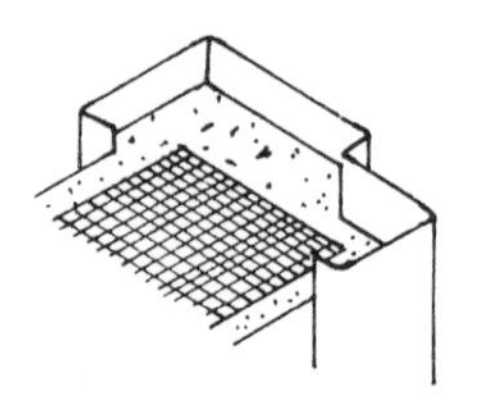

FIG 11 METAL FRAME

Figs 8 & 9 illustrate alternative arrangements of frames. Fig 10 shows the use of linings and Fig 11 shows a metal frame.

PARTITIONS 2

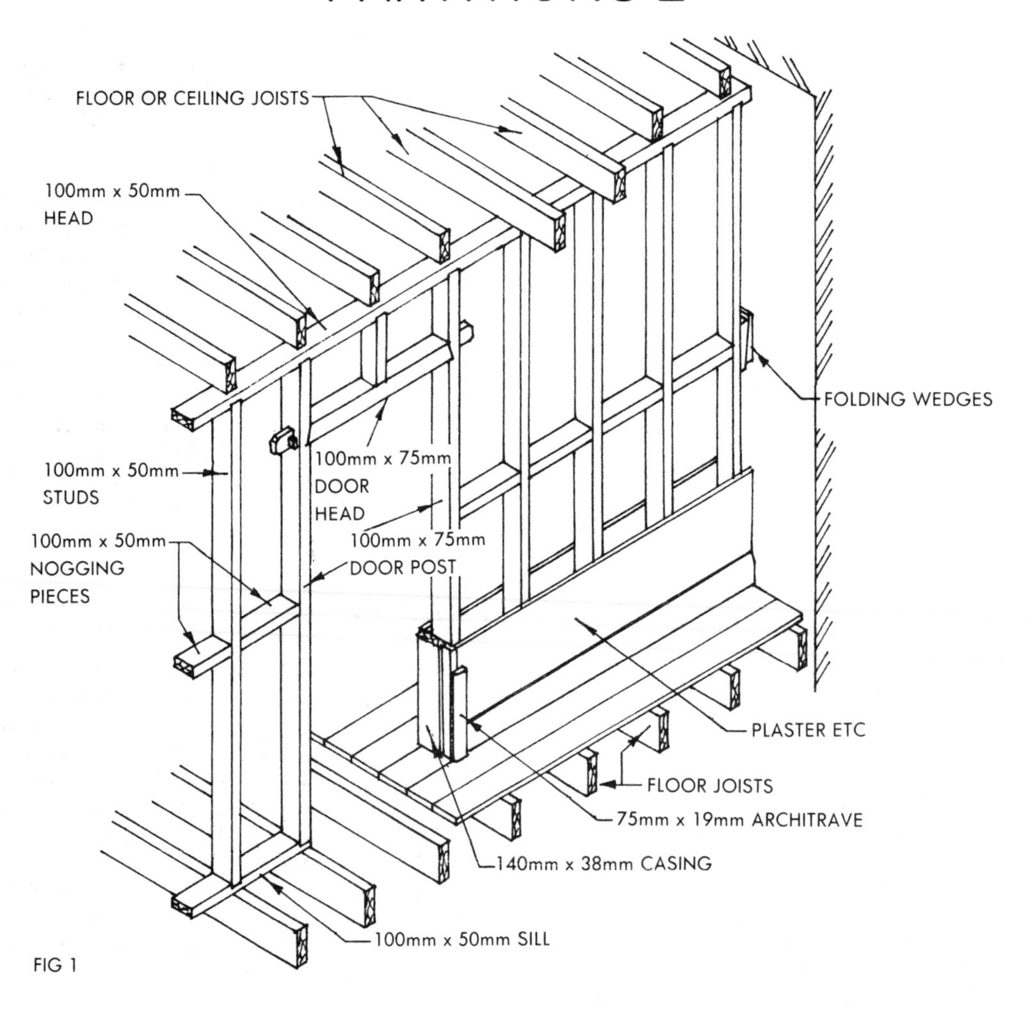

FIG 1

TIMBER STUD PARTITIONS

Also known as stoothed partitions.

The framing of the partition consists of timber studs fixed between a timber sill and head which are nailed to the joists Fig I. For partitions up to 2.400m high 75mm x 50mm studs may be used, but above this height and up to 3.000m, 100mm x 50mm studs should be used. Sill and head timbers are usually either 75mm x 50mm or 100mm x 50mm. To ensure rigidity, stiffen the studs and keep them from bowing, timber noggings are nailed tightly between the studs as shown. These noggings are at intervals in height of 1.000m approximately. The partition should be firmly wedged at each end as shown, the wedges being placed opposite the noggings.

The studs may be simply nailed to the head and sill, but it is better if they are housed into the sill and head Fig 2, or secured by a stub tenon Fig 3.

The partition may be faced with plaster on metal lathing, or covered with plasterboard, wallboard, plastic boards etc, and the spacing of the studs will largely depend upon the method of facing. Where plasterboard is used, studs should be spaced at 406mm c/c for 9.5mm thick boards and 610mm c/c for 12.7mm thick boards. If the partition is to be boarded or panelled the studs may be spaced at up to 600mm c/c.

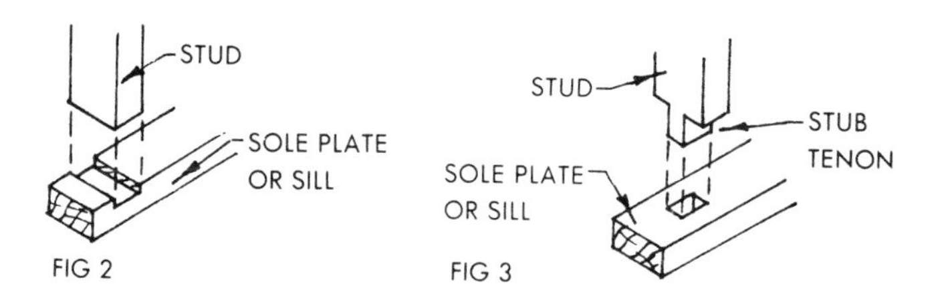

FIG 2 FIG 3

DOOR OPENING

A door opening in a stud partition is formed by framing around the opening in stouter timbers and casing around these timbers as Shown Fig I. 100mm x 75mm timbers would be suitable and the head and jamb posts should be rigidly secured with a wedged mortices and tenoned joint Fig 4.

N.B. If partition is parallel with the joists it shouild be sited immediately over a joist or a pair of joists placed together.

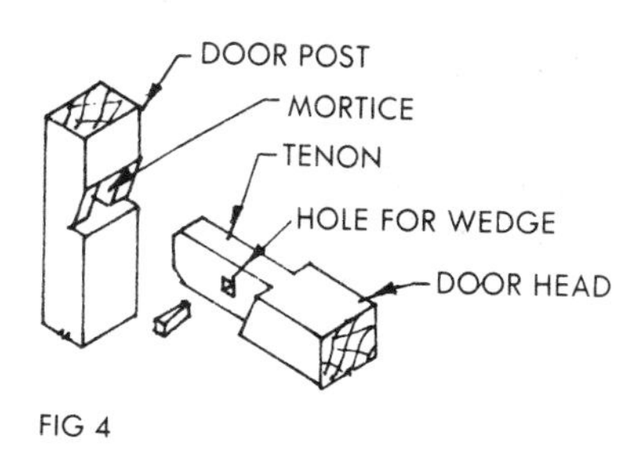

FIG 4

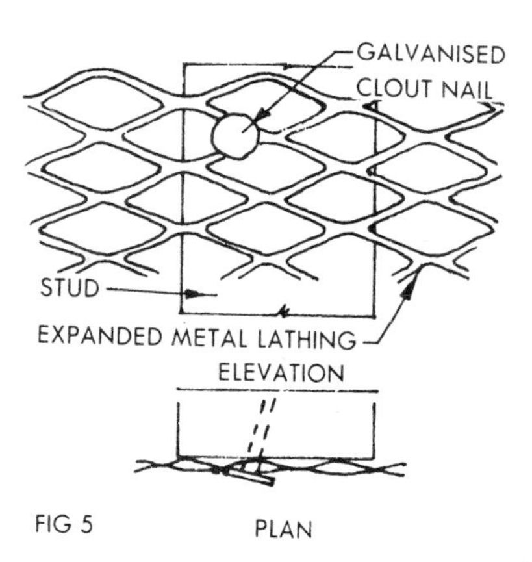

FIG 5

METAL LATHING

Metal lathing should comply with BS 1369 for partitions heavy gauge (24 S.W.G) expanded metal lathing should be used and should be protected by black asphaltum paint or be galvanised.

The lathing should be fixed with the long way of the mesh running across the supports (Fig 5) and be secured by 38mm galvanised clout nails at 100mm intervals or with galvanised staples. End laps should be not less than 25mm when the lap is on a bearer, and not less than 50mm between bearers. Side laps should be not less than 25mm and laps should be tied together with 18 gauge soft tying wire at 150mm intervals. The wire should be as taut as possible and this is best achieved by nailing in the centre and extending throughout its length. Cut ends of wire should be bent inwards away from the finishing coat.

PARTITIONS 3

PLASTERING ON METAL LATHING

The metal lathing should be taut and firmly secured Metal lathing plaster is recommended, this does not shrink and contains a rust inhibitor, which providing the plaster coverage of cut ends is adequate will ensure that rust stains do not occur.

Three coat work should be used and the renderinig coat of I part Thistle metal lathing plaster to 1½ parts clean sharp sand by volume should contain a suitable additive, rayon fibre being commonly used today instead of hair. The first coat should be scratched as it stiffens to provide a good key when the rendering coat has set hard. The floating coat of I part Thistle metal lathing to 2 parts clean sharp sand by volume should be applied, ruled and scratched to provide a key for the finishing coat. The finishing coat may consist of Thistle finish applied neat or with the addition of well slaked lime.

Where a cement mix is required, the following are alternatives:- 1:2:8-9 cement, lime, sand; or 1:8 cement, sand with mortar plasticiser; or 1:6 masonry cement, sand.

PLASTERBOARD
Consists of an aerated gypsum core encased in specially prepared paper liners. Plasterboard should comply with the requirements of BS 1230 'Gypsum plasterboard', and is available in plain or insulating grades, the latter having a veneer of polished aluminium on one side. There are four basic types of plasterboard:-

GYPSUM WALLBOARD
Available with tapered, square or bevelled edges as shown on left. The boards have one grey or foil surface and one ivory surface for direct decoration, and are available in two thicknesses 9.5mm and 12.7mm. Sheet widths are 600, 900 and 1200mm and lengths are 1800, 1350, 2400, 2700 and 3000 and 3600mm.
Framing:- Movement of timber framing due to moisture fluctuations is a common cause of cracking at the joints of the plasterboard, and properly seasoned timber complying with the recommendations of CP112 'The structural use of timber in building' should be used. the following table gives the spacing for timbers:-

TAPERED EDGE FOR SMOOTH SEAMLESS JOINTING

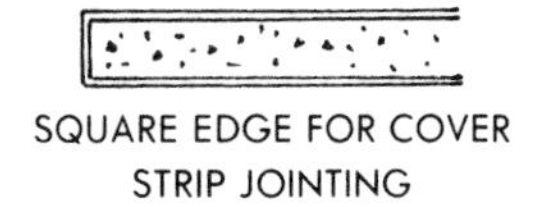
SQUARE EDGE FOR COVER STRIP JOINTING

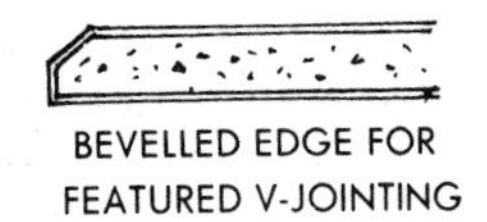
BEVELLED EDGE FOR FEATURED V-JOINTING

'A':- Covers situations where timbers are not precise centres and unlikely to become distorted.
'B':- Covers situations where timbers have to be at maximum indicated or where conditions under 'A' cannot be maintained.
*Noggings desirable. **Noggins essential.

The timber supports to which two paper bound edges of plasterboard are to be fixed should permit nailing of the plasterboard not nearer than 13mm from its edges and allow nails to be driven in straight. Where two cut edges of board are fixed to a timber section, the timber support should permit a 3mm gap between boards. Boards should be nailed to every support of 150mm c/c commencing from the centre of the boards and working outwards. The board layout should be arranged so that as far as possible cut edges occur at internal angles. Joints directly above door jambs and above and below window jambs should be avoided.

When lining all internal surfaces, fix ceiling boards first and then follow by fixing to walls and partitions.
Nails:- Wallboard nails are galvanised 14 SWG with small flat heads and smooth shanks. 32mm for fixing 9.5mm boards and 38mm for 12.7 mm boards. Nails should be driven home without fracturing the paper surface and the final hammber blow should leave a shallow depression to facilitate spotting. All nails should be spotted with Gyproc joint filler and joint finish.
Joint treatment:- Smooth, seamless jointing is possible with tapered edge boards. Joints are filled with a suitable filler, e.g. 'Gyproc joint filler' and reinforced with joint tape. When the filler has set a thin layer of Gyproc joint finish can be applied. A slurry of gyproc joint finish can then be applied over the whole board, using a jointing sponge, to given an even texture.
Decoration:- Before decorating a single coat of wallboard primer should be applied to the whole surface. Any form of decoration can then be used, normally applying two coats. N.B. A primer is essential before paper hanging.

A subdued panel effect is obtained using square edged boards and appropriate paper cover strips, wood, plastic, or metal strips give a more pronounced panel effect.

FRAMING MEMBERS	BOARD THICKNESS	'A' RECOMMENDED CENTRES	'B' RECOMMENDED CENTRES
Joists	9.5mm 12.7mm	406mm* 457mm*	457mm** 610mm**
Studs or furring	9.5mm 12.7mm	406mm 610mm	457mm 610mm

PARTITIONS 4

GYPSUM LATH
A narrow width plasterboard which is less expensive than wallboard and has been designed as a base for Gypsum plaster. The edges are rounded as shown. Fig 1. There is no need to reinforce the joints with scrim except at ceilings and wall angles. Thicknesses are 9.5mm and 12.7mm, width of sheets is 406mm and lengths are 1200, 1219 and 1372mm.
Fixing:- Gypsum lath is fixed across the studs with joints staggered as shown. Fig 2. Each lath should be secured to each support with at least four nails equally spaced across the width and driven not closer than 13mm to its edges. Any nut end should be located over a support. Noggings are not required to support edges, but provision must be made to support the ends and edges of the lath at the perimeter of a ceiling.
N.B. Approx' 3.0 kg (6.6lb) of 32mm nails, or 3.3 kg (7.2lb) of 38mm nails are required to fix 100m² of Gypsum wallboard and approx' 6.8 kg (15lb) of nails for 100m² of Gypsum wallboard.
Plastering:- It is possible to obtain a good finish with a single coat of board finish plaster 5mm thick, but it is essential in this case that the boards are fixed to a true, plane surface. If this is

not possible and there are any variations in the surface, two-coat work should be used.
Two-coat work :- For traditional work a floating coat of sanded, Class 'B' plaster (e.g. 1 part Thistle haired plaster to 1½ parts sand by volume) approx' 8mm thick will be satisfactory. This coat should be lightly scratched to form a key. Any normal finishing coat may be used, e.g. Thistle finish plaster applied neat or with the addition of well slaked lime putty (not exceeding 20%) may be used. Alternatively Sirapite may be applied neat.
N.B. (i) Sirapite plaster should not be applied direct to plaster boards.
(ii) Lime should not be added to any plaster in direct contact with plasterboard.
(iii) Plasterboards should not be wetted before plastering.

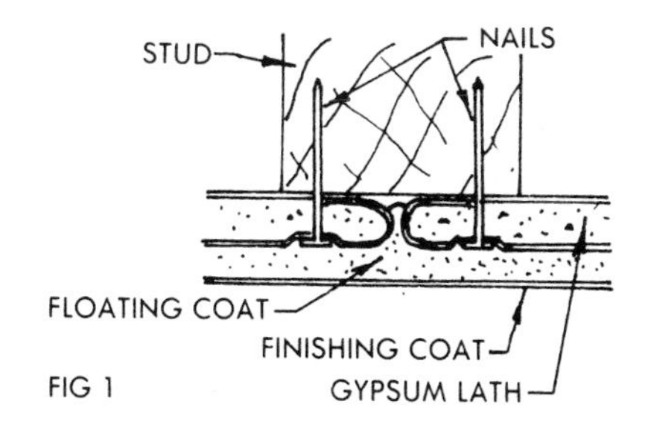

FIG 1

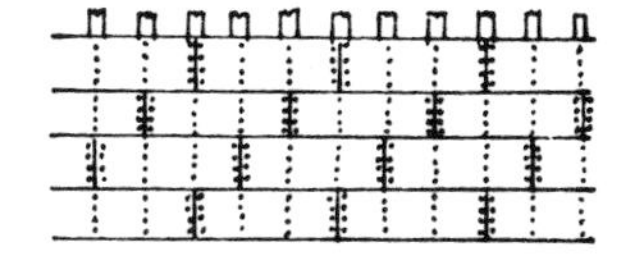
FIG 2

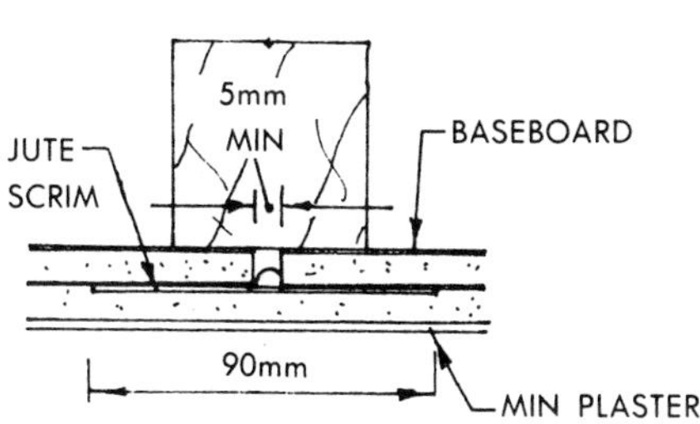

FIG 3

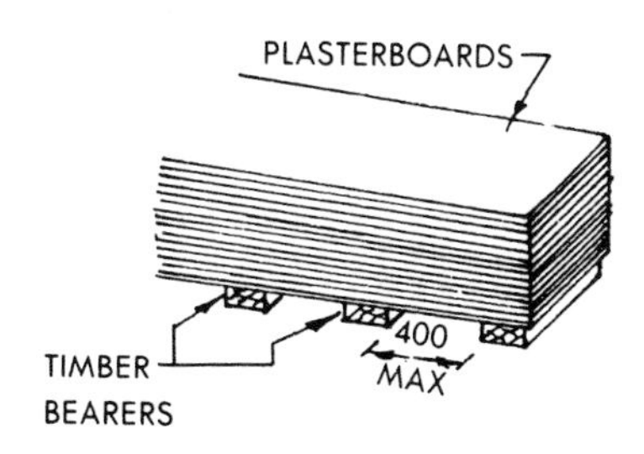

FIG 4

An alternative method is to use pre-mixed lightweight Gypsum plaster, e.g. a floating coat of Carlite bonding coat plaster, followed by a finishing coat of Carlite finish plaster. This will increase fire resistance and reduce condensation.
Gypsum Baseboard :- A square edged plasterboard designed as a base for Gypsum plaster, supplied in relatively small sheets 914mm wide and 1200, 1219 and 1372mm long. Framing is similar to that previously described and centres must not exceed 457mm. A gap not exceeding 5mm should be left between all edges of adjoining boards and all joints and angles should be reinforced with jute scrim at least 90mm wide. Fig 3.
Gypsum Plank :- A narrow width board of greater thickness than any other type of plasterboard. Thickness 19mm, width 600mm, lengths 2350, 2400, 2700 and 3000mm. Obtainable either with grey surfaces designed as a plaster base, or with one ivory surface for direct decoration. May be used for dry lining any framed structure when a high standard of fire protection is required without plastering, and provides a very high fire resistance for column and beam encasement when plastered. Framing must be accurately spaced at centres not exceeding 914mm for ceiling lining or 760mm for partition and wall lining. 57mm wallboard nails should be used for fixing.
Storage :- Plasterboards should be stacked flat in a dry place on a level surface and properly protected from rising damp and bad weather. They should preferably be stacked on 100-125mm wide timber bearers, of length not less than the width of the boards, which should be spaced not more than 400mm apart. Where plasterboards for direct decoration are used, the ivory surfaces should be together. (These boards are usually supplied in pairs with the ivory surfaces in contact and should not be separated until they are required). Where boards must be stacked outside, the top and sides of the stack must be completely covered with waterproof sheets securely anchored, and the space between bearers should be ventilated to stop ground moisture rising and being trapped inside.
Boards should be decorated or plastered as soon as possible after fixing, so that dust, dirt or grease will not accumulate on the surface and interfere with the adhesion of the plaster. Angles may be strengthened with 'corner tape' — a strong paper tape reinforced with nylon strips. When maximum protection at external angles is required, galvanised steel angle bead is recommended.

BUILDING BOARDS

Fibre building boards may be used for stud partitions, wall linings and ceilings. Framing battens for wall linings should be at least 38mm wide, and partition studs at least 50mm wide. Boards should be securely fixed to supports in the middle as well as at the edges, noggings provided as required, and at not more than 1200mm c/c. Studs should be spaced according to the width and type of board being used.

Storage:- Boards should be stored in a cool, dry place, with edges and surfaces protected against damage. Insulating boards should be 'conditioned' by stacking loosely on edge in the room in which they are to be used for at least 24 hrs. Hardboards may be conditioned by sponging the back (a litre of water to 6m² of board) and then stacking back to back in the room where they are to be fixed for at least 24 hrs, or preferably 48hrs.

Conditioning:- The moisture content of board will vary with the condition of the atmosphere and slight movement of the boards will result from these variations. Fixing of the boards should preferably be delayed until the building has been allowed to dry, or at least until the building is glazed and external doors fixed.

Joints:- A panel effect may be achieved by using cover strips of wood, metal, plastic or strips of fibre board. Alternatively the edges of the boards may be bevelled or wooden beading inserted in the joints. For flush joints the edges of boards can be bevelled and the joints filled with scrim and a suitable filler.

Nailing:- Rust resisting nails must be used. Round wire nails with flat heads, 10 gauge, 38mm long for insulating board and 17 gauge 19 or 25mm long for hardboard. If joints are not to be covered use lost head nails punched down below surface of board. There should be a 3mm gap between boards (boards should never be forced into position) nails being spaced at 100mm intervals around perimeter of boards and 200mm elsewhere.

In some cases it will be necessary to treat the boards in order to control the rate of surface spread of flame, this may be achieved by painting the surface with a suitable flame retardant paint. Boards are also available impregnated with flame retardant chemicals.

Asbestos substitute boards (insulating and wallboard) and plastic boards (expanded polystyrene and foamed polyurethane) are suitable for plastering. The latter have low resistance to impact and when plastered should be fully bonded to a firm background. Joints between boards, and all angles should be reinforced with jute scrim, or galvanised wire scrim to prevent cracking for single coat work class 'B' board finish plaster is suitable, and for two coat work an undercoat of 1:1½ class 'B' gypsum plaster and sand (if type 1 sand is used or 1:1 if type 2 sand) followed by a final coat of neat gypsum plaster. Alternatively two-coat premixed lightweight gypsum plaster may be used.

N.B. BS 1198 'Sands for internal plastering with gypsum plasters' stresses the importance of good grading and refers to 'Type One' and 'Type Two' sand giving the sieve analysis for each. Where 'Type Two' sand is used in a mix instead of Type One the proportion of sand should be reduced by one-third.

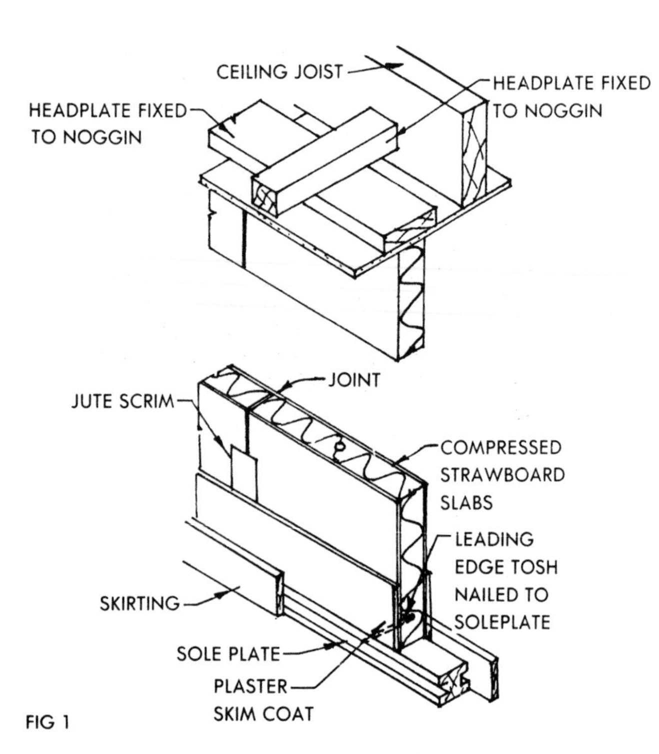

FIG 1

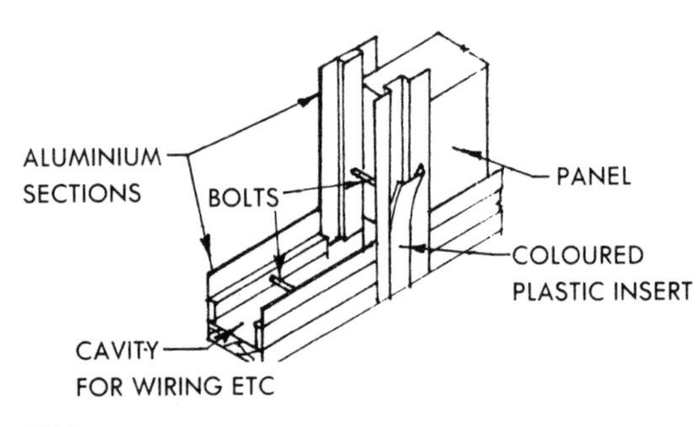

FIG 2

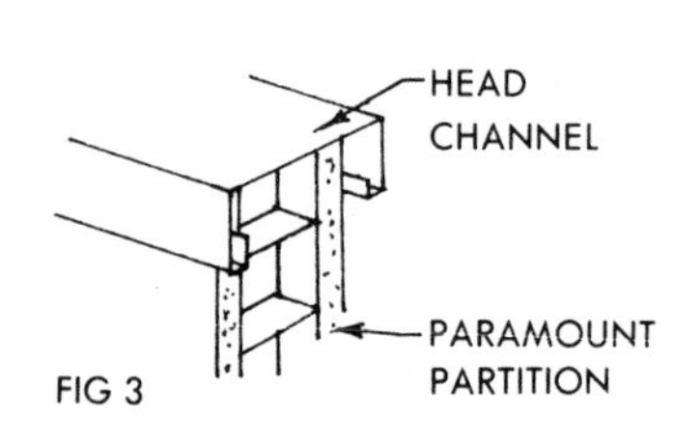

FIG 3

LIGHTWEIGHT AND DEMOUNTABLE PARTITIONS

A wide range of patent lightweight, demountable and folding partitions are available. The construction is usually based on metal, wood, or plastic sections or trim with various infill panels in a variety of materials, strawboard, insulation board, chipboard etc. Provision is normally made for the housing of electrical services. Most demountable partitions are fixed by pressure between the floor and ceiling.

Fig 1 shows a partitioning system, based on compressed strawboard slabs cut to size and jointed together with adhesive. Holes are provided in the slabs for conduit or wiring and the partitioning is designed for receiving a skim coat, board plaster finish. Slabs are 50mm thick, 1200mm wide and 2200, 2300 and 2400mm long. A variety of partitions are available and Fig 2 illustrates partitioning in adjustable metal sections.

Paramount demountable partitioning is shown in Fig 3. This is a gyproc partition with a cellular core fixed into light aluminium framing.

SHORING I

Shoring may be needed to give temporary support to walls and floors during alteration work, for demolition work, during underpinning operations, and as a safety measure where a structure has become unsound.

DEAD SHORING

The purpose of dead shoring is to support dead and superimposed loads of the building. It is often necessary to strut existing floors and roofs to relieve the weight from the walls during alteration and repair work, and in this case the shoring may be of a lighter construction than the main shoring supporting the walls. Fig I shows a typical example.

While timber shores as shown may be quite satisfactory, and it may be possible to cut costs by using sound second-hand timber, and possibly building up heavier shores from lighter sections, (e.g. using three 225mm x 75mm to build up a 225mm x 225mm shore). The use of adjustable steel props, Fig 2 is to be recommended. These are very strong, quick and easy to fit and adjust, and may be hired at reasonable rates.

An example of the use of needles and dead shores to support the wall above while a new opening is being formed is shown in Fig 3.

Needles :- May be of timber, or alternatively steel sections may be used. Size will depend on spacing of needles, distance apart of supports and loading.

Timber shores :- Must be of adequate strength and free from major flaws. A rough guide to size, is to see that the height does not exceed approx' 24 times the least width of the timber. Shores must be upright and should be securely fixed to baseplates and head, preferably using dogs. Fig 3.

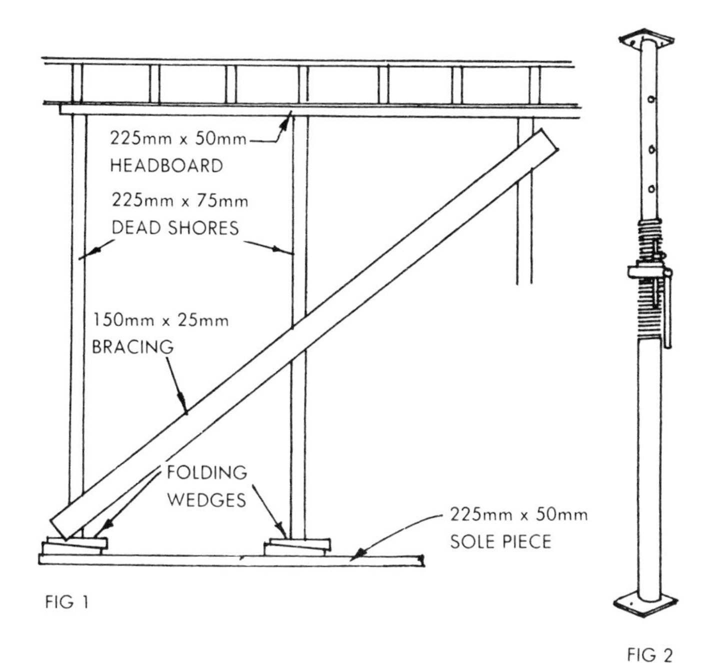

FIG 1

FIG 2

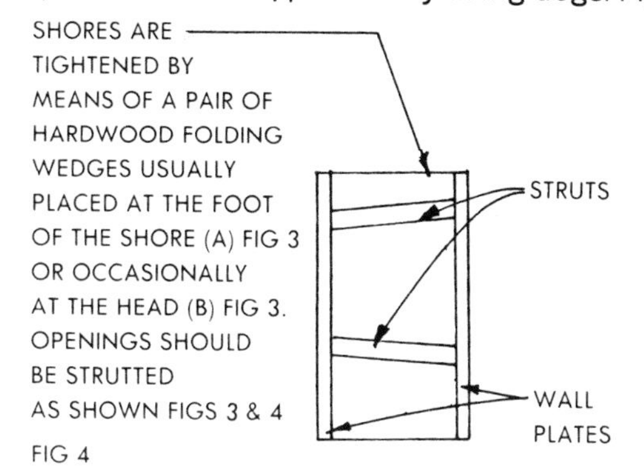

FIG 4

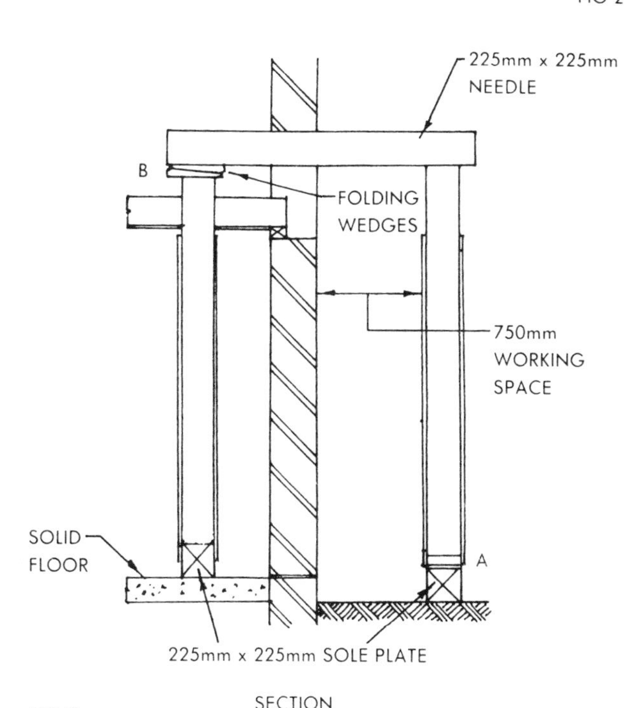

SECTION

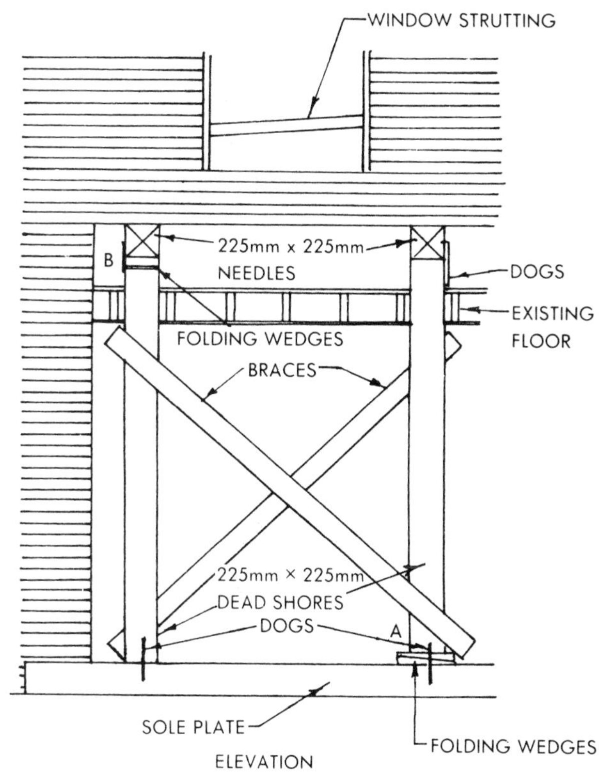

ELEVATION

FIG 3

Spacing:-Needles should not be placed beneath window openings. Brickwork in good condition in cement mortar will only require supports every 1.5m to 1.8m. Brickwork in lime mortar however will require needles every 0.9m to 1.2m. Needles will normally have to be long enough to allow for working space of at least 750mm on at least one side of the wall, and possibly on both sides. Fig. 3.
Bracings:- Should be fixed as near as possible at an angle of 45 deg and securely nailed in position. 225mm x 25mm boarding may be used.

Sole plates:- Must be on firm ground. Where the sole plate or base plate is placed on an existing floor care must be taken to see that the floor is strong enough to take the imposed load. Sole plates should be continuous and at right angles to the joists. It may be advisable to take the shores through the floor and place the sole plate on the over site concrete.

On a restricted site it may be necessary to place a new beam or lintel against the wall at the base, before placing the shoring in order to manoeuvre the beam into place.

SHORING 2

RAKING SHORES

Raking shores may be required to provide temporary support to a wall which has become defective and in danger of collapsing, or as a precautionary measure to ensure adequate support to the existing structure during alteration work.

The design and arrangement of the shoring will depend upon the height of the building, loads to be carried, openings in the building and space available on the adjacent ground. Raking shoring may consist of a single raker or a compound system of shores Fig 1.

The sole piece usually 75mm to 100mm thick should be at an angle of 85 deg (approx') to the top raker, and if necessary a grillage provided Fig 1. A crowbar can be used to lever the raker into position Fig 4. (A heavy hammer or maul should not be used when adjusting a raker as this can result in damage and possible movement of the wall). Finally a cleat is nailed in position. Fig 1.

All members should be fixed with dogs, cleats and braces as shown in Fig 1.

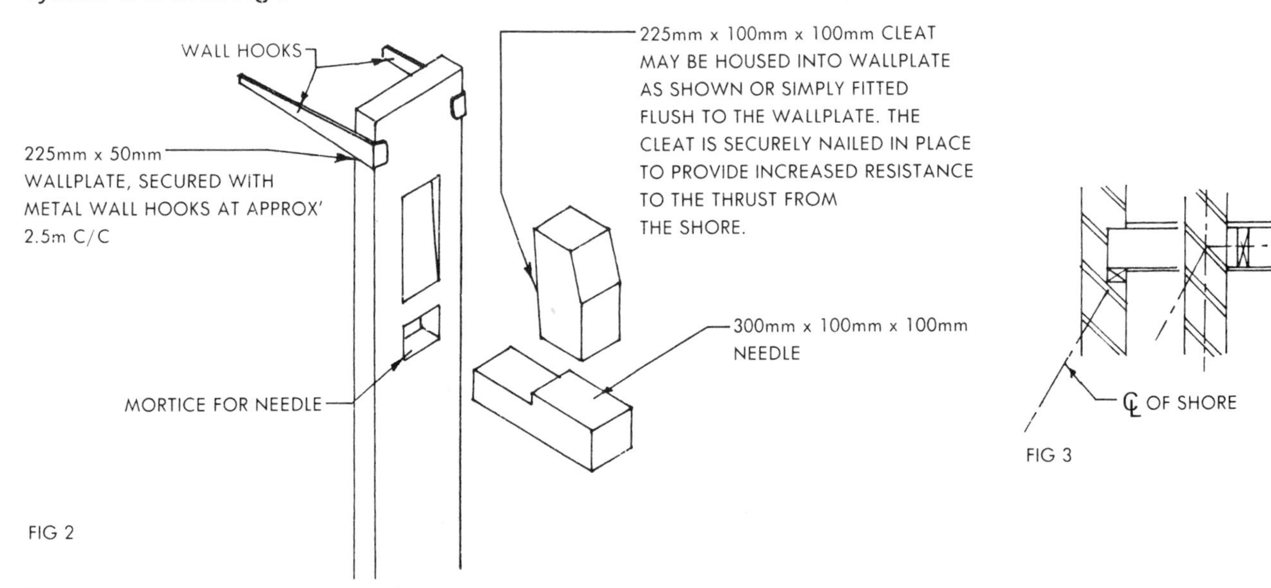

FIG 2

FIG 3

The rake of the shoring should be such, that as far as possible, the outside shore is not less than 60 deg nor more than 75 deg to the horizontal. The distance apart of the shores will depend upon prevailing conditions and may vary from 2.5m to 5.0m. It is advisable to place shores at each end of the wall to be supported.

The centre line of each raker should intersect the wall plate or centre line of the floor. Fig 3.

A hole is cut into the wall in the required position so that a needle may be inserted at the head of the raker. The wallplate, which gives an even distribution of the load is secured in position by wall hooks as shown, the needle then inserted and strengthened by the provision of a cleat (which may be housed into position) nailed to the wall piece. Fig 2.

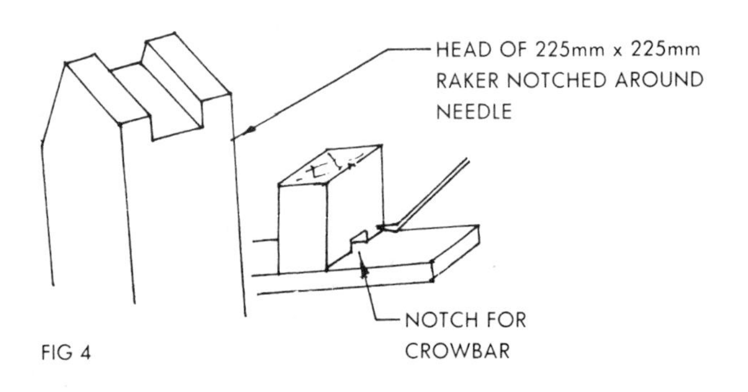

FIG 4

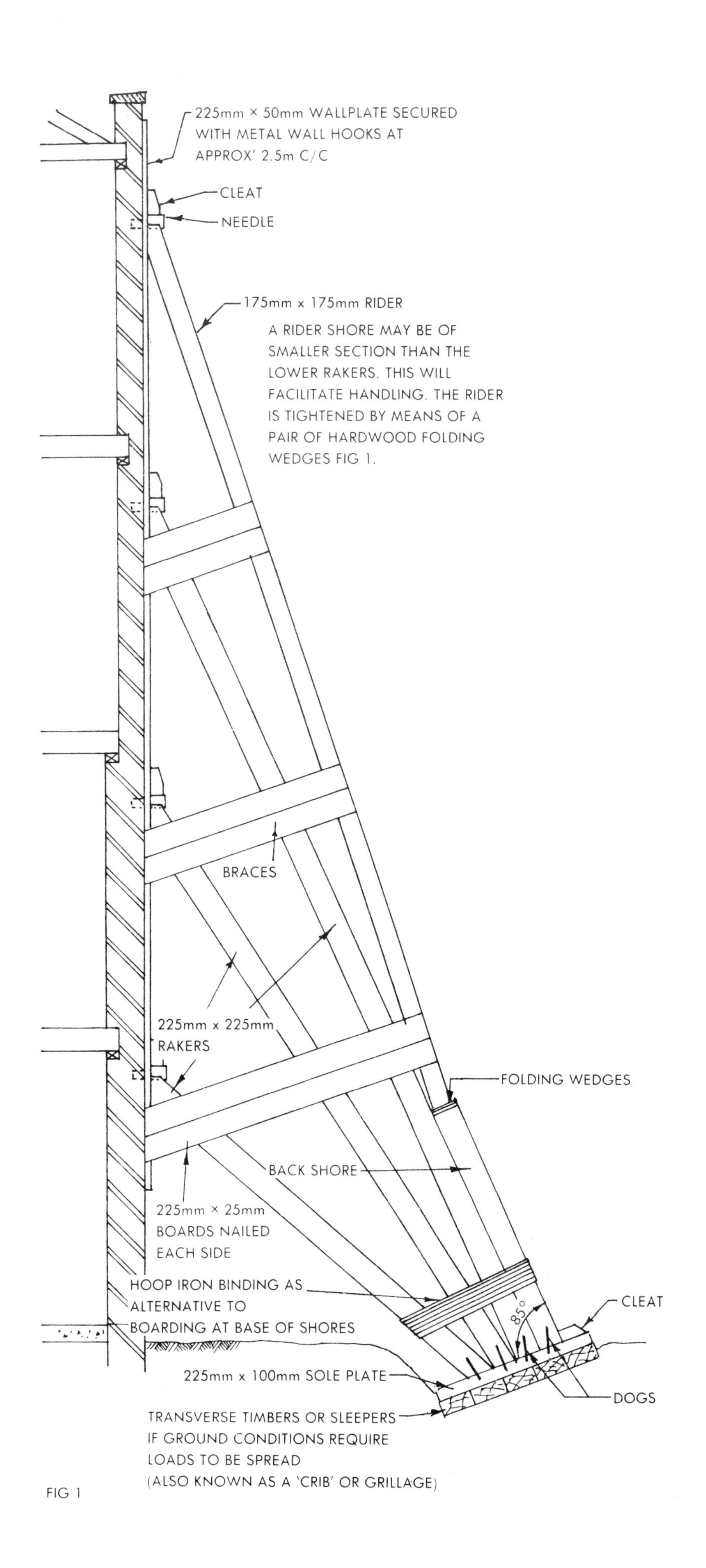

225mm × 50mm WALLPLATE SECURED WITH METAL WALL HOOKS AT APPROX' 2.5m C/C
CLEAT
NEEDLE
175mm x 175mm RIDER
A RIDER SHORE MAY BE OF SMALLER SECTION THAN THE LOWER RAKERS. THIS WILL FACILITATE HANDLING. THE RIDER IS TIGHTENED BY MEANS OF A PAIR OF HARDWOOD FOLDING WEDGES FIG 1.
BRACES
225mm x 225mm RAKERS
FOLDING WEDGES
BACK SHORE
225mm × 25mm BOARDS NAILED EACH SIDE
HOOP IRON BINDING AS ALTERNATIVE TO BOARDING AT BASE OF SHORES
85°
CLEAT
225mm x 100mm SOLE PLATE
DOGS
TRANSVERSE TIMBERS OR SLEEPERS IF GROUND CONDITIONS REQUIRE LOADS TO BE SPREAD (ALSO KNOWN AS A 'CRIB' OR GRILLAGE)

FIG 1

SHORING 3

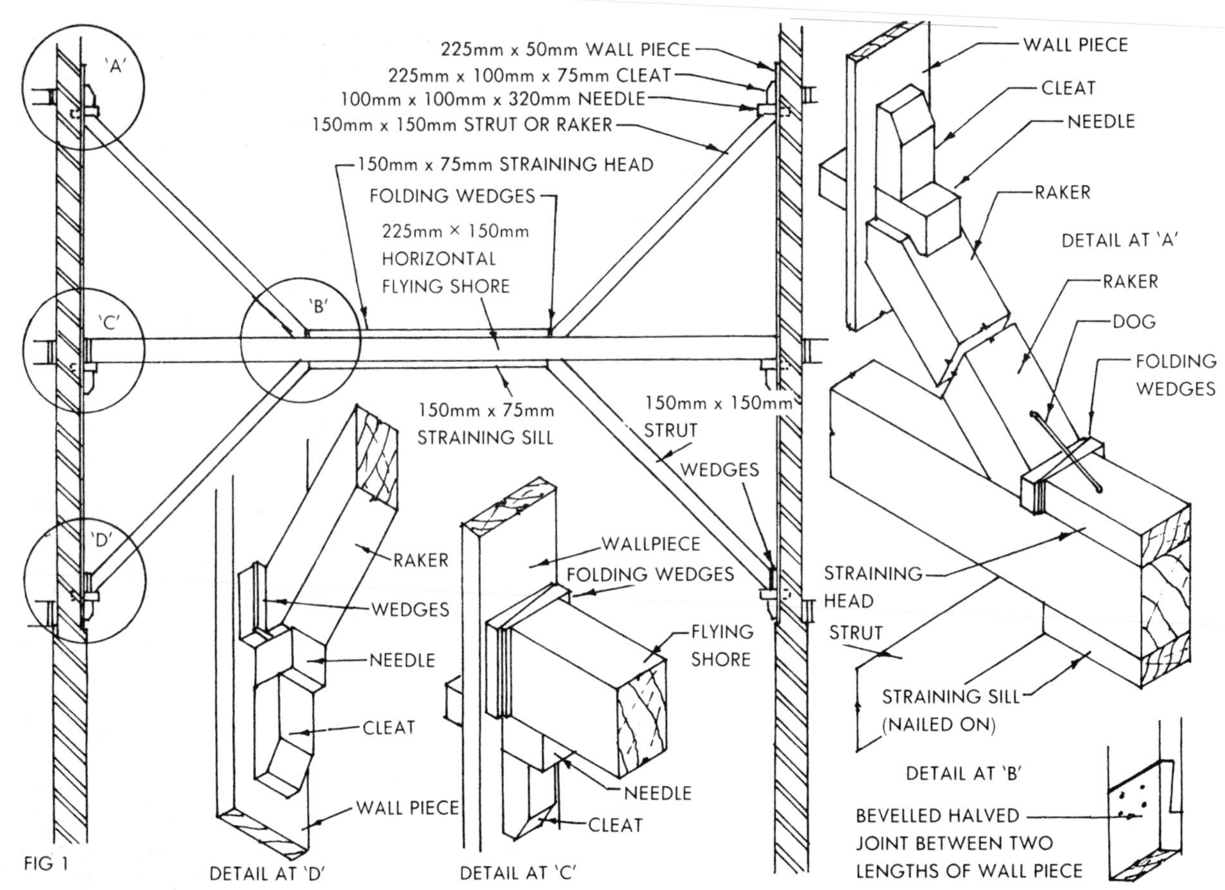

FIG 1 DETAIL AT 'D' DETAIL AT 'C'

FLYING SHORES

Flying shores are used to provide horizontal restraint between parallel walls, flying shores are often used when one of a terrace of buildings has been removed and the adjoining buildings are thus deprived of support.

N.B. All shoring and strutting should be designed or approved by an experienced person.

Where the distance between two buildings does not exceed about 9m, timber flying shores may be used, but for greater spans it is advisable to use structural steel members or a combination of timber and steel.

Flying shores have the advantage of not obstructing the ground below.

Fig I shows an example of flying shores over a span of 7m. Shores should be needled into the wall as described . Rakers should be inclined as near as possible at 45 deg and the system securely wedged and dogged. Fig 2 shows an example when one building is higher than the opposite building. Where the floor levels of two buildings do not coincide it is usual to select the floor levels of one building for the horizontal members, and provide a stouter wall piece at the other end to transmit the thrust to the floors of that building.

TUBULAR SCAFFOLDING

Shores may be built up of tubular scaffolding provided they are *properly designed and braced.* Fig 3 indicates a raking shore and Fig 4 a flying shore.

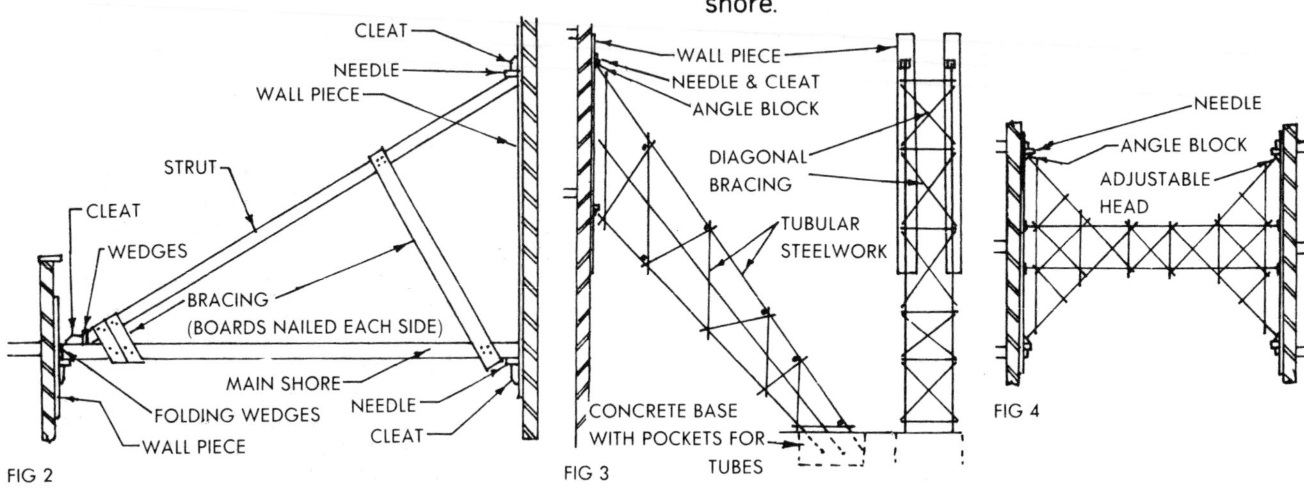

FIG 2 FIG 3 FIG 4

SHORING 4

For alteration work, particularly when forming a new opening in a wall, both dead shoring and raking shoring may be necessary. Caution must be observed, a careful survey should be made to ascertain the condition of the structure, and the work supervised by an experienced person. A sketch showing the combined use of the various shores is given in Fig I.

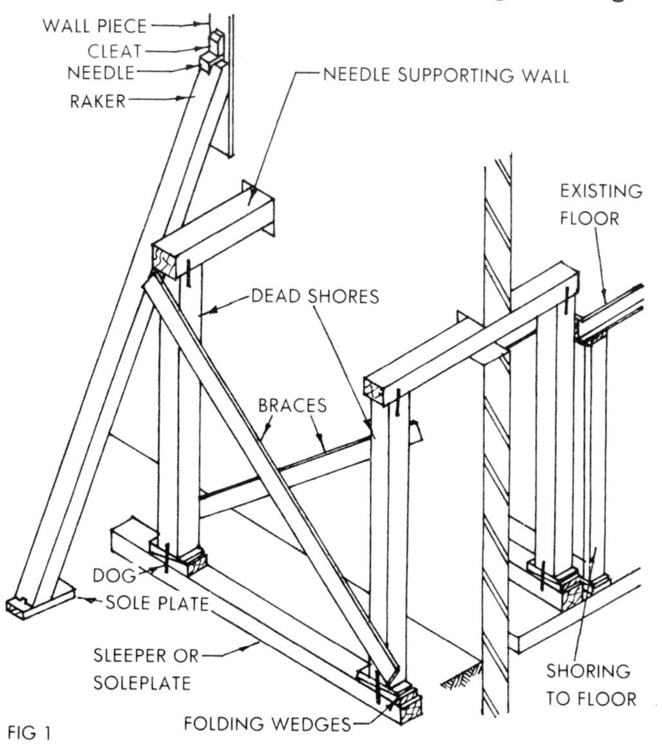

FIG 1

Sequence of operations:- The window openings should be securely strutted as shown in as a precaution against any deformation taking place. If conditions warrant the use of raking shores, they are secured in position as described in . The procedure is then as follows:- holes are cut through the wall and the needles inserted. These may be of timber or steel sections, and must be above the level of the new lintel or beam to the opening. A clearance should be left between the top of the needle and the wall which it is to support, so that a bed of cement mortar can be rammed in to provide full area contact between the needle and the wall. Where a wall is heavily loaded it may be necessary to insert padstones between the needle and the wall to distribute the weight. Fig 2. These padstones should be wider than the needle. The sole plate should be firmly bedded in position and the dead shores tightened into place. Dead shores may be of timber, or adjustable steel props may be used. Heavy duty adjustable steel support trestles capable of a safe working load of 20 tonnes are available.

When the shoring has been completed, the wall below the level of the needles is removed. The new beam or lintel is lifted into place and the brickwork above carefully pinned up to the wall above. A week should be allowed for the new work to harden before any shoring is struck.

When striking the shoring needles are removed first, then the window strutting, followed by the floor strutting and lastly the raking shores. It is good practice to allow a day or two between each operation so that the loads are gradually transferred to the new supports.

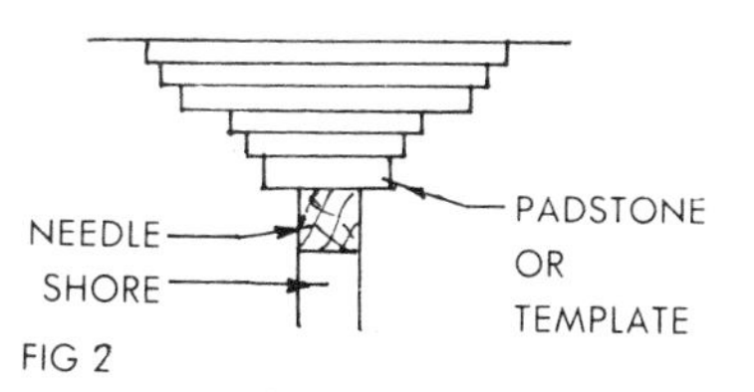

FIG 2

DOUBLE FLYING SHORE

It is usually considered that the limit for single flying shores is a span of about 9m. Should flying shores be required for spans between 9m and 12m then a double flying shore, Fig 3, is used. The arrangement for wall pieces, needles, cleats and rakers is similar to that for single flying shores.

The following table indicates suitable sizes for flying shores of various spans.

Span	Horizontal Shore	Struts
up to 4.5m	150mm x 100mm	100mm x 100mm
up to 7.0m	225mm x 150mm	150mm x 100mm
up to 12.0m	225mm x 225mm	150mm x 150mm

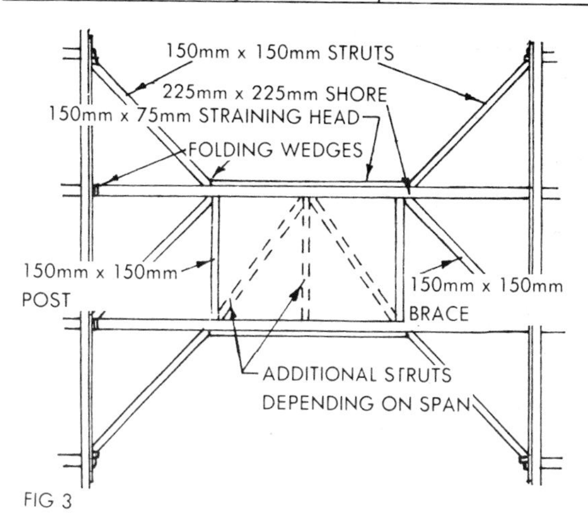

FIG 3

Where a series of shores is erected in a continuous line, diagonal bracing may be fixed between each set of shores.

Maintenance of shoring:- Timbering in shoring may swell or shrink, and means of adjustment should be provided by hardwood wedges, which should be readily accessible. All shoring should be regularly inspected by a competent person. Particular attention should be paid to the effects of freezing and thawing of the ground, any softening of the soil, movement of soil due to adjacent excavation, effects of traffic vibration, and any damage by impact. Care must be taken to see that no unauthorised person interferes with the shoring in any way.

UNDERPINNING I

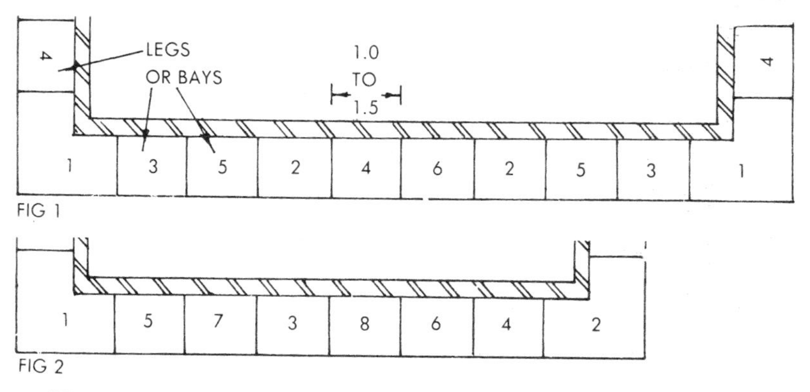

FIG 1

FIG 2

The object of underpinning is to transfer the load carried by an existing foundation to a new foundation at a lower depth. Underpinning may be necessary for a number of reasons:-

1. Settlement of the existing foundations has occurred. This may have been due to uneven loading, movement of the sub-foundation due to moisture movement, action of tree roots etc, or possible deterioration of the foundation concrete by sulphate action etc.
2. To increase the load-bearing capacity of the foundation. Either to allow for the building-on of additional storeys or to allow for an increase in the superimposed loads on the structure.
3. To allow the level of the adjacent ground to be lowered, e.g. when constructing a new basement.

Before underpinning commences a careful survey of the structure should be carried out, notice given to adjoining owners, their consent obtained and agreement reached on the work to be carried out, temporary shoring required etc. 'Tell-tales' should be fixed where there are existing cracks, to indicate any further movement.

Permission should be obtained to stop up any adjoining flues and fireplaces to prevent damage by falling materials and nuisance from falling soot.

GROUND CONDITIONS

An investigation of the subsoil beneath the building should be carried out before underpinning work is started to ascertain:-

1. The conditions responsible for any settlement.
2. The general nature of the ground, particularly if the adjacent ground level is to be lowered.
3. The load bearing capacity of the ground on which the new foundation is to be founded.
   N.B. Underpinning should not be rushed, care is necessary and the work should be supervised by an experienced and competent person.

INITIAL PRECAUTIONS

The loading on the structure should be reduced as much as possible, the adjoining owners permission and co-operation may be required. Superimposed floor loads should be removed. It may be necessary to strut floors and beams, so that the loading is transferred to a solid bearing clear of the underpinning.

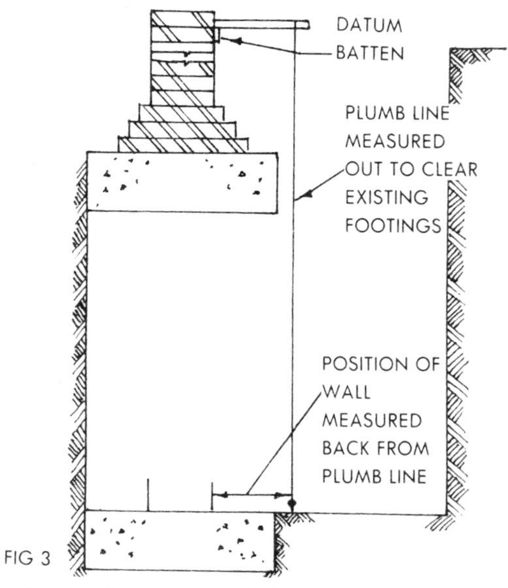

FIG 3

Periodic checks for any possible movement may be made by taking readings of heavy plumb bobs suspended from high parts of the structure and at the corners of the structure. Additional checks may be made by level readings taken from a reliable datum situated away from the work.

PROCEDURE

Underpinning is usually carried out in a series of bays or 'legs'. Generally in brickwork each leg would be lm to 1.5m long, and extend out from the wall line to give sufficient working room. The sequence of work should be such that the wall above is amply supported at all times. Each leg should be completed, and the wall 'pinned up' before the next is commenced. No two adjacent legs should be excavated at the same time and the work should be set out so that the sequence of operations avoids this. Should it be necessary for any reason to work on legs adjacent to each other, then one leg should be completed and 'pinned up' before work is started on the next one. The sum total of unsupported lengths should never exceed one quarter of the wall length, and for heavily loaded walls less still, say one-fifth to one-sixth of the length.

Fig I shows an example of the order of excavations for a wall 14m long. In this case the wall is long enough to permit more than one section to be worked on at the same time, so that both sections No I could be carried out simultaneously, then sections No 2 and so on. Fig 2 shows a shorter wall 10m long, and in this case the legs would probably be worked in single sequence as shown.

Excavation:- A datum batten is spiked to the wall as shown and the levels of the new foundation concrete and the brickwork gauged down from this Fig 3. All excavations should be securely timbered and strutted to prevent any movement of the adjacent ground. In some cases chemical consolidation of the ground or pressure grouting, may be used to ensure the stability of the surrounding ground.

UNDERPINNING 2

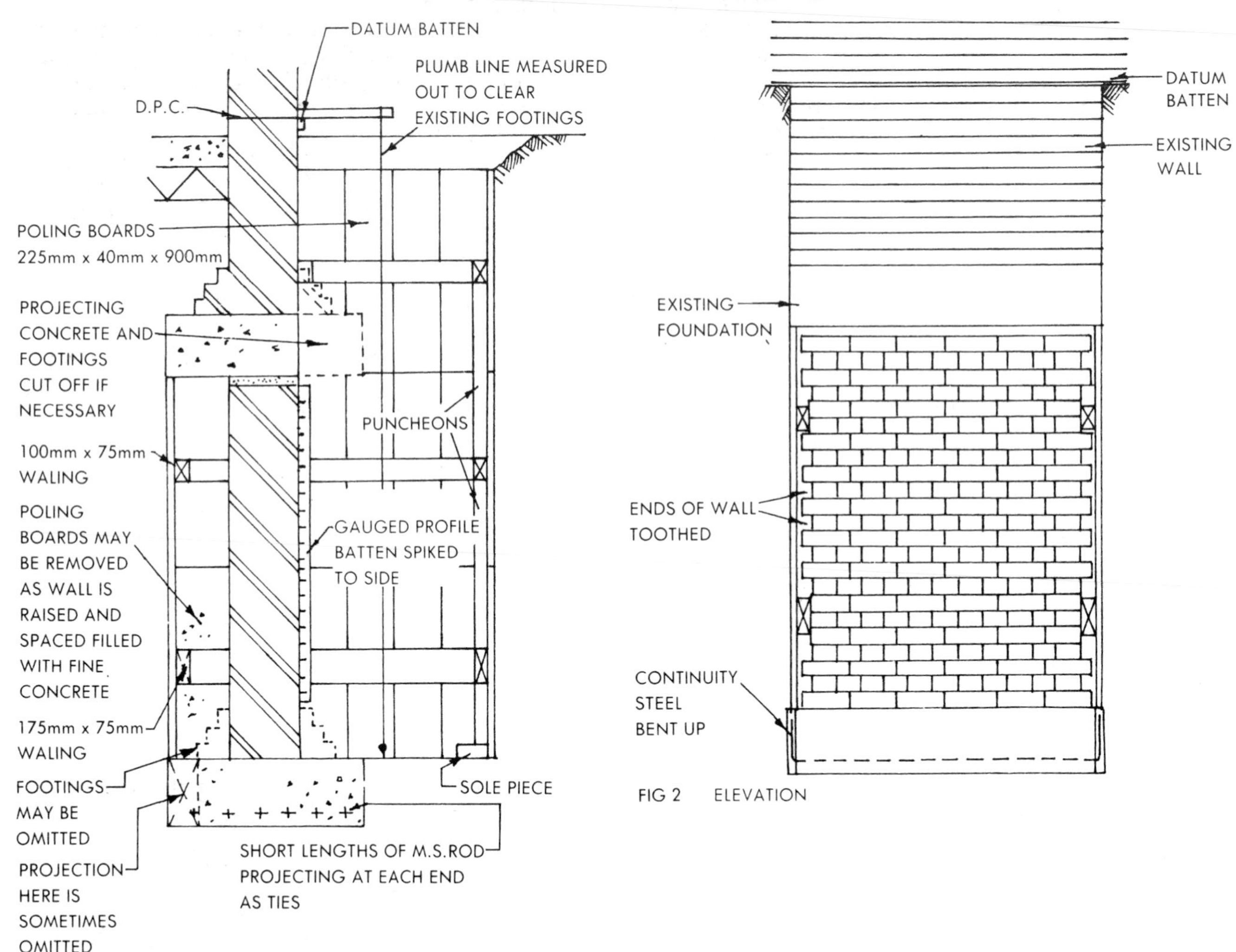

FIG 1 SECTION THRO' UNDERPINNING

FIG 2 ELEVATION

EXISTING FOUNDATIONS

The existing foundation concrete may be cut away, and if so, this is usually done before excavating below the foundation. This facilitates plumbing down and gauging for the new foundation and wall, and since the underside of the old work is often uneven is more convenient when pinning up.

If the existing foundation is in good condition it may only be necessary to cut-back any projecting concrete and footings flush to the wall line. Fig I.

However, prevailing conditions will govern the method adopted, and the concrete foundation may be excavated with the particular leg, and the projecting footings cut off after pinning up has been completed and hardened.

WORK SEQUENCE

The position of the new foundation and wall is measured back from a plumb line as shown, the levels being gauged down from a datum battened spiked to the wall above Fig I. The ends of each section are toothed (Figs 2 & 3) to enable subsequent sections to be properly bonded on, in due course. Particular care must be taken in levelling and alignment of each section, to ensure that the wall is properly continued and 'lined up'.

Plumbing is usually dispensed with, and a line strained between gauged profiles spiked to the timbering at each end of the length of wall Fig I.

The new foundation concrete may be reinforced, and in any case should have short lengths of mild steel rods left projecting at each end to act as ties with the adjacent sections. Fig I. If the foundation concrete is reinforced, the steel rods should be bent up at the ends, and then straightened when the next section is continued, to ensure continuity. Fig 2.

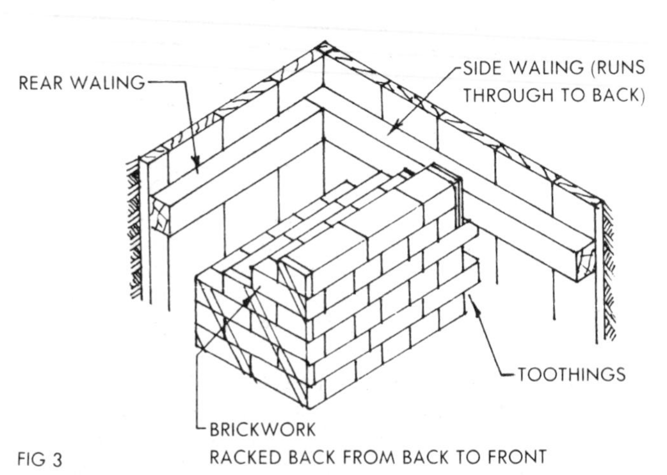

FIG 3

TIMBERING
If it is decided that the rear poling boards are to be left in position and not withdrawn the use of concrete poling boards is recommended these would normally be 900mm x 300mm x 50mm thick, and may have holes left in them to permit grout to be subsequently pumped through, thus filling any spaces or voids between the soil behind and the back of the polings, and preventing subsequent movement. The space between the back of the wall and the poling boards is filled with fine concrete as the work proceeds. Fig I.

If it is decided to remove the rear poling boards, one method is as follows:- when the height of the walings is reached, the brickwork is racked back as shown. Fig. 3. The face brickwork is then wedged tightly up to the side walings as shown, thus acting as a strut and allowing the rear waling and boards to be removed. The rear brickwork is then built up level, and the space behind filled with concrete.

UNDERPINNING 3

PINNING UP
If the underside of the existing foundation concrete has a sound, level, even surface it may be possible to pin up to this, providing it is thoroughly cleaned and no clay-type material is adhering to it. If however the existing concrete is in rough condition it should be removed (usually when excavating) and the wall pinned up to the footings. The projection of the footings can be cut off after pinning up is completed, thus avoiding disturbing the existing wall as much as possible.

The actual pinning up is carried out in the order shown in Fig I, working from the back to the front. The brickwork is racked from front to back as shown, the bricks being laid in the sequence indicated. The brickwork is gauged so that a gap of at least 25mm is left between the top of the new work and the underside of the old. When brick No 3 has been laid it is firmly pinned up, then No 4 and so on. A fairly dry strong cement mortar is used (say I:2), thrown in with a trowel and well rammed home with a piece of 25mm thick wood and a club hammer.

STOOL AND BEAM METHOD
This technique uses special 'stools' to support the building whilst a new beam and any additional construction is prepared. Fig 2 shows a typical concrete stool. A range of stools is available to cover various situations. For most two storey buildings (where loads up to approx' 2 tonnes per 300mm run are concerned) standard concrete stools may be used, and steel stools for heavier walls.

Sequence of work :-

1. Cut holes in wall, insert stools and pin up Fig 3.
2. Cut away remaining walling between stools, building now supported by the stools.
3. Long lengths of reinforcement are threaded through the stools and together with stirrups form a continuous reinforcing cage. Fig 4.
4. Shutter out and pour concrete. Strike when hardened, and pin up walling above beam.

The new beam may in itself be sufficient to spread the load evenly, and provided that the gross bearing capacity of the site is adequate, may provide a satisfactory remedy for differential settlement.

Alternatively the new beam may be used in conjunction with new bases, such as spreader pads, concrete piers, piles etc. The beams are generally constructed before the bases. Thus they support the building during excavation to a sound underlying stratum and then, after pinning up, they carry the building loads to these bases.

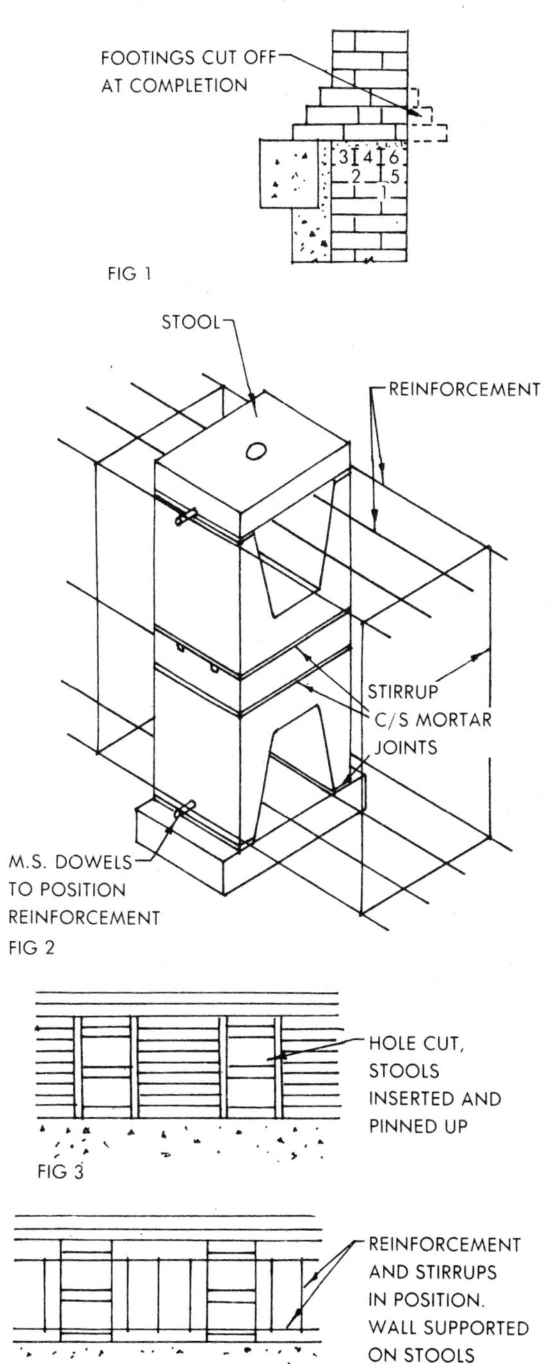

PILED UNDERPINNING

Where a satisfactory load bearing strata lies at a considerable depth below the existing foundation, piling may be resorted to. In order to avoid excessive vibration and noise, bored or jacked-in piles are commonly used.

This method is carried out as follows:-sufficient space is excavated under the foundation to allow for the insertion of the first section of the pile, the bearing plate and hydraulic jack Fig 5. The first length of pile has a steel cap, and the sections are holed, for inserting steel continuity rods or dowels. The first section of the pile is driven down by the action of the hydraulic jack, operated by the pump as shown. Above excavation level, the jack is removed and a steel dowel is inserted into the hole and grouted into position to provide an effective joint between the sections. The next section of pile is then placed in position, the jack replaced and the process continued. A pressure gauge is used to check when adequate bearing resistance has been achieved. The jack is then removed and the space between the head of the pile and the foundation pinned up.

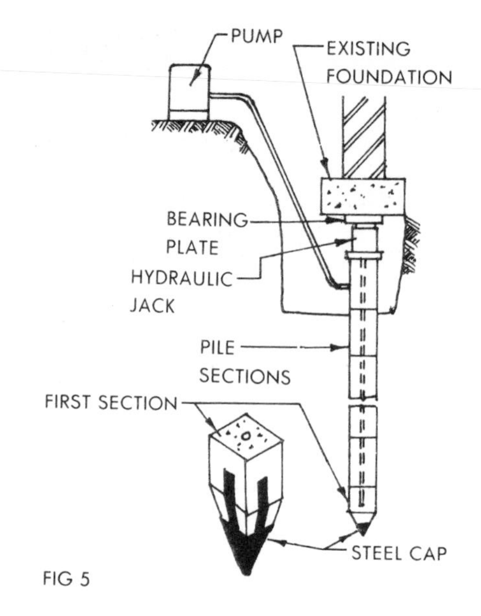

FIG 5

WAFFLE FLOORS I

WAFFLE FLOOR
A two-way spanning in-situ reinforced concrete floor. The intersecting ribs form a grid as shown Fig I, and this coffered grid, reminiscent of a waffle iron gives the floor its name.

A series of polypropolene or glass fibre moulds Fig 2 are supported on decking, or on timber or metal supporting beams. The spaces between them are filled with concrete to form the ribs, the tops of the moulds providing the soffit support for the main floor.

Fig 3 shows a typical arrangement. The ribs are reinforced with M.S. rods and stirrups, as shown, and the main floor slab reinforced with a mesh fabric to BS 4483 'Steel fabric for the reinforcement of concrete'.

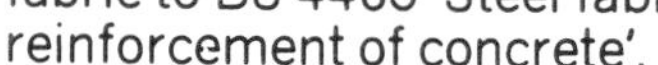

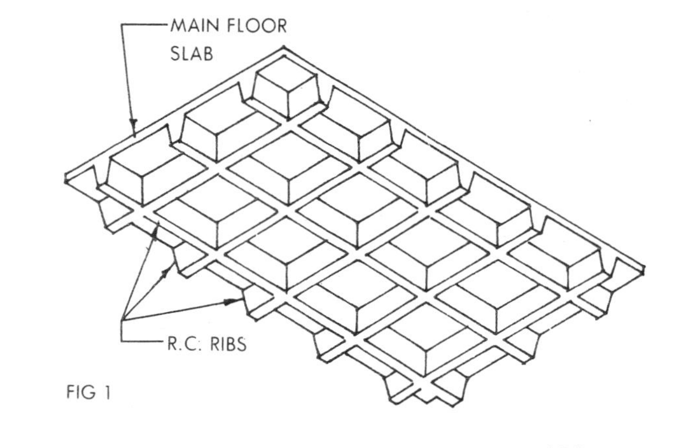

FIG 1

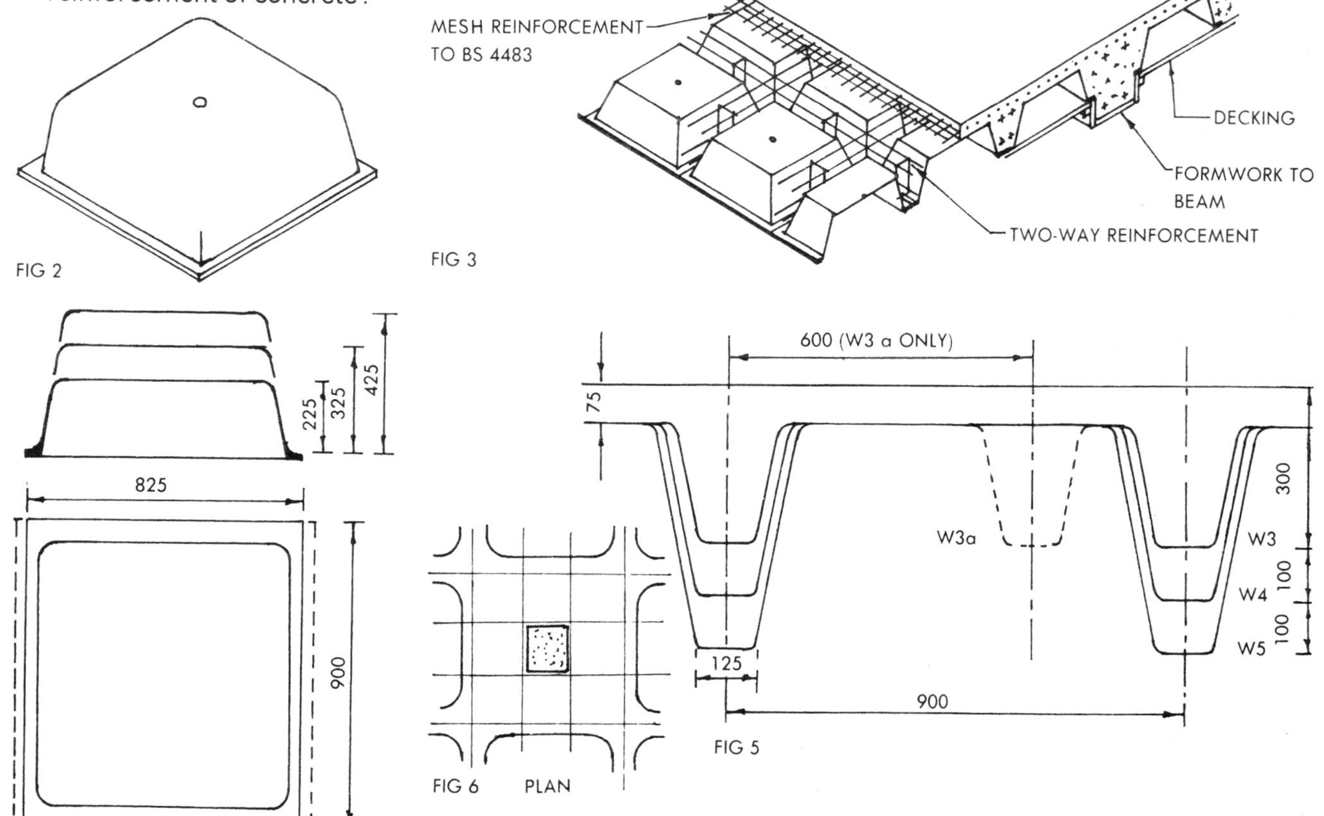

FIG 2

FIG 3

FIG 4

FIG 5

FIG 6

MOULDS
To facilitate striking, the sides of the moulds are tapered about 10 deg from the vertical. A Schrader valve is applied to the central hole in the top of the mould, and the mould easily removed by compressed air. A range of moulds is available and they may be purchased or hired.

Fig 4 shows details of the 'beemoulds'. These glass fibre moulds have been designed to meet the requirements recommended by a working party set up by the Concrete Society, for the standardisation of waffle floors. The 900mm x 825mm moulds are designed for use in conjunction with a Quickstrip system, and when used with a 75mm timber or metal supporting beam will produce a minimum rib width of 125mm.

Moulds are available in three depths as shown, and when used with a main floor slab of 75mm depth, overall depths of 300mm, 400mm and 500mm are produced, complying with the proposals of BS 4330 'Recommendations for the co-ordination of dimensions in building controlling dimensions' Fig 5. Intermediate depths of floor between those shown can be achieved by adjusting the depth of the topping, where modular depths have no relevance.

When required 900mm x 900mm moulds (Fig 4) are available, these being used with a full deck supporting system.

The W3d mould (shown dotted Fig 5) is available should a 600mm grid be required, the floor depth is 300mm as for the W3 mould. When columns are used the floor slab around the column is solid to resist shear. Fig 6 shows a possible arrangement on a 300mm basic grid.

WAFFLE FLOORS 2

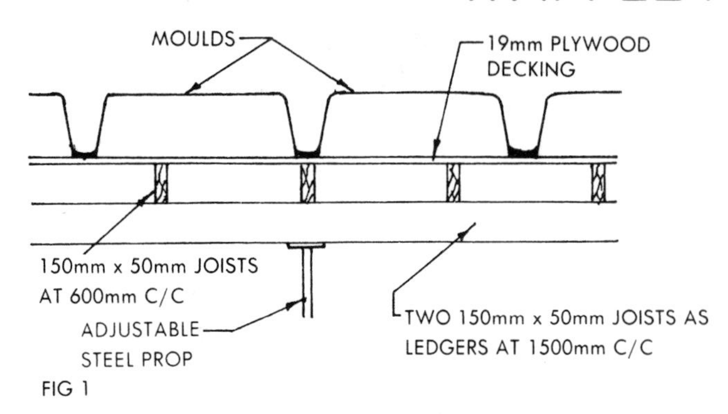

FIG 1

METHODS OF SUPPORT

One method of supporting the moulds and floor slab during construction is to provide a level decking of boarding, plywood or steel. Figs 1 & 2 show a typical formwork arrangement. This type of decking is suitable for square moulds
The 900mm x 825mm moulds.
may be used with a Skeletal or Quickstrip formwork. Fig 3. This will give a minimum rib width of 125mm and a rib spacing of 900mm c/c in both directions. The cleats, secured by double headed nails as shown, may be struck after about three days and the moulds removed for re-use, the supporting joists remaining in position, to be struck at a later stage.

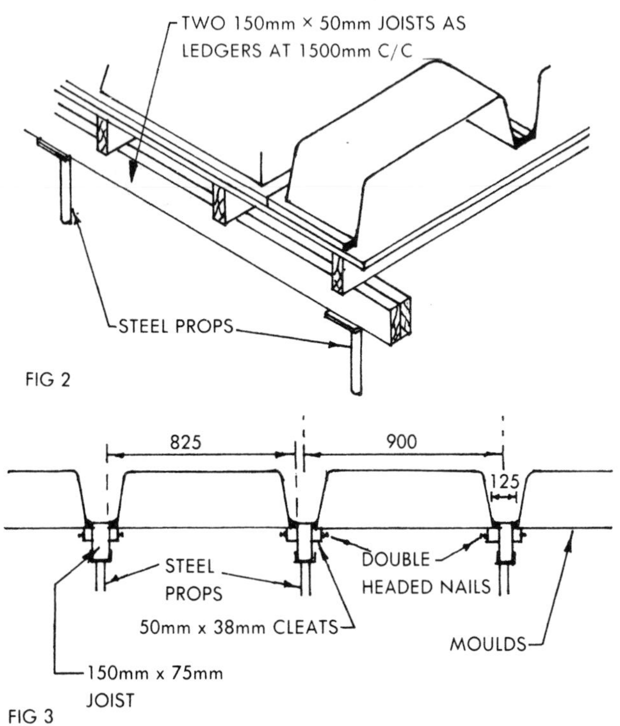

FIG 2

FIG 3

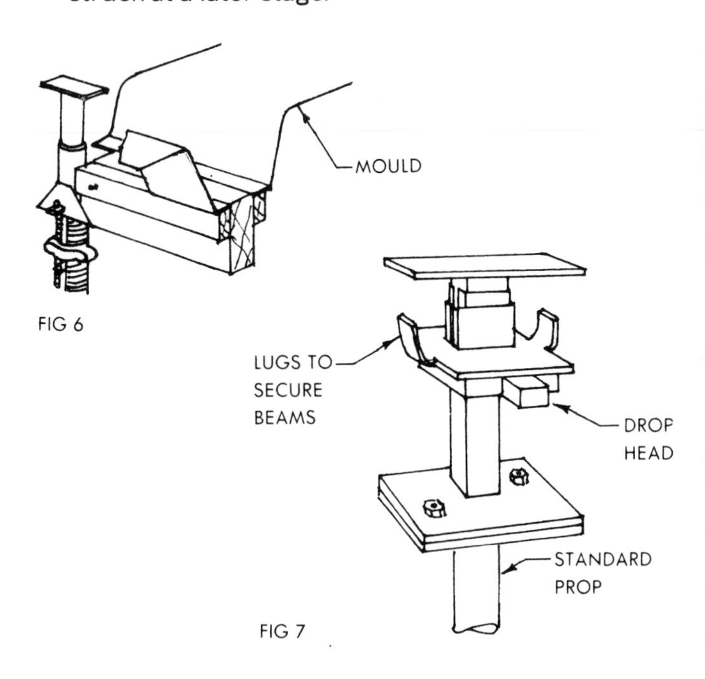

FIG 6

FIG 7

PROPRIETARY SYSTEMS

A number of proprietary systems of formwork are available. Designed for general use the equipment may be purchased or hired. One such system is illustrated in Figs 4, 5 and 6. The special prop Fig 4, is fitted with an independent sleeve with a 'U'-shaped saddle to carry the support beams. The plate at the prop head maintains a continuous smooth surface between the beams, at the same time providing support for the slab after the moulds and support beams have been struck. Timber support beams may be used as Fig 6 where grids of non-standard dimensions are used. For standard grids, steel support beams Fig 5 are available. Lightweight panels of steel and marine ply are available for decking or wall forms. Another system uses a quick release drop head, which is bolted to a standard prop Fig 7. The drop head allows the panels and beams to be struck and re-used after three days or so, leaving the props only in position for the remainder of the slab curing period. The drop head fitting increases the height of props by 210mm, thus frequently enabling smaller props to be used. Lightweight steel beams, Fig 8 are used between props.

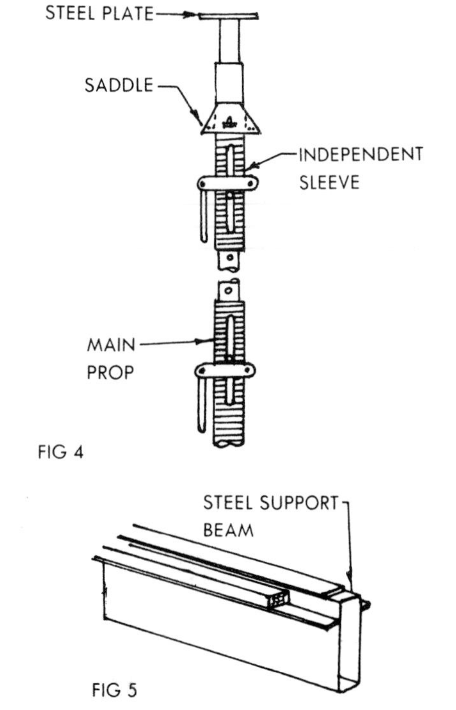

FIG 4

FIG 5

TROUGH FLOORS

Where column grids are square or nearly so, waffle floors are economic. If however the column grid is rectangular, trough floors, with the ribs running mainly one way may be more economical. The troughs span one way between main beams or perimeter walls. The end moulds must be closed across the end. Moulds may be of hardboard polypropylene or glass fibre and Figs 9 & 10 show a range of glass fibre 'T' units. The units are supplied in four lengths and depths as shown. Any two end units and a combination of mid units give a wide range of void lengths.

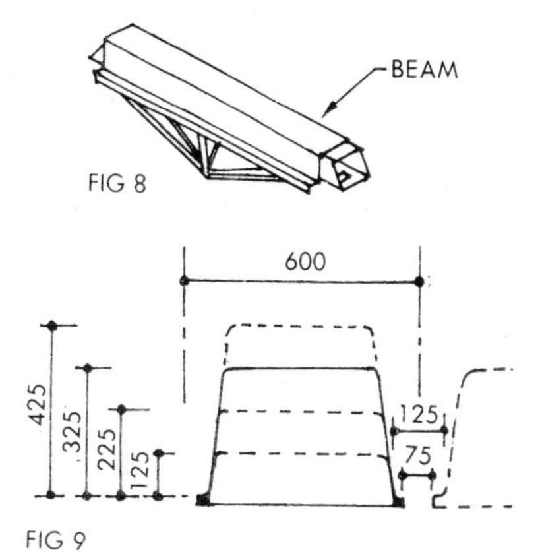

FIG 8

FIG 9

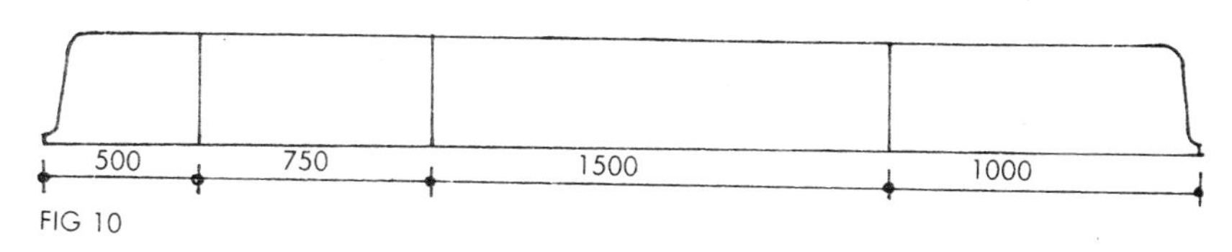

FIG 10

CHIPBOARD FLOORING I

Chipboard (wood particle board) consists of wood residues compressed together with urea formaldahyde or other adhesive. Although more stable than solid timber and much more tolerant of the dryness of central heating, chipboard is affected by atmospheric changes and will swell on becoming damp.

Chipboard may be worked with normal hand or machine tools and is now widely used for a variety of building purposes. Although the strength is adequate for many purposes it is lower than for solid timber and it is most important that the correct grade of board is used, particularly for the purpose of flooring.

SUSPENDED FLOORS

It is essential to specify a board which has the requisite strength properties for flooring. These may be described by the manufacturer. As a flooring board, flooring grade or similar, e.g. Weyroc flooring boards. Thicknesses are usually 18mm or 19mm, or thicker boards usually 21mm or 22mm. A range of sizes is available but sheets 2440mm x 1200mm or 1220mm are commonly used.

Manufacturers specify joists spacing and Fig 1 illustrates the use of Weyroc flooring boards with joists spacing of up to 450mm for 18mm thick boards and 610mm for 22mm thick boards. Square edge boards should be laid with the joists, the short ends being staggered and supported on noggins. Figs 1 and 2.

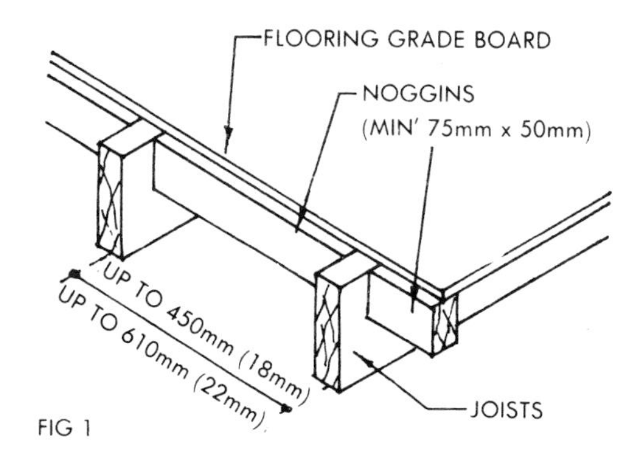

FIG 1

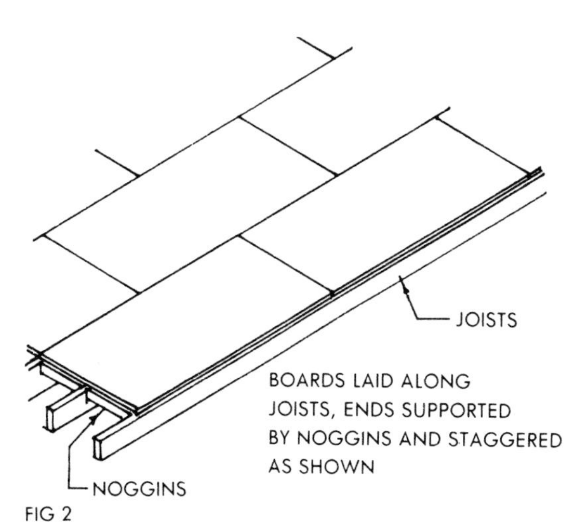

FIG 2

TONGUED AND GROOVED BOARDS

Boards are available grooved all round with loose plywood tongues supplied and all joints glued, but boards with integral tongue or groove on all four edges are now widely used.

Weyroc 'Teegee 4' is tongued on one long edge and one short edge and grooved on the other long and short edges Fig 3. Each panel is 2440mm x 600mm laid measure with the tongues projecting beyond this dimension Fig 4. The grooved edges have a projecting lower shoulder which protects the joint in transit, provides a convenient locator for the tongue when assembling and gives added strength to the joint in service.

Tongue and grooved boards are normally laid across the joists, and the cross joints should be staggered. With T&G boards noggins are not required, but the short edges should be supported by a joist Fig 5. The jointing of boards parallel to the joists should be avoided.

Because the boards will expand or contract slightly with chnages in humidity, it is necessary to provide an expansion gap around the perimeter of the room. An allowance of 2.5mm per metre run should be made. Thus for a run of 4.0m a gap of 10mm is required. This will be concealed by the skirting. For very large buildings intermediate expansion gaps may be required — maximum run of 6.0m approximately.

Although not always essential the strength and rigidity of the floor may be improved by glueing the joints before nailing and this is to be advocated when 18mm boards are used. The glue used should be of a PVA type.

Ventilation:- On joisted ground floors good ventilation is essential. Air bricks should be placed around the walls so as to provide $30cm^2$ ($3000mm^2$) of open area per metre run of perimeter wall. It is presumed that adequate D.P.Cs and honey-comb sleeper walls to permit adequate through ventilation will be provided.

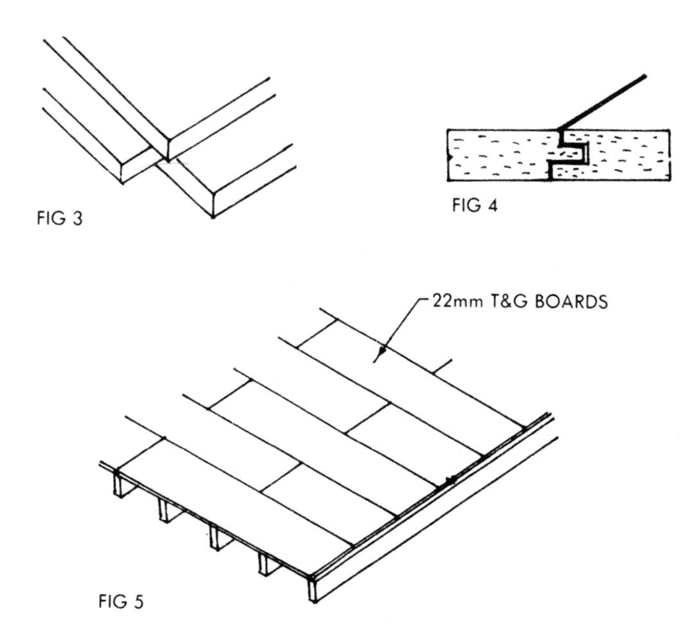

FIG 3

FIG 4

FIG 5

CHIPBOARD FLOORING 2

Boards should not be laid until the building has dried out, and before fixing the boards should be conditioned within the room or building for as long as possible.

Nailing:- Chipboard flooring boards should be nailed along all supports at spacings not greater than 200mm. Nails should be inserted at an angle to the vertical as in normal practice, with not less than 38mm of nail penetrating the board. With T & G boards some secret nailing is possible. Ordinary lost head nails may be used, but where coverings such as PVC tiles are to be laid on the surface, nails with better holding power, eg annular ring nails or screw nails should be used or alternatively wood screws give good results.

Joists:- It is important that joists be kept in dry condition and that they be allowed to dry out thoroughly, particularly if treated with a water-borne preservative, before laying the floor boarding. Failure to ensure this may result in a squeeky floor, some loss of rigidity and the possibility of fungal attack.

STRESSED SKIN FLOORS

A method that has proved very effective and has been used successfully for industrialised building is to preform floor sections by both nailing and gluing flooring boards to timber joists Fig I. By this means it is possible to either reduce timber sizes or alternatively, to increase the length of span that a given joist can cover. By this means joist depth may be reduced approximately 25% e.g. 150mm x 50mm joists may be used instead of 200mm x 50mm or alternatively the span could be increased by approximately 12%. A futher increase in stiffness is gained if ceiling panels of 9mm standard grade chipboard are glued and nailed to the underside of the joists Fig 2.

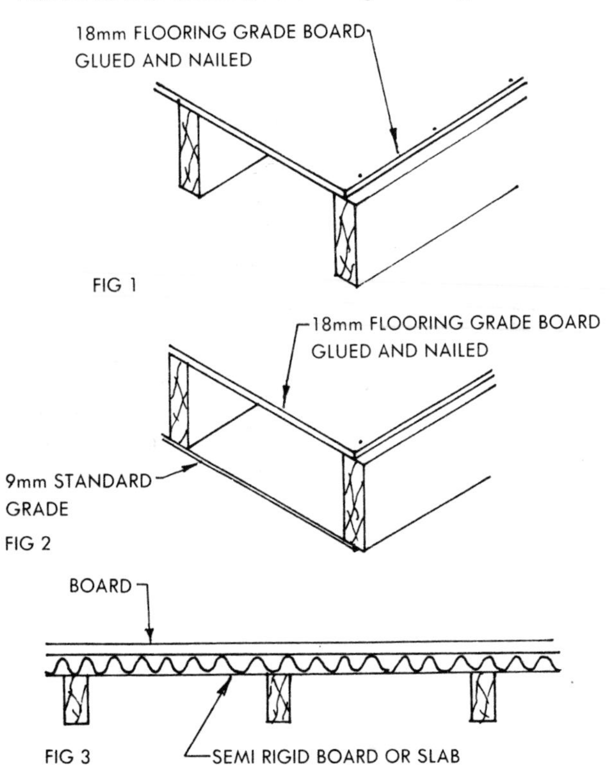

INSULATION

Increased thermal insulation for a suspended timber ground floor may be achieved by means of a continuous layer of semi-rigid material (Fig 3) or flexible material (Fig 4) laid over the joists. Alternatively semi-rigid insulation may be provided between the joists (Fig 5).

Resilient layers:- Materials for use as resilient layers or slabs beneath chipboard floors include, glass wool quilts of the long fibre type, (should be 13mm minimum thickness before compression), glass wool resin bonded slabs, mineral wool quilts of the long fibre type, mineral wood slabs, expanded polystyrene slabs (150 grade — available in self-extinguishing type). Further information may be obtained from the various manufactuers and from BRS Digest 145 'Heat loss through ground floors' and from BRS Digests Nos 102 and 103 'Sound insulation of traditional dwellings'.

ACCESS TRAPS

Traps to give access to services beneath the floor are best incorporated into the layout at the design stage, and such traps should be supported at the ends by noggins and the panels screwed down.

N.B. Screwing provides the strongest fixing, and if the screws are dipped in glue, their holding power is increased.

If a trap is required subsequent to the floor being completed, the access space may be cut using a pad saw and the floorboard replaced by noggin back.

Support should be provided to boards at perimeter walls, as Fig 6.

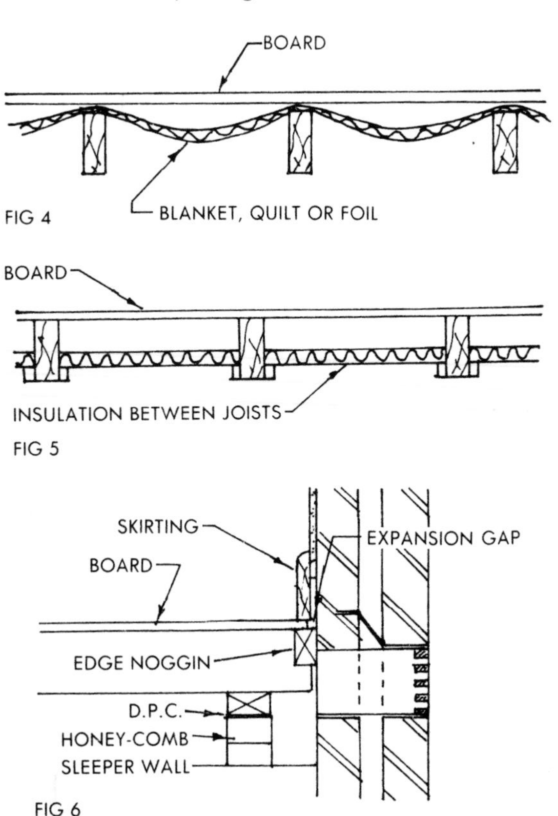

CHIPBOARD FLOORING 3

FLOATING FLOOR CONSTRUCTIONS

These forms of construction have good thermal properties and can provide good sound insulation. They are based on the principle that impact sound insulation is best provided by the complete separation of the floor finish from the structure, and that airborne sound insulation is directly related to the mass of the structure. Any partitioning should be fixed to the structural floor so that the floating floor is self contained within each room, the floor being held in position by its own weight and by the skirting fixed to the perimeter walls.

The simplest form of this type of floor, a floating floor without battens, is shown in Fig I. In this assembly only tongued and grooved boards are used. Floors of this type are suitable for in-situ or precast slabs, when screeding should not be necessary. The following recommendations should be carefully observed:-

1. The concrete base must be given sufficient time to thoroughly dry out before laying the floor.
2. On the concrete sub-floor pre-compressed expanded polystyrene 16 kg/m^3 density is laid at least 25mm thick. Thickness will depend on space available, need to accommodate services and degree of insulation required. Channels are easily cut in the polystyrene for services. Alternative materials are available.
3. A vapour barrier of 1000 gauge polythene laid over the polystyrene. The vapour barrier must be turned up above the level of the boarding around the perimeter (Figs I & 2). Joints lapped over 150mm and taped on the upper side. Alternatively the resilient layer may have a vapour barrier factory bonded to it. In this case joints must also be taped.
4. All joints should be securely glued with a suitable PVA adhesive and the boards should be cramped while the adhesive cures. This can be achieved by wedging the boards from the perimeter walls and partition sole plates. In this case edge insulation should be fixed after curing. Wedges *must* be removed to allow for possible moisture movement, before the skirting is fixed. This is essential.
5. A gap must be allowed around all perimeter walls for expansion, a minimum of 2.5mm per metre run of floor being allowed.
6. The jointing of boards should be staggered to avoid the four corners abutting
7. For maximum sound insulation there must be no contact between the wall and the floorboards. A resilient strip should be inserted beneath the skirting, and the resilient slab may be turned up around the perimeter or a vertical strip of the resilient material placed in position. Fig 2 shows an example of this technique. In this case a dry lining of plasterboard is shown.
8. If it is required to place a partition on the flooring, then to avoid crushing the resilient layer, the floor should be supported under the boarding by the insertion of a batten, preferably foam-backed to give maximum sound insulation.
9. Threshold pieces should be screwed or nailed to battens insulated from the slab.

Protection:- The building should normally be weathertight before conditioning and fixing boards, and boards should always be protected on site to avoid any increase in the moisture content. Boards can be further protected by the application of one coat of polyurethane to both faces and edges before delivery to site.

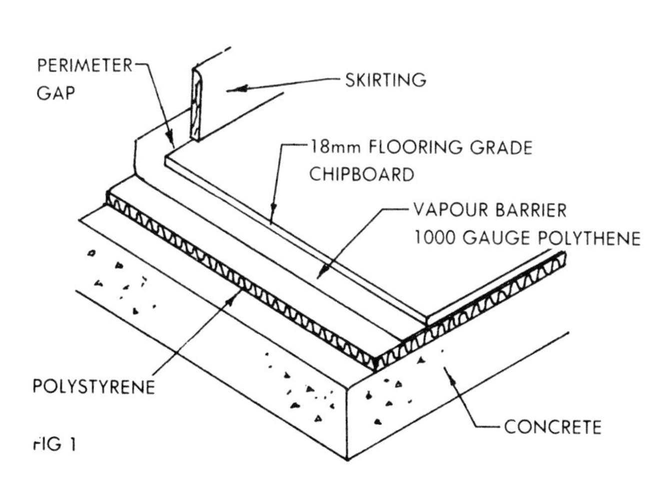

FIG 1

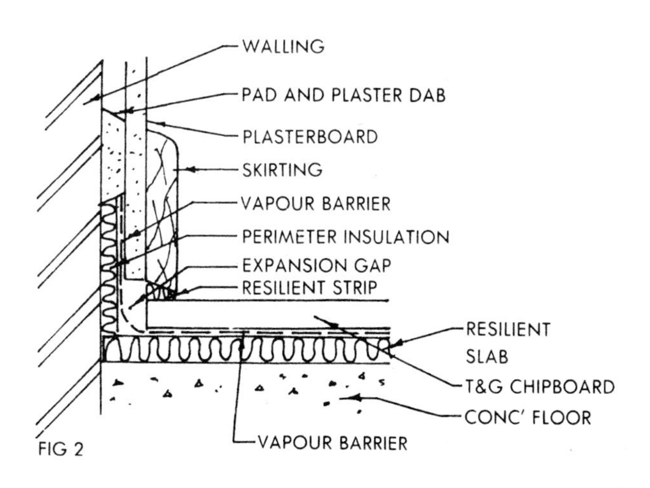

FIG 2

CHIPBOARD FLOORING 4

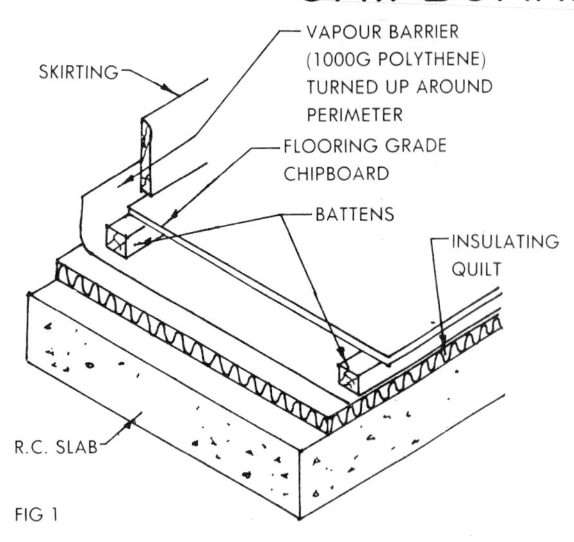

FIG 1

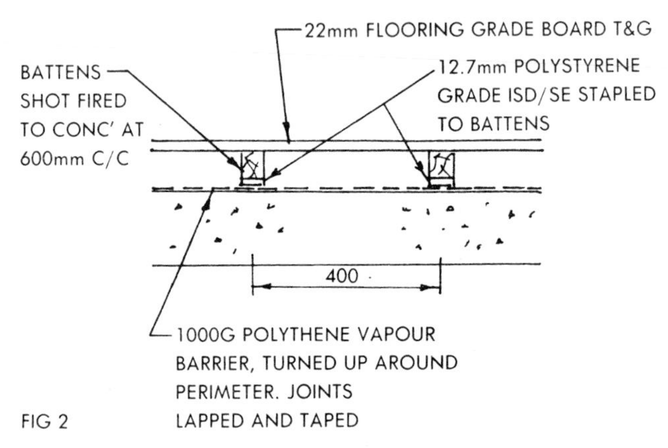

FIG 2

BATTENED FLOATING FLOORS

A floating floor of this type (Fig I) may be employed over any concrete sub-floor. Such a floor provides good insulation and is recommended where low impact sound transmission is of critical importance. Services may be accommodated between battens, which are notched for services running at right angles to them.

The recommendations laid down should be carefully observed.

A suitable resilient quilt is laid over the concrete, e.g. 12mm paper faced glass fibre or mineral wool quilt Fig I. Care must be taken to ensure that the whole floor is covered, and the quilt may be turned up around the edges.
Battens :- Thicknesses vary and 25mm (approx) are sometimes used, although some authorities advocate they should not be less than 40mm deep. Certainly where 50mm nails are used the battens should not be less than 50mm deep. In all cases the battens should not be less than 50mm wide.

Boards should be supported, and the battens spaced at the same centres as for suspended floors. Boards should be nailed to battens as described. The battens must be conditioned to low moisture content, particularly important where they have been pressure treated with a water-borne preservative.

INDUSTRIAL FLOORS

Flooring arrangements vary with prevailing circumstances and requirements. Fig 2 shows a construction suitable for a light industrial floor. Suitable polystyrene is stapled to the underside of the battens as shown. The usual expansion gap will be required around the perimeter and at intervals in large areas, bay should not normally exceed 10m square without expansion joints. For heavier duty floors the arrangement as Fig 3 is suitable, providing the base is reasonably smooth. If the sub-floor is rough or uneven then the alternative heavy duty grid system, Fig 4 may be used. This method is particularly suitable for gymnasia, recreation halls etc, as it provides a resilient floor, while avoiding impact fracture by limiting deflection. N.B. Where dry sand is used a continuous covering sheet of polythene (500 gauge) overlapped and taped should be used to prevent the 'puffing action' of the dry sand.

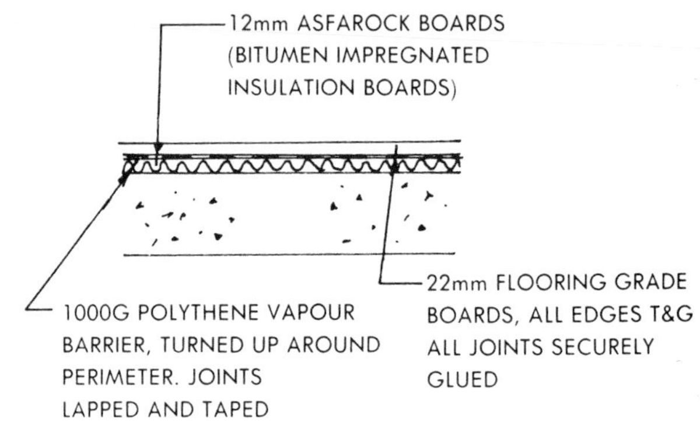

FIG 3

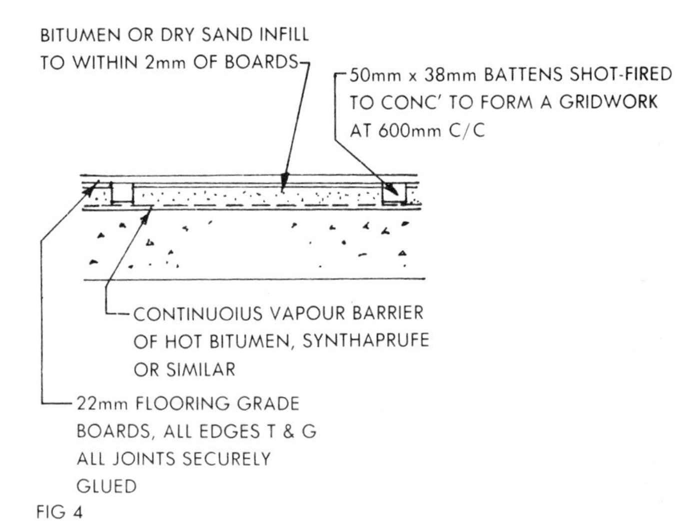

FIG 4

MOISTURE RESISTANT BOARD

A chipboard which has a high degree of resistance to moisture is suitable for all domestic flooring applications, for joists up to 450mm c/c. It is resistant to adverse site conditions, e.g. where floors are left exposed to weather before the building is made weathertight. Since the moisture resistance is an integral property of the board, no protection is necessary to the edges if on-site cutting is required. This type of board has many other uses, in roofing, shuttering etc.

STORAGE ON SITE

Boards should be close pile flat with all four edges flush, and stacked on a level surface. For temporary storage, timber or chipboard bearers about 100mm wide and of equal thickness may be used. A clear space of 400mm to 500mm is suggested, to prevent sagging Fig. 5. Incorrect stacking, Fig 6, may result in permanently buckled boards. Boards stacked for long periods in a partially completed building or unheated stores, should be covered with polythene sheeting, drawn tightly round them.

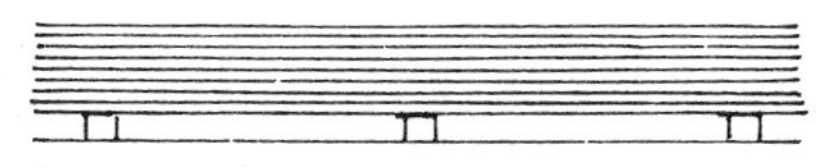

FIG 5 CORRECT STACKING

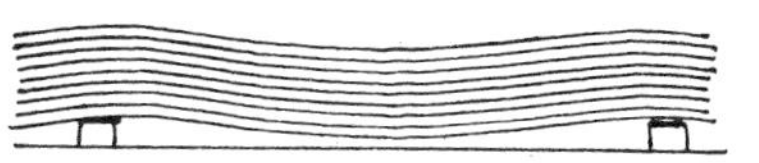

FIG 6 INCORRECT STACKING

CLADDING I

In a framed structure where the loads are carried by the reinforced concrete frame and not by the external walls, the use of thin concrete or stone units to cover the structural frame is referred to as cladding. Brickwork may also be used as a cladding medium.

When considering cladding, certain precautions must be taken. Allowances must be made for permanent and temporary movement of the structure due to shrinkage and elastic deformation under load, and for thermal movement. Considerable care is required to ensure permanent fixing and to prevent or control the ingress of water and allow for condensation.

BRICK CLADDING

The main walling is constructed as a normal cavity wall with brick outer skin and brick or block inner skin, the walling being supported at each floor level. Problems arise where the brickwork serves as a facing to beams and columns. If a full half-brick is allowed in front of the structural frame the walling may be secured by means of dovetail anchors which fit into dovetail slots. Fig I.

BRICK SLIPS

Where brick slips (also called brick tiles) are used to cover the face of columns, floor slabs or beams, particular care is necessary. In recent years there have been a number of failures due to courses of slips buckling, usually attributed to the squeezing of cladding by vertical shrinkage and creep of the concrete frame. (Creep, is the gradual compression of the structure due to sustained stress and this may take place over a long period of time). It is important that movement joints (or pressure relieving joints) be provided at storey height intervals. Fig 2.

Where the walling is supported by the structural frame, any overhang of the brickwork must not exceed one-third the width of the brick. Fig 3.

Fixing of brick slips :- Considerable research has been carried out in this field and the BDA have produced a Technical Note (Vol I No 4) dealing with the problem.

The use of conventional cementitious mortars and brick fixing processes, have in a number of cases provided inadequate for holding brick slips, even where support was afforded beneath them by the top course of a brick infill. The use of special mortars or some type of mechanical support and restraint therefore is recommended.

Brick slips, which are usually 25-37mm thick present a special problem through being porous. (The following comments do not apply to the use of ceramic tiles which pose a separate problem).

The ideal adhesive system should have a high strength, be impermeable to water and be free of voids in which water could collect.

Horizontal stresses may arise due to differences in the coefficients of thermal and moisture expansion of the concrete and the brick slips. Where slips are bonded to the toe of a floor slab vertical stresses may occur due to creep and drying shrinkage to the concrete, as well as long term vertical expansion of the brick infill. This suggests the necessity of providing expansion joints to accommodate these movements.

Backing surface :- It is important that the toe or nib of the floor slab is accurately cast so that the edge is vertical and plumb with the floors above and below. The concrete surface must be clean, free from dust, grease and any mould oil. Any laitance should be removed and preferably the aggregate exposed. A retarder may be used on the shuttering to achieve this result, or a key may be obtained by using ribbed shuttering. Alternatively the aggregate may be exposed by mechanical means.

The face of the concrete should not be treated with a bituminous D.P.C, as this will seriously affect the final bond. The adhesive system itself is largely impermeable and will act as a D.P.C.

Epoxy resin mortars :- These actually consist of a liquid resin, a liquid hardener and selected sand. Epoxy resin systems have a working life of 2-3 hours, with full cure developing in 24 hrs. They should not be used when the ambiant temperature is below 4 deg C. It is most important that manufacturers instructions are strictly complied with. The use of epoxy resin mortars is expensive, but the risk of failure of bond and the high cost of remedial measures warrants the cost.

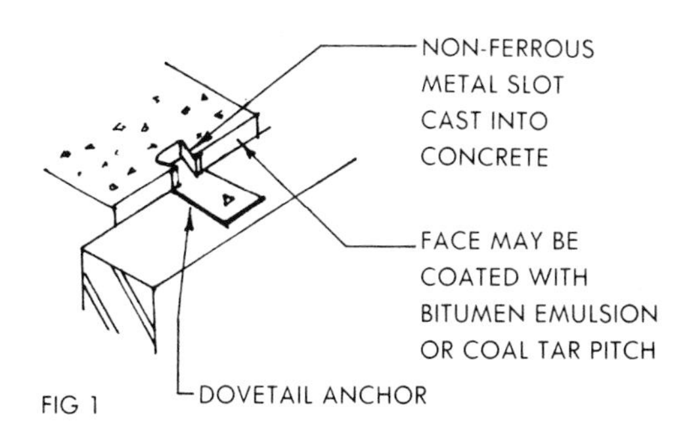

FIG 1

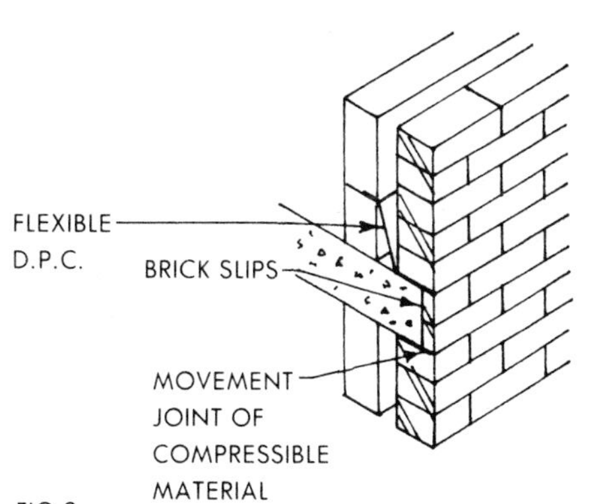

FIG 2

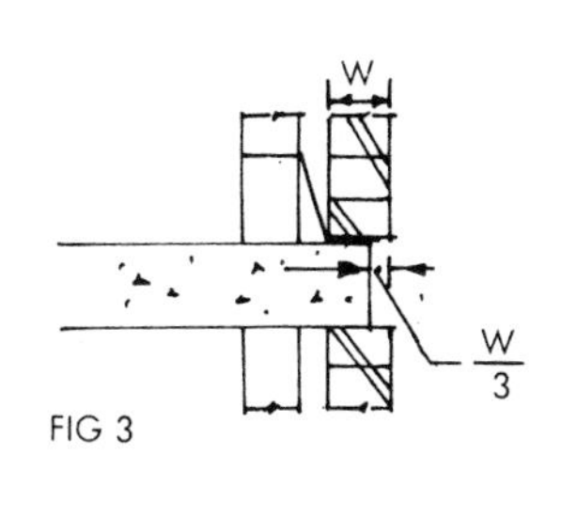

FIG 3

CLADDING 2

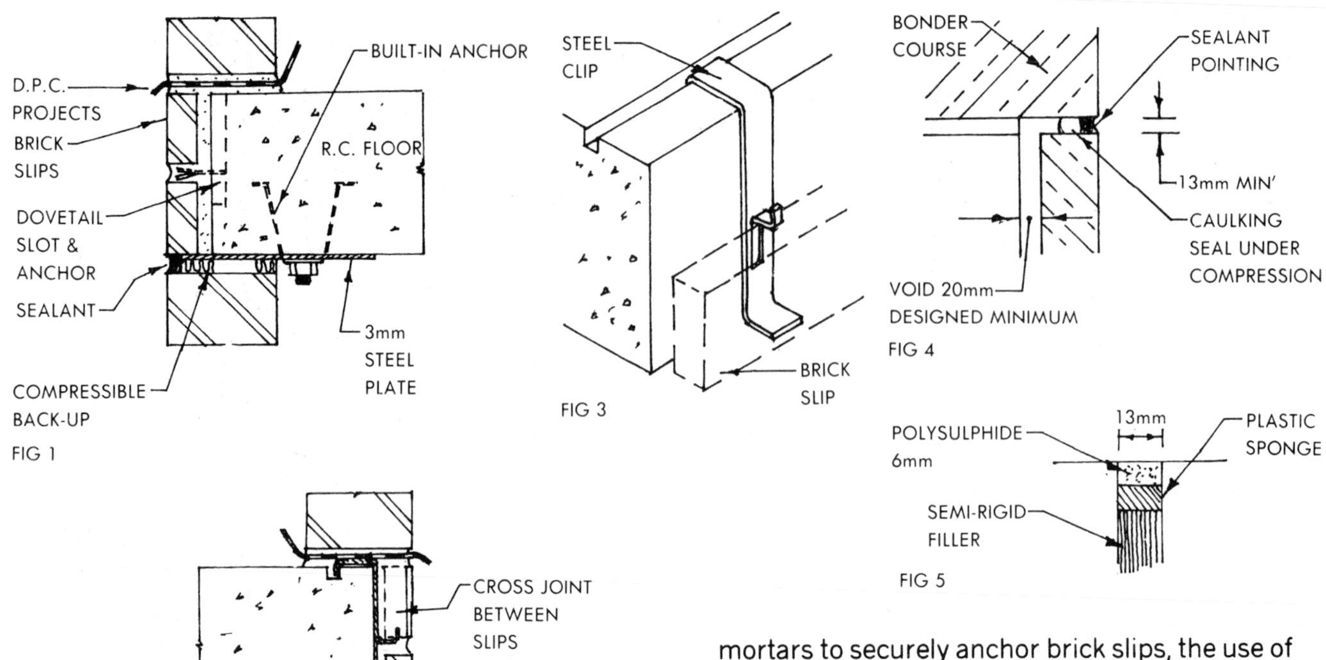

FIG 1

FIG 3

FIG 4

FIG 5

FIG 2

Polyester resin systems:- These consist of a resin in a solvent together with a hardener. A graded silicious aggregate is frequently added, both as an extender and to give the correct degree of plasticity and workability.

Both polyester resin and epoxy resin adhesives properly mixed and applied rapidly develop a high bond strength which is resistant to water and most chemicals.

Cementitious mortars with styrene/butadiene rubber (SBR):- A number of SBR emulsions are available for improving the bond of mortars, renderings and floor screeds. One part P.C. (by weight) to 2½- 3 parts of selected sand, with the gauging water wholly or partly replaced by SBR. Manufacturers' instructions must be strictly observed. The adhered surfaces are prepared by coating with a slurry of cement and SBR. Whilst this is still tacky, the mortar is buttered on to the grouted surfaces and the slips pressed into place.,

Some form of temporary wooden battening is usually necessary above openings to provide support until the adhesive is sufficiently cured.

Applications in the form of dabs of adhesive is not recommended as this will not produce an impermeable coating and problems of freezing may arise.

Mechanical support:- The use of a soft compression joint at each floor level means that problems arise in providing support and restraint to the brickwork. As well as the use of special mortars to securely anchor brick slips, the use of some form of mechanical support has been adopted in a number of cases.

One method is to use a projecting continuous steel plate, (stainless steel or galvanised) bolted to the floor slab, upon which the weight of the units can be sustained. Fig I. High adhesive mortar may be used in bedding the slips back to the face of the concrete, as an extra precaution.

An alternative method is to use special stainless steel (or galvanised) clips Fig 2. These provide both vertical support and lateral restraint. One such clip would be required for each brick of the bottom course of slips Fig 3. The initial shrinkage of the structural concrete and deformation under load can cause spalling of the joints of the lower stones. The effect may be further aggravated by moisture movement and thermal movement. Dimensional changes may result from a combination of these factors. But the change most likely to have the most serious effect on cladding, is the drying shrinkage of a building, particularly where the concrete is cast in-situ. Where possible it is recommended that cladding should not be fixed until the basic structure is completed.

The timing for the insertion of the compression joints will depend upon the design of the units and the materials used, but the joints should be left open as long as possible so that the main shrinkage may take place before the joint is filled.

Compression joints should be provided at each floor level immediately under the support for the cladding, and along the full length. The width of the joint should allow for maximum movement of the frame, so that pressure is not transmitted. A 13mm wide joint is the accepted minimum per floor. Fig 4.

Compression joints are formed of a soft elastic material. Butyl rubber compounds and preformed bitumised polyurethane sections may be used or polysulphide compounds applied by gun. Fig. 5.

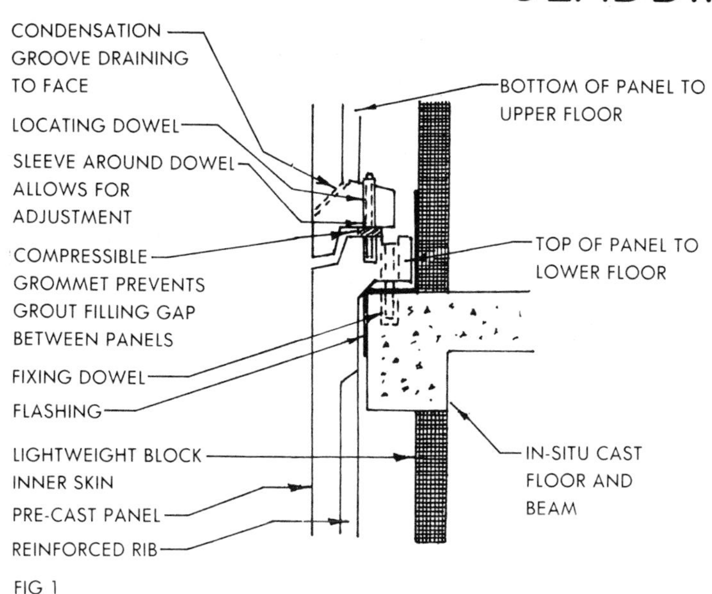

FIG 1

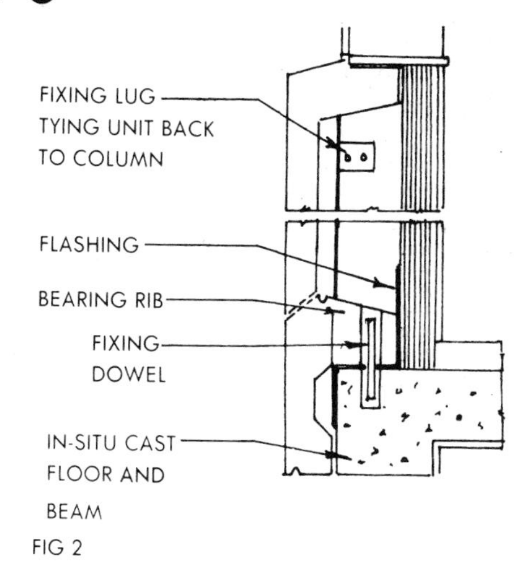

FIG 2

PRECAST CONCRETE CLADDING

May be storey height panels attached to the structural frame, as an applied facing to a solid background of brick or concrete, or as permanent formwork held mechanically to a background of in-situ concrete.

Size of units :- If 20mm aggregate is used for the concrete, units of uniform thickness should not be less than 65mm thick, and such units should not exceed 1070mm in any direction. (For larger aggregate the units would be thicker). Prestressed units may be larger. The maximum superficial area of plain units should not exceed 1.14m². For framed units, if the maximum size of coarse aggregate does not exceed 10mm and the reinforcement is confined to the ribs, the thickness can be reduced to 40mm, provided the spacing of the ribs in any one direction does not exceed 900mm. N.B. The maximum weight two men can handle and place in position is about 55 kg, above this weight a lifting appliance is required.

Where the cladding is supported by the structural frame units should have a bearing of at least 50mm. They should only be supported at one bearing level which may be positioned at the top (hung unit) Fig 1, at the bottom (standing unit) Fig 2, or at any convenient intermediate level in the height of the unit. A support should be provided at each storey height. Designs vary but Fig 1 showing hung storey height panels, and Fig 2 showing a standing apron panel are typical examples and indicate general principles.

Finishes :- A wide variety of surface finishes to concrete units are available (i) exposed aggregate (ii) textured finishes (iii) cast stone — can be given a range of masonry and other tooled finishes (iv) mosaics — only mosaics known to be frost resistant should be used externally (v) tiles — these and brick slips should be frost resistant and have well formed grooves to provide a good key to the backing (vi) stone facing veneer to precast concrete.

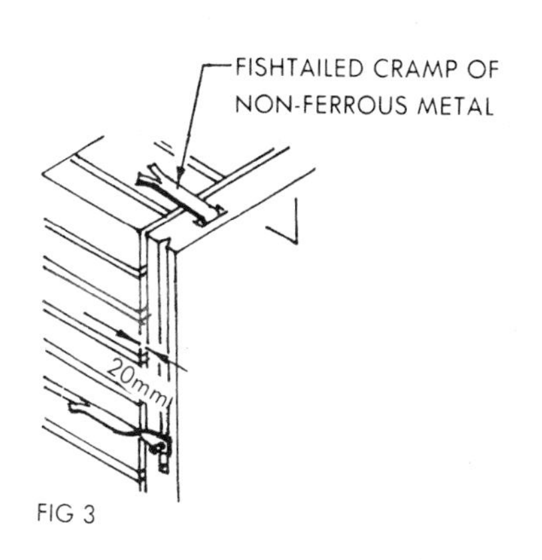

FIG 3

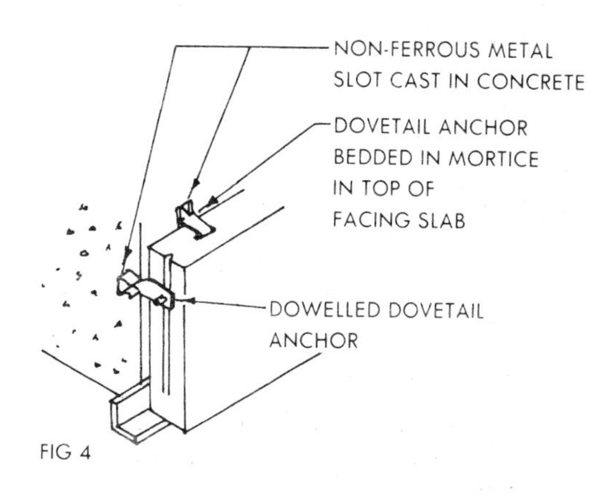

FIG 4

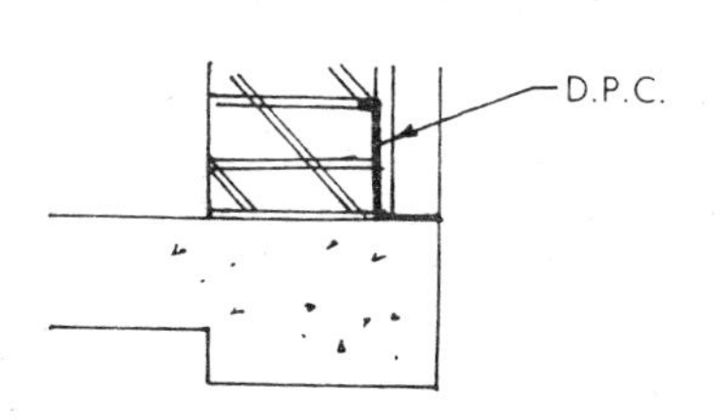

FIG 5 CONCRETE CLADDING AS APPLIED FACING TO BRICK OR CONCRETE

Units should preferably be finished with a chamfered edge, this will not only help to mask any irregularities in alignments, but there is less chance of damage to edges. Sharp edges and thin projections should be avoided as far as possible, as they are prone to chipping and breakage in transit and handling.

Concrete cladding as applied facing to brick or concrete :- A 20mm cavity should be left between the back of the units and the face of the background Fig. 3.

Cladding may be tied back to brickwork by means of fixing cramps built into the joints of the walling as shown Fig 3. The units may be supported by the structural frame Fig 5, or by means of corbel plates.

Fig 4 shows ways of securing cladding to a concrete backing. Over an opening the cladding may be supported on a non-ferrous angle bolted or screwed to the structural concrete.

Concrete cladding as permanent formwork :- The units and the inner formwork must be securely fixed to withstand the pressure of the wet concrete and the vibration when placing. The ties must be spaced close enough to ensure adequate strength — say 600mm apart vertically and horizontally. An example is shown Fig 6.

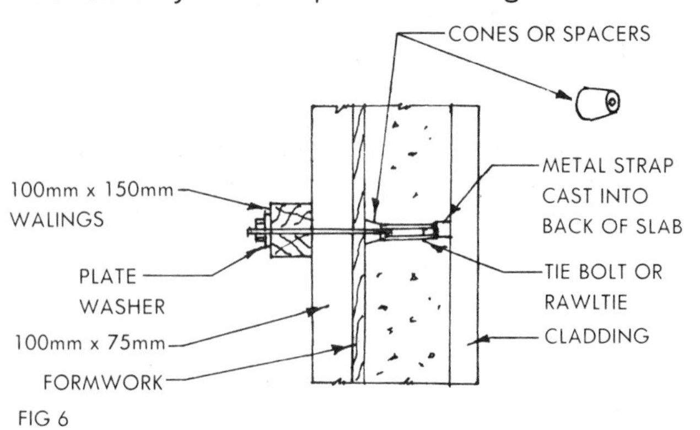

FIG 6

CLADDING 4

	Minimum thickness of cladding				Minimum thickness of material behind the cramp mortice			
	Granite mm	Marble mm	Slate* mm	Limestone & Sandstone mm	Granite mm	Marble mm	Slate mm	Limestone & Sandstone mm
Pilasters, stall risers and similar below first floor level (not exceeding 3.7m)	20	20	40	75	7†	7†	20	37 or half the overall thickness if over 75
Fascia and soffits below first floor level (not exceeding 3.7m)	35	35	40	75	13	20	20	37 or half the overall thickness if over 75
All cladding above first floor (exceeding 3.7m)	40‡	40‡	40	75	20	25‡	25	37 or half the overall thickness if over 75

With riven material such as slate, slabs should be selected to achieve the required thickness behind the cramp mortice and to leave not less than 9mm of material thickness in front of the cramp mortice

†It may be necessary in certain circumstances to increase this thickness

‡This thickness of marble and granite may be reduced to 35mm minimum for cladding and 20mm behind cramp mortice where the load of each unit is accepted by a continuous structural member and where each unit is adequately tied back near the top of the unit (e.g. window aprons, balcony units etc).

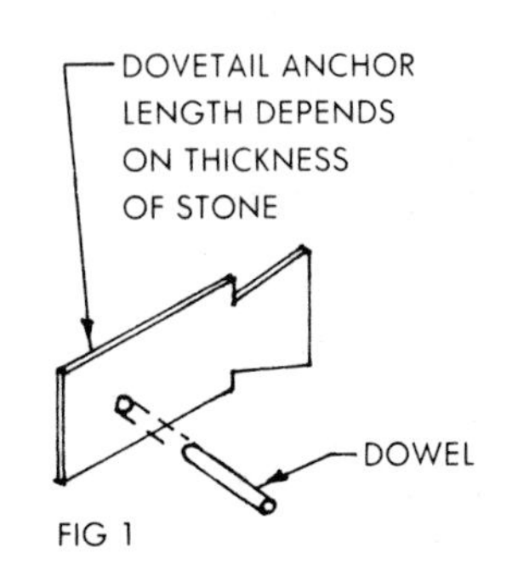

FIG 1

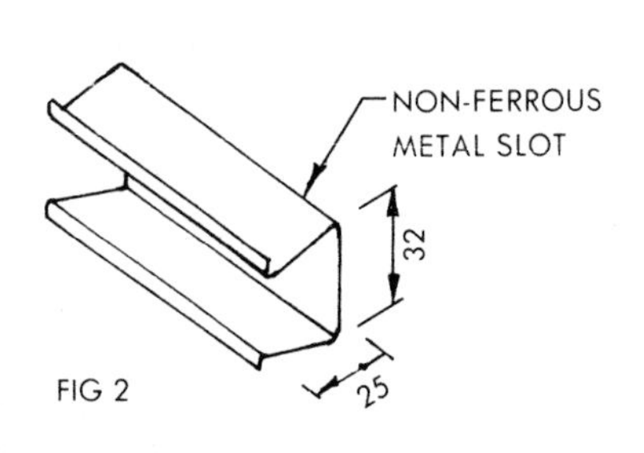

FIG 2

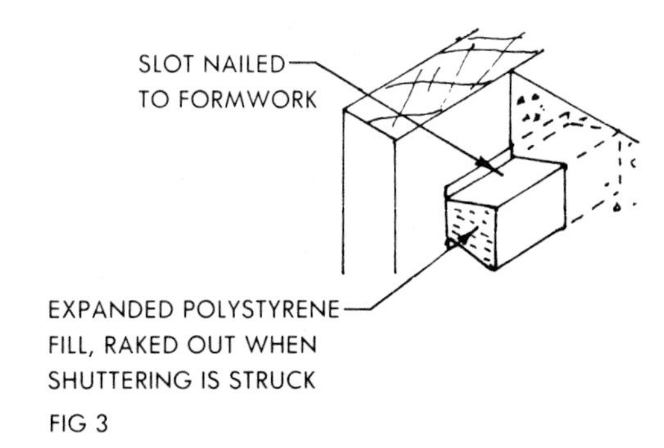

FIG 3

STONE CLADDING

Limestone, standstone, granite, marble and slate may all be used for cladding. It is important to maintain a minimum thickness for stone units, and mortices must not penetrate too deeply. The table is reproduced from CP298: 1972.

CHOICE OF STONE

If thin cladding, e.g. granite, marble and particularly slate, which is normally face bedded are to be used, the desirability or otherwise of using face fixings back to the structure should be considered at the design stage. This is particularly relevant on highrise buildings or wherever a high degree of vibration, wind loading or movement of one type or another may occur. (e.g. shrinkage, elastic-deformation-under-load, temperature movement etc). The fixing of soffits also requires careful consideration.

The likelihood of some rain penetration and/or condensation within the cavity between the cladding and the inner leaf or back up wall must be allowed for, and adequate provision made for drainage and for damp proofing over openings. Particular care must be taken when dense cast units are used and typical precautions were shown in Figs 1 & 2.

N.B. Certain stones are incompatible and it is essential that cast stone made with limestone is not used in conjunction with sandstone, if water from the former can drain onto the latter.

FIXING CLADDING

There are two important considerations (i) supporting the stone (ii) tying back the slabs. Cramps and fixings should be of suitable non-ferrous metal e.g. copper, aluminium bronze, phosphor bronze, gunmetal, or stainless steel. For any particular job, fixings and anchor slots should preferably be of the same metal, thus obviating failure from galvanic action. Iron or steel fixings should not be used even if coated. Protective coatings are vulnerable and damage to the coating will result in corrosion and ultimate failure of the fixing.

METHODS OF TYING BACK

Dowelled dovetailed anchors which fit into dovetail slots (Figs 1, 2 and 3) are commonly used. The slots are tacked to the formwork and cast in position. A temporary filling is needed during concreting and expanded polystyrene is useful for this purpose as it is easily removed. Four anchors are normally required for each block.

CLADDING 5

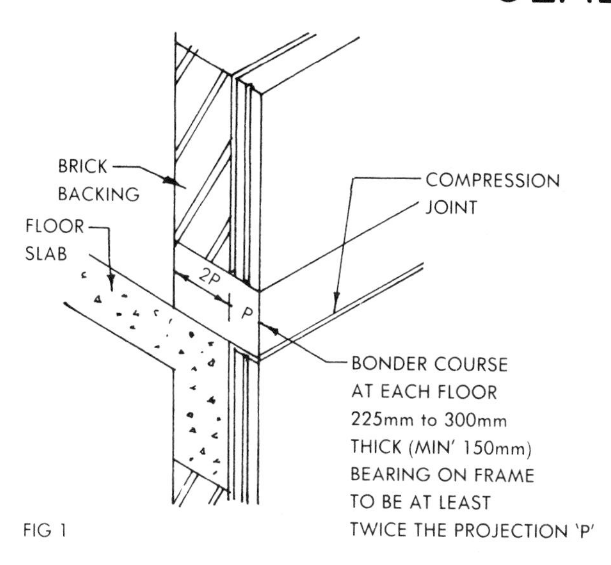

FIG 1

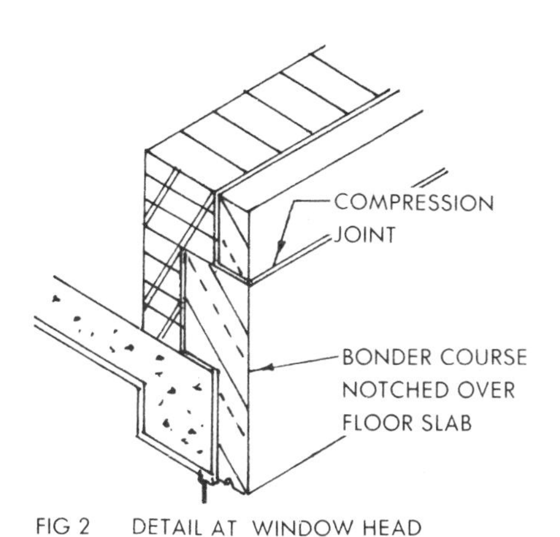

FIG 2 DETAIL AT WINDOW HEAD

METHODS OF PROVIDING SUPPORT

The weight of the cladding should be carried by the structural frame. Methods of support vary, Figs 1 and 2 show examples of the use of bonder courses and Fig 3 shows the use of a concrete nib to provide support.

Where these methods are adopted the positional accuracy of the structural supports is most important, and these should be constructed with a permissible deviation not exceeding ± 7mm.

Where nibs or structural corbel courses which provide direct support for the units are used, it is important to ensure that between half and two-thirds of the unit thickness bears direct upon the support.

Metal fixings :- Supports may be provided by means of non-ferrous metal corbel plates possibly cast integrally with the structural concrete, or more commonly grouted into a mortice formed in the concrete Fig 4. Alternatively, adjustable corbel plates may be bolted to anchor slots, as illustrated in Fig 5.

Another method of support is to use a suitable non-ferrous metal angle bolted to the structure Fig 6.

Where fixings or inserts of any kind are cast integrally with the structural concrete, it is essential that these are placed accurately: Should the fixing not co-incide with slab unit fixing position, a satisfactory new fixing, securely anchored to the concrete must be provided.

Where fixings are secured to the structure (as shown Fig 4), a strong mortar 1:2 PC (or R.P.C.)/clean sharp sand should be used. The mortar should be mixed to a fairly dry consistency, well tamped around the fixing and allowed to mature for at least 48hrs before being subjected to stress, unless the cladding is temporarily supported on head trees, when the time may be halved. Certain epoxy or polyester resin mixes may also be used for securing fixings, but the manufacturers instructions must be strictly observed. The positions of corbels are indicated in Fig 7.

Compression joints :-

These joints should be provided at each floor level immediately under the stone which is transmitting the load from the storey above onto the structure, and along the full length (As Fig 4). The joints must be watertight. Sealants may be butyl, oleoresinous, acrylic, polysulphide, silicones or polyurethanes. Wider joints may be built up, allowing sealant to set between applications. Back-up material may be fibre-board, rope free of oil and bitumen, expanded plastics or rubber.

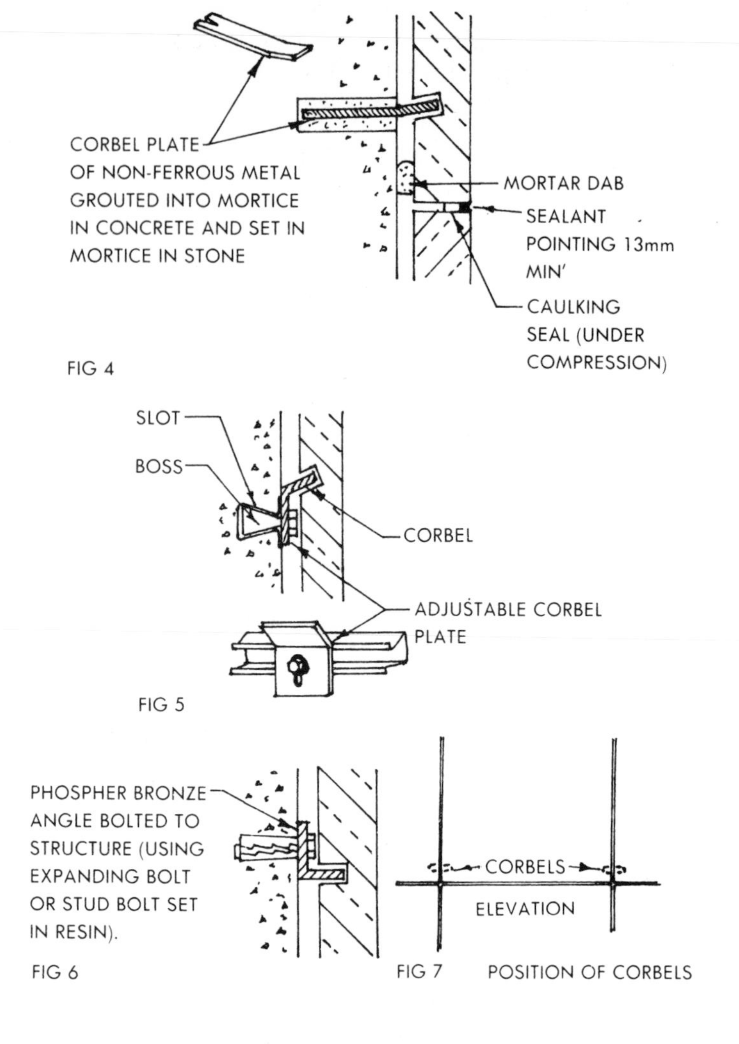

FIG 4

FIG 5

FIG 6

FIG 7 POSITION OF CORBELS

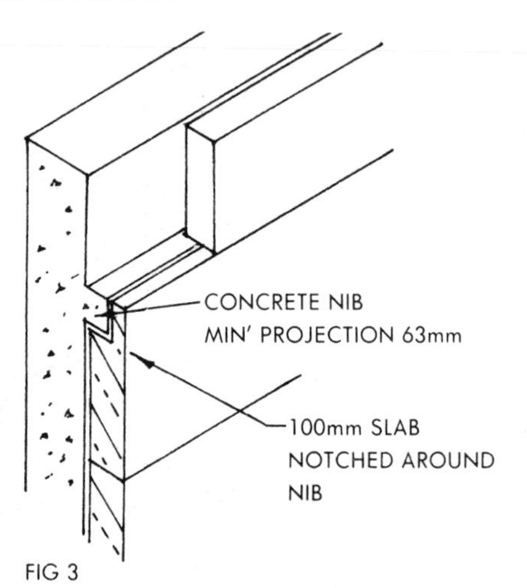

FIG 3

CLADDING 6

ADAMANTINE CLADDINGS
(GRANITE, MARBLE, QUARTZITE, SLATE)

The thickness of the material is critical and the minimum thickness laid down in the table, should be carefully observed. Thin slabs should be positioned against dabs of weak mortar or lime putty applied to the backing. Copper or brass cramps are caulked into mortices in the structural frame, and set in mortices drilled in the top or side of the slabs. 'S' hooks are set in mortices drilled in the bottom edge of the upper slab and clipped behind the top edge of the lower slab. Over openings they may be set in the sides of slabs. N.B. 'S' hooks and tie cramps should preferably not be more than 25mm apart.

The locations of the various fixings are shown in Fig 3 and details of individual fixings are shown in Figs l and 2.

N.B. Adamantine facing slabs are sometimes used for internal wall linings and in the case the limitation of overall height makes the use of corbels unnecessary.

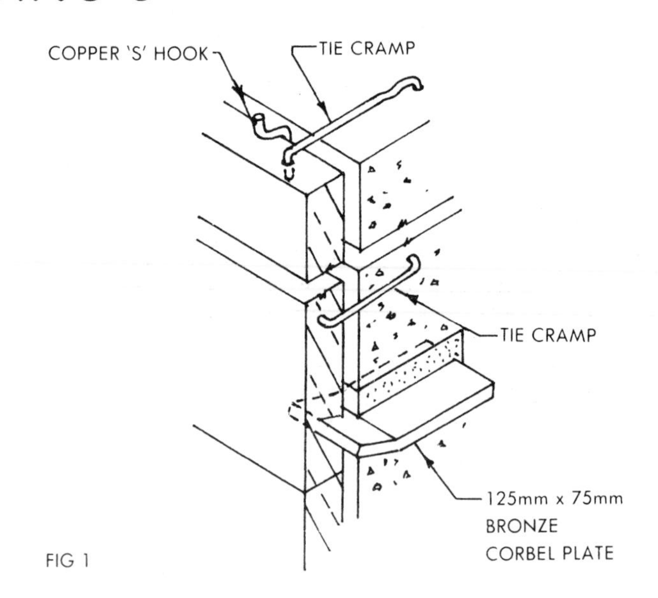

FIG 1

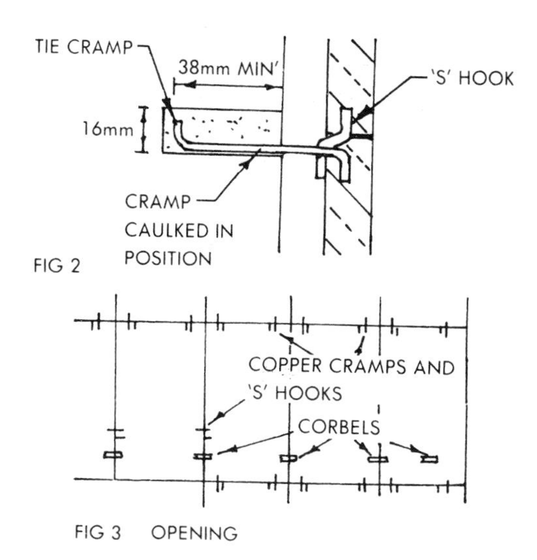

FIG 2

FIG 3 OPENING

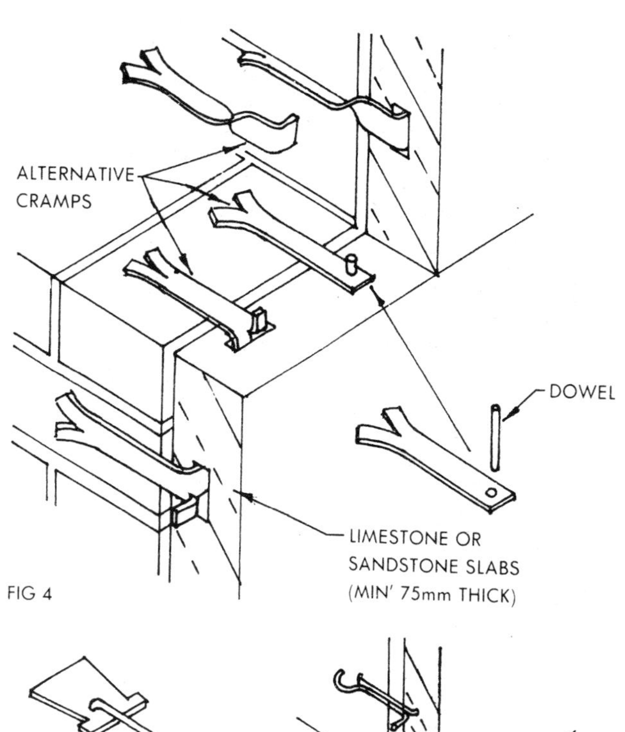

FIG 4

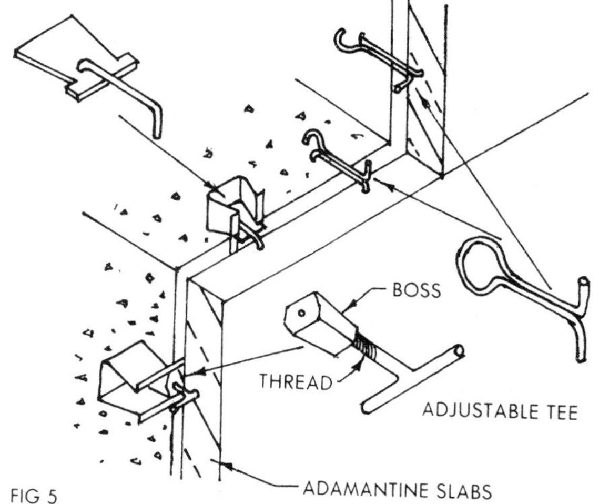

FIG 5

VARIOUS FIXINGS

A wide variety of fixings are available to suit various situations, e.g. fixing Adamantine, limestone or sandstone slabs to the structural backing of brickwork or concrete, securing slabs at the heads of openings etc.

It is important that fixing mortices and holes in the stone should coincide with the anchor slots in the structural frame, and particularly when using limestone and sandstone it may be advisable to cut the mortices or dowel holes on site, but this is not really practical for granite and slate and when these are being fixed it is essential that particular care is exercised when siting the corbels and cramps. Four anchors are normally required for each block. Figs. 4 and 5 show a range of typical fixings.

Joints :- Because of the possibility of movement due to the initial shrinkage of the structural concrete and deformation under load, which can result in spalling of the joints of the lower stones, very tight joints should be avoided. Ideally a maximum width of 13mm for mortar filled joints is desirable and the minimum width should be:-(i) limestone 5mm (ii) marble or slate 3mm (iii) granite 5mm (iv) slate (Riven finish) 7mm.

CLADDING 7

SOFFIT CLADDING

The attachment of cladding to soffits presents special problems, the usual method is to suspend the slabs from bolts or hangers which slide into soffit anchorages cast into the concrete. Figs 1 & 2.

The minimum thickness for stone units as recommended in the table may not be sufficient for soffit cladding and should be increased if the size of units and method of fixing require it.

MORTAR

A number of factors must be considered when selecting the mortar to be used. The type of unit, the size of the units, the surface finish of the cladding and the degree of exposure to severe weather conditions. For natural stone ashlar work mixes of from 1:5:7 to 1:2:8 cement:lime:stonedust may be used, the stonedust should pass through a 1.4mm sieve. Where sand is used in the jointing mortar stonedust may be used for the pointing mix. White cement is sometimes used to give a particular effect.

N.B. Mixes other than those given above are used at times e.g. a 1:3:12, cement:lime:fine crushed stone, or even 2:5:7, cement:lime:fine crushed stone. This latter mix although very plastic and having good workability, may possibly result in a high shrinkage which might cause cracking. Because of the effects of structural movements it is important that the mortar should not be so strong as to transmit stresses and cause damage and spalling to the arrises of the stone units. For this reason the leaner mixes given above and recommended by CP298: 1972 are preferable.

For the jointing of granite, slate or similar units, a 1:4 cement:sand mix may be used.

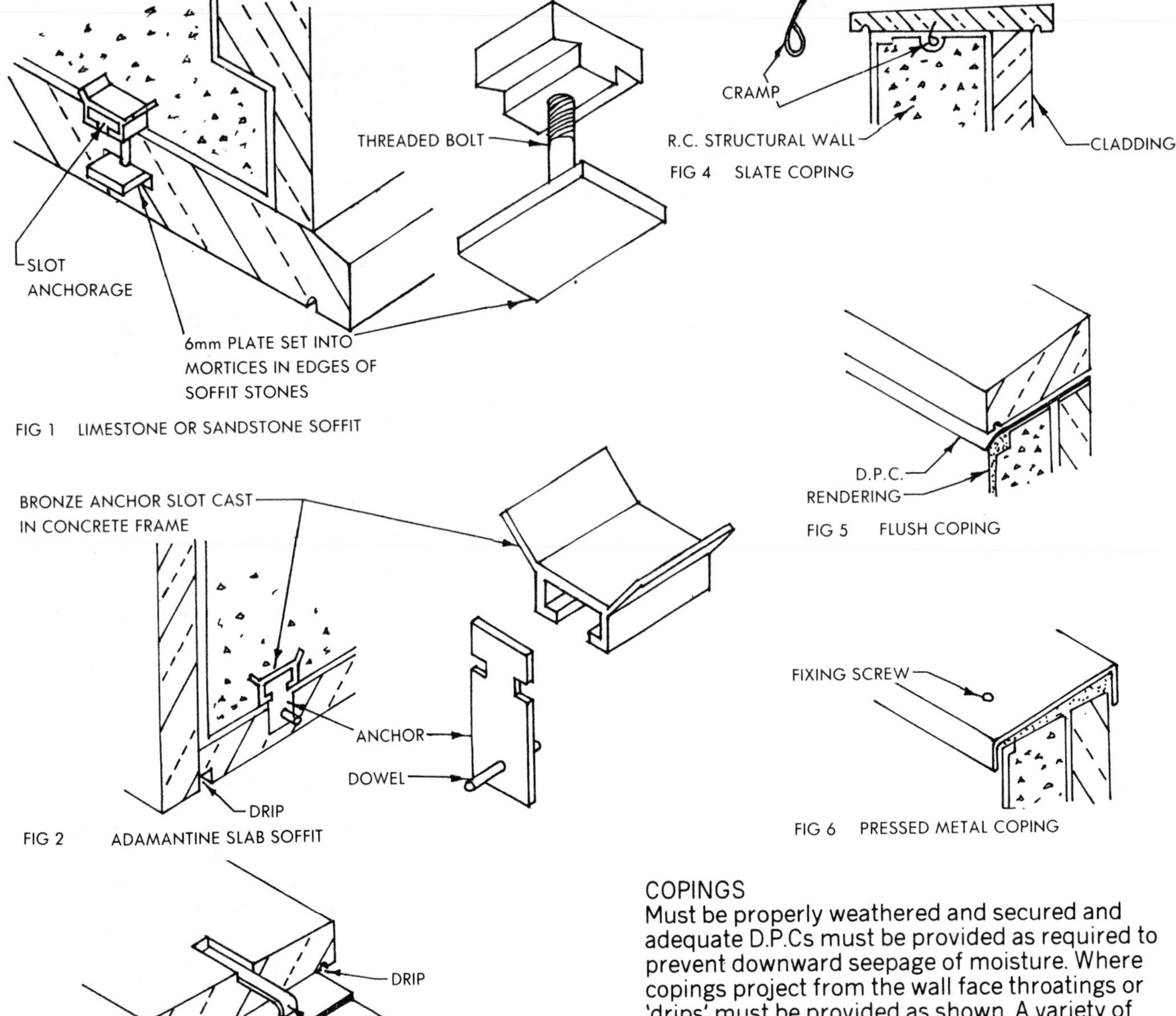

FIG 1 LIMESTONE OR SANDSTONE SOFFIT

FIG 4 SLATE COPING

FIG 5 FLUSH COPING

FIG 2 ADAMANTINE SLAB SOFFIT

FIG 6 PRESSED METAL COPING

COPINGS
Must be properly weathered and secured and adequate D.P.Cs must be provided as required to prevent downward seepage of moisture. Where copings project from the wall face throatings or 'drips' must be provided as shown. A variety of copings are illustrated in Figs 3, 4, 5 and 6.
N.B. It is recommended that adequate movement joints are provided in the cladding to parapets and copings, as being very exposed they are more affected than other parts of the building.

FIG 3 FEATHER-EDGE COPING

CLADDING 8

MOVEMENT JOINTS
Any construction movement joints in the main building structure must be repeated through the cladding and effective movement joints must be incorporated in the facing work.
These may take the form of a void between stones, or are filled with a compressible filling material and caulked and pointed with a mastic sealant.

Vertical expansion joints 6mm to 10mm wide to allow for differences in thermal movement between cladding and structure are normally provided at approximately 6m intervals in areas where the stonework is continuously horizontal.
Polysulphide compounds, bitumenised foamed polyurethane strip and sprung copper strip are commonly used.

PROTECTION OF FINISHED WORK
A stiff slurry of stonedust and lime not stronger than 12:1 mixed with water and brushed liberally over the face of limestone units affords protection from staining. The slurry should be kept clear of adhesive joints or sealants. Slurry may be removed by a light rub with a flexible disc or fine grinding wheel. Further protection may be afforded by using timber strips and heavy duty polythene sheet.

STONE FACED CAVITY WALLS

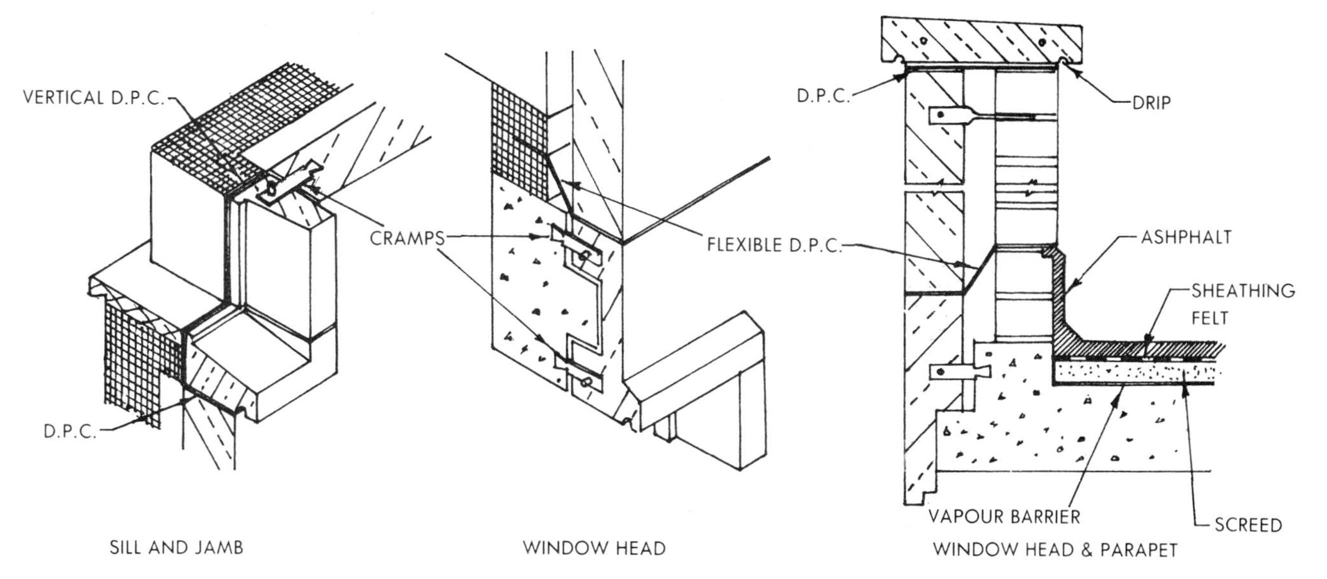

TYPICAL CAVITY WALL DETAILS

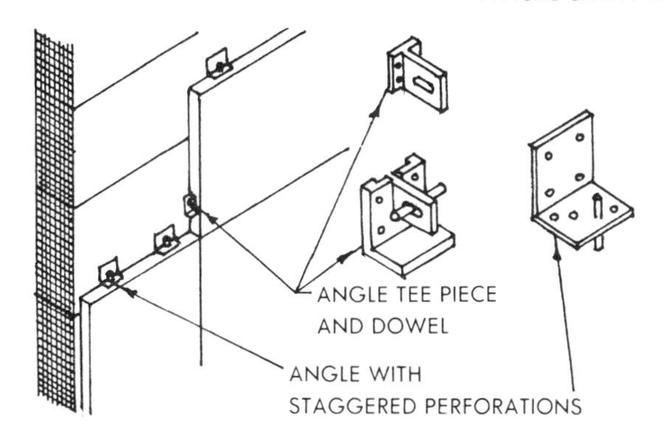

THINWALL CLADDING TO LIGHTWEIGHT BLOCKS

Bronze angles as shown fixed either by non-ferrous metal wood screws to lightweight blocks (Celcon, Thermalite or similar) or by expanding bolts.

HANDLING AND STORING

Handle stone with normal care to avoid damage to edges and corners. Store under cover stacked on battens. (chamfered edges, particularly for cast blocks are less likely to be damaged and are useful for masking any irregularities in alignment).

GLASS REINFORCED PLASTIC CLADDING

Panels of hot moulded glass reinforced plastic have been developed for building cladding. They are extremely lightweight and a variety of finishes can be applied, ranging from smooth paint to light or medium aggregate. A recommended finish is 3 coat polyurethane comprising, undercoat, 'spatter' coat and top coat. This gives a hard reflecting surface broken with a light stippled effect. Panels are available without backing (where this is not required, e.g. for fixing to an existing wall or for use on their own) or with backing to form a complete wall ready for internal decoration.

STAIRS I

Approved Document K of the Building Regulations 1985 gives provisions for stairways which are in a building or form part of the structure of the buildings.

BS 585: 1984 Wood Stairs, relates to the quality and construction of interior stairs, in wood with close strings for use in houses.

Common stairway:- An internal or external stairway of steps with straight nosings on plan which form part of a building and is intended for common use in connection with two or more dwellings.

Private stairway:- A similar stairway either within a dwelling or intended for use solely in connection with one dwelling.

DESIGN DETAILS

Going:- The going of a step shall be measured on plan between the nosing of its tread and the nosing of the step or landing next above it (Fig I). (BS 585: 1984 states:- going is measured from face of riser to face of riser).

The going of a step shall be not less than 220mm for a private stairway, and not less than 230mm for a common stairway.

Between consecutive floors there is an equal going for every parallel step. (BS 585 states:- Treads shall be not less than 235mm wide).

Parallel step:- A step of which the nosing is parallel to the nosing of the step or landing next above it.

Nosings:- The nosings on treads usually project 32mm (or the thickness of the tread) from the face of the riser below, and should in any case project not less than 15mm, and the profile should be rounded. Fig I.

The nosing of the tread of any step or landing which has no riser below it, (e.g. an open tread stair) overlaps on plan the back edge of the step below it by not less than 16mm Fig 2.

Pitch:- The angle between the pitch line and the horizontal ('o' Fig I) shall not exceed 42 deg for a private stairway and 38 deg for a common stairway.

It is recommended that the pitch should not be less than 25 deg. A stairway at a shallow angle occupies more space and is tiring to ascend.

Pitch line:- A line drawn from the floor or landing below a stairway to connect the nosings of all the treads in a flight of stairs (Figs I&3)

Rise:- Measured from top of tread to top of tread. (Figs I & 3). Between consecutive floors there must be an equal rise for every step or landing.

The rise of a step must be not more than 220mm for a private stairway and not more than 190mm for a common stairway.

Part I. 14 of Approved Document K of the Building Regulations 1985 states the number of risers in a flight should be limited to 16 if a stairway serves an area used for shop or assembly purposes. Part I.15 of Approved Document K states that stairways of more than 36 risers in consecutive flights should make at least one change in direction between flights of at least 30 deg.

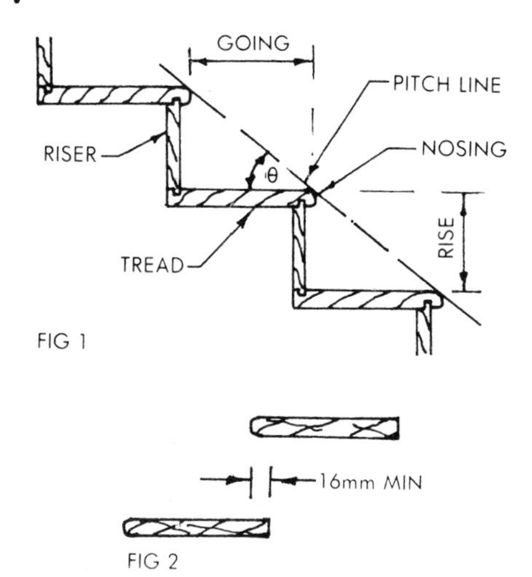

FIG 1

FIG 2

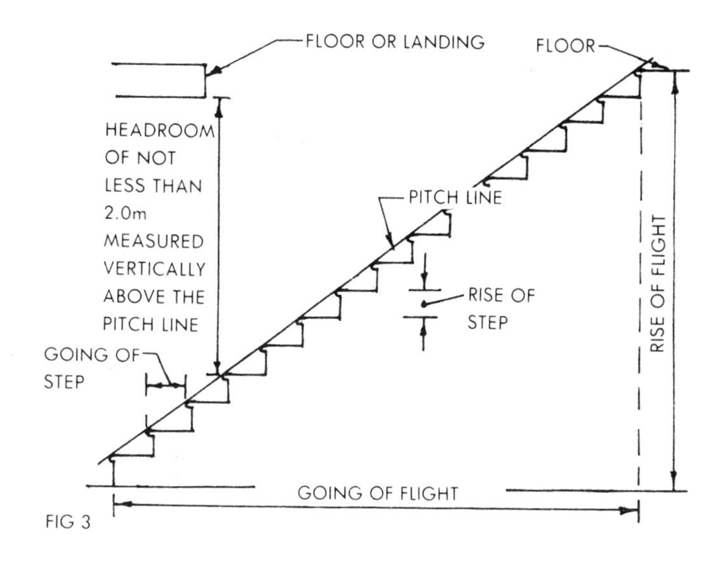

FIG 3

Flight:- An uninterrupted series of steps between floors or between floor and landing.

The sum of the going of a parallel step plus twice its rise must be not less than 550mm and not more than 700mm. (BS 585 states the following rule:- 2R + G = 575mm to 630mm. Where 'R' = rise and 'G' = Going.

Design:- The number of steps is governed by the floor to floor height which is usually fixed, and the flight going space available. (Fig 3). Alteration work may present difficulties, but in new work the flight going space is usually flexible.

Example:- A straight flight stairway is required. Floor to floor is 2.7m and going space is unrestricted. (A rise of between 170mm and 200mm, and going of 220mm to 250mm would normally be satisfactory). Assume risers of 175mm. No of risers = 2700/175 = $15\frac{3}{7}$. Since a whole number is required, try 15. The actual rise of each step will be 2700/15 = 180mm. If going of each step is 250mm, then 2R + G + 2 x 180 + 250 = 610mm. Since this is between 575 and 630 it meets both the requirements of the Building Regulations and BS 585. The Building Regulations regarding headroom are shown in Fig 3.

STAIRS 2

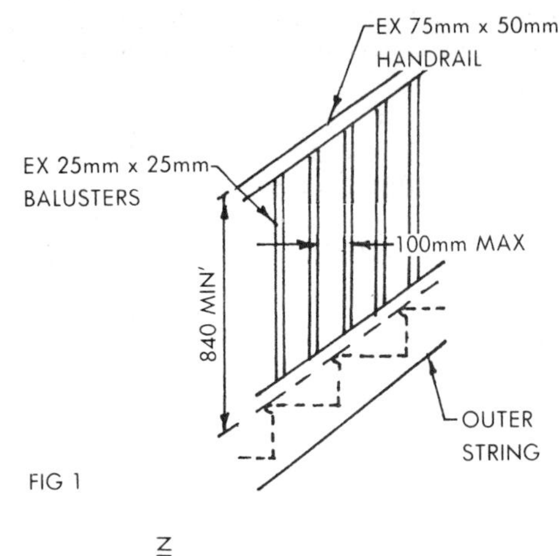

FIG 1

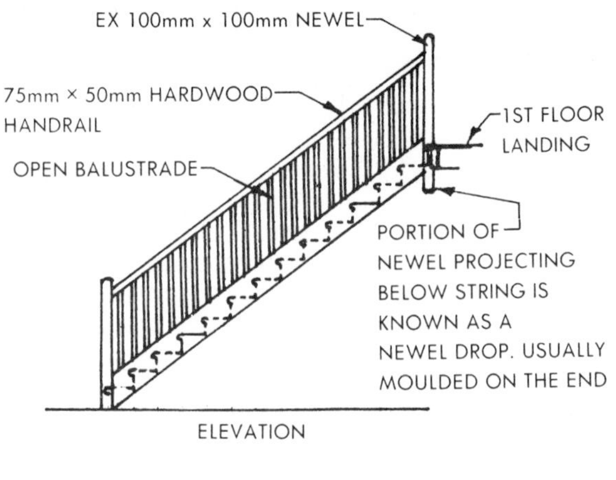

ELEVATION

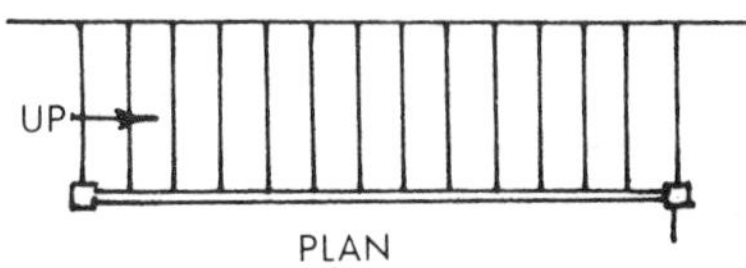

PLAN

FIG 2 STRAIGHT FLIGHT STAIR

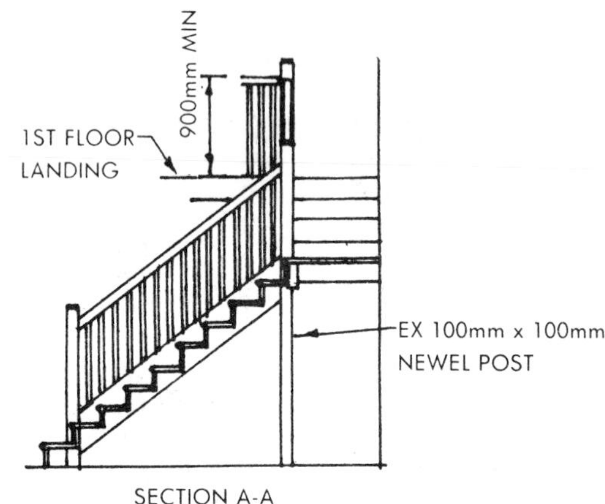

SECTION A-A

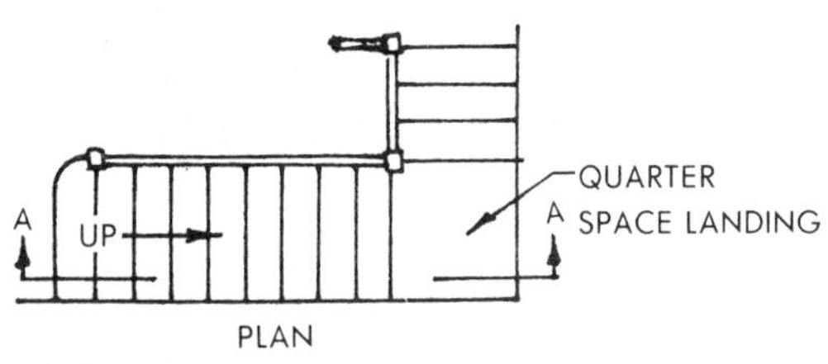

PLAN

FIG 3 QUARTER TURN STAIR

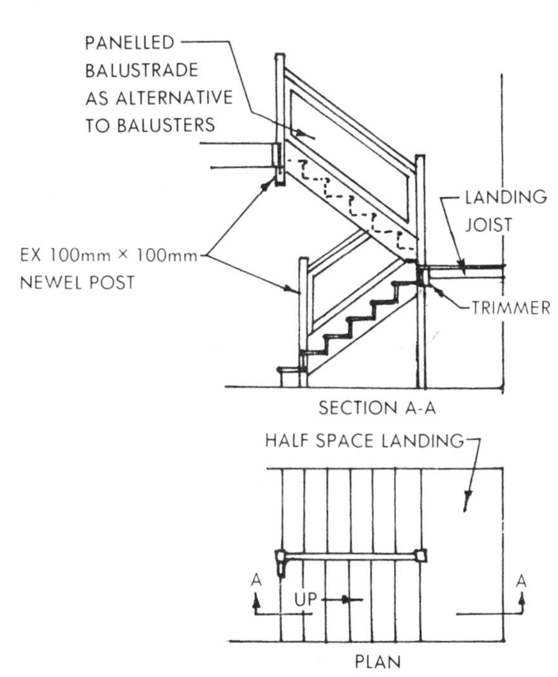

FIG 4 HALF-TURN (DOG-LEG) STAIR

Width of stairs :- Items 1.12 and 1.13 of Approved Document K of the Building Regulations 1985 give maximum widths of stairs which will satisfy the requirements of Paragraph K1 — depending on the type of stair. The widths are given in table 2 of the Approved Document. An exception to this is where a staircase forms part of a means of escape to meet paragraph B1 of schedule 1. The widths may need to increase to comply with the 'Mandatory rules for means of escape in case of fire'. The widths given in the table are unobstructed widths and are not measured in the same way as the widths given in the British Standards referred to in the 'Mandatory rules for means of escape in case of fire'.

BS 585 recommends that the width of the stairs measured overall the strings shall be not less than 860mm.

Height of handrails :- Approved Document K of the Building Regulations 1985 states that flights should have a handrail on at least one side if they are less than 1m wide — and they should have a handrail on both sides if they are wider. There is no requirement for handrails to be provided beside the bottom two steps of a stairway. Handrails should be at a height of between 840mm and 1000mm and give firm support.

The vertical height from floor to the top of a horizontal handrail (e.g. in the case of a landing) shall be not less than 900mm for a private stairway, (Fig 3) and not less than 1m for a common stairway.

Building Regulations states:- No opening in any balustrade shall be such as to permit the passage of a sphere of more than 100mm diameter. The underside of the handrail shall be grooved to receive the balustrade.

TYPES OF STAIR

A variety of designs are to be found and some of these are illustrated above. The straight flight (Fig

2), the quarter turn (Fig 3) and the half turn, also known as a 'dog leg stair' (because of its appearance in section — Fig 4), are among the more commmon.

Straight flight stairs :- A straight flight runs from floor to floor in one direction, and where the height between floors is excessive the 'run' may be broken by a landing.

Quarter turn stair :- Usually consists of two flights with a quarter space landing in between, the staircase changing direction 90 deg to the right or the left. The change of direction is sometimes achieved by using winders instead of a landing and occasionally a quarter turn geometrical stair is used.

Half turn stair :- This type of staircase turns through an angle of 180 deg. The change in direction may be achieved by a half space landing, Fig 4, or two quarter space landings, or by using winders either with or without landings.

STAIRS 3

TYPES OF STAIRS

Open well stairs :- Where there is a space between the flights of a staircase it may be described as an open well stairway. Fig 1 shows a half turn stair with a half space landing and an open well. A number of designs however may be used which incorporate an open well, and Fig 2 shows the plan of an alternative open well stair, with either quarter space landings as at 'A' or using winders as at 'B'.

Width of stairs:- Items 1.12 and 1.13 of Approved Document K of the Building Regulations 1985 give minimum widths of stairs which will satisfy the requirements of paragraph K1 of the regulations, depending on the type of stair. The widths are given in Table 2 of the Approved Document.

An exception to this is where a staircase forms part of a means of escape to meet paragraph B1 of Schedule 1. The widths may need to increase to comply with the 'Mandatory rules for means of escape in case of fire'.

The widths given in the table are unobstructed widths and are not measured in the same way as the widths given in the British Standards referred to in the 'Mandatory rules for means of escape in case of fire'.

Height of handrails :- Approved Document K of the Building Regulations 1985 states that flights should have a handrail on at least one side if they are less than 1m wide — they should have a handrail on both sides if they are wider. There is no requirement for handrails to be provided beside the bottom two steps of a stairway.

Handrails should be at a height of between 840mm and 1000mm and give firm support.

Width of stairway :- Approved Document K of the Building Regulations 1985 shows that, subject to the provisions of paragraph B1 of Schedule 1, the width of a stairway shall be measured clear between the handrail on one side and the other side (Fig 3) or, if there is no handrail — the surface of the wall, screen or balustrade facing the stairway or railing (Fig 4).

Paragraph 1.16 of Approved Document K deals with the requirements for tapered treads.

Guarding of landings :- The requirements of the Building Regulations 1985 in respect to height of handrails was dealt with in Sheet 107. Approved Document K also describes the requirements for landings.

Landings should be provided at the top and bottom of every flight. The width and depth of the landing should be at least as great as the smallest width of the appropriate stairway (See Table 2 of the Approved Document).

Handrails can form the top of the guarding if the heights can be matched (see paragraph 1.28 and Table 3 of the Approved Document).

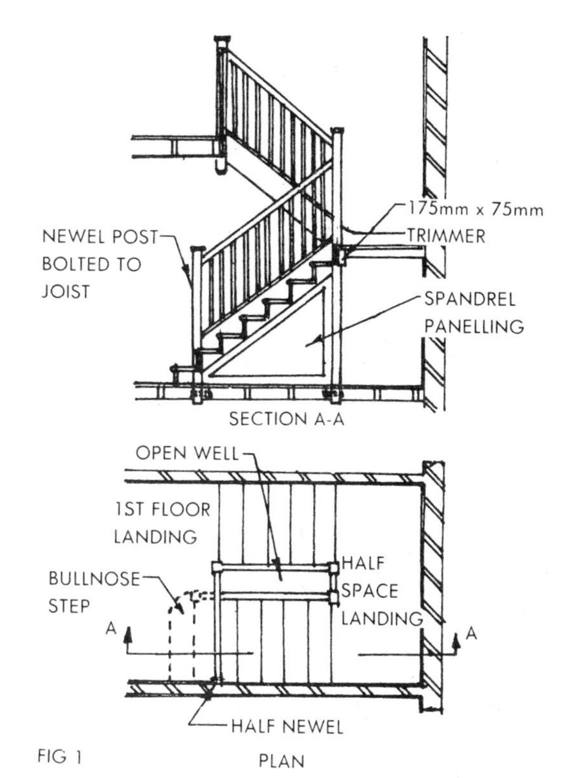

FIG 1

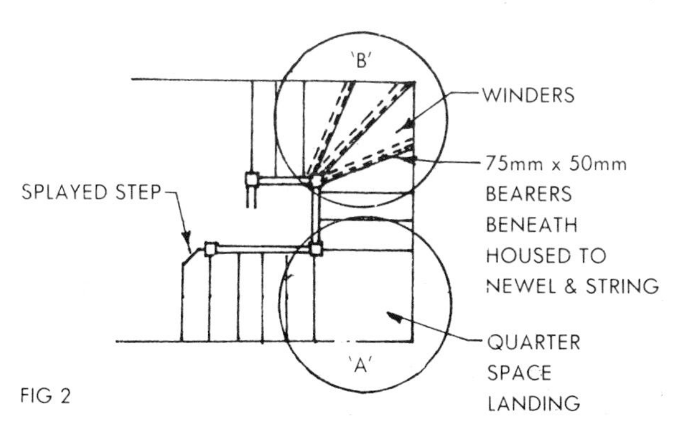

FIG 2

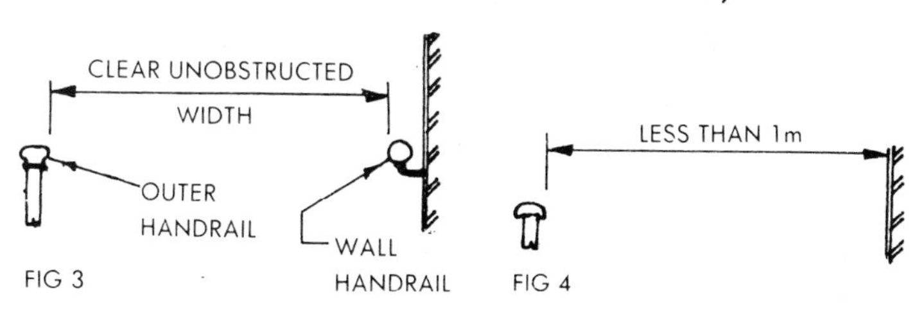

FIG 3

FIG 4

TAPERED STEPS

In situations where space is restricted, tapered steps or winders (Fig 2) are sometimes used. The introduction of winders should be avoided as far as possible as they can be hazardous, particularly to small children and the elderly. While tapered steps are permitted providing they comply with the Regulations, the use of kite winders where four treads are fixed to one 100mm square newel is not now accepted.

Going and pitch :- Whatever the width, the rise and going should be within the limits of Table I of Approved Document K of the Building Regulations 1985, and the tread should measure at least 50mm at the narrower end.

For steps with tapered treads, the going should be measured as follows:-

a) if the flight is narrower than 1m, measure in the middle, and
b) if the flight is 1m or wider, measure 270mm from each side.

Consecutive tapered treads should have the same taper.

Geometrical stair :- Has a continuous outer string around a shaped open well. The handrail forms a continuous line and newels may be used at the top and bottom of this type of stair, but the use of newels is not essential. (Fig 6).

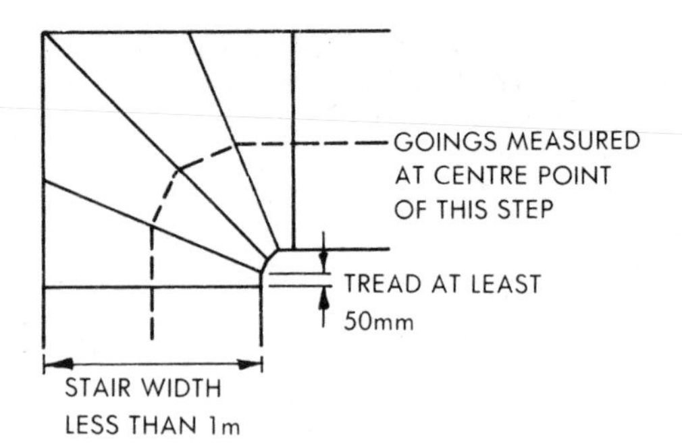

FIG 5

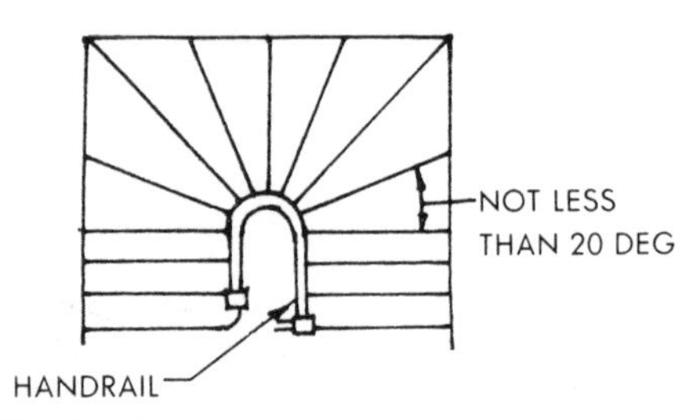

THE GREATEST AND LEAST GOINGS OF CONSECUTIVE TAPERED STEPS TO BE UNIFORM

FIG 6

STAIRS 4

CONSTRUCTION OF WOOD STAIRS

Thickness of members :- The following table showing the minimum finished thickness of members is reproduced from BS 585: 1972.

Width of Stair (overall strings)	Tread Thickness	Riser Thickness Wood	Riser Thickness Plywood	String Thickness
	mm	mm	mm	mm
Up to and including 990mm	20	14	9	27
Exceeding 990mm and not exceeding 1220mm	27	14	9	27

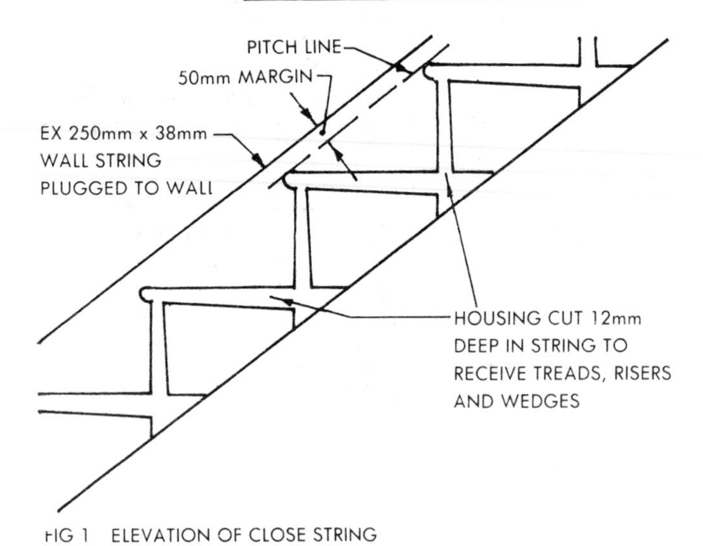

FIG 1 ELEVATION OF CLOSE STRING

In traditional construction the staircase usually consists of a wall string fixed against a wall and an outer string secured to newel posts at the foot and the head of each flight. The strings support the treads and risers which are housed into the strings, and securely glued and wedged in position. (Figs 1 & 2).

Wall strings are usually close strings having top and bottom edges parallel, while outer strings may also be close strings, or sometimes are cut strings in which the lower edge is parallel to the pitch line of the stair while the upper edge is cut to accommodate the treads and risers (Fig 5). BS 585: 1972 states:- Treads shall be not less than 235mm wide. Their nosings shall project beyond the face of the risers by not less than 15mm and the profile shall be rounded. (Fig 3).

The top of the risers shall be tongued, or housed 6mm deep into the underside of the treads. Figs 2 & 3.

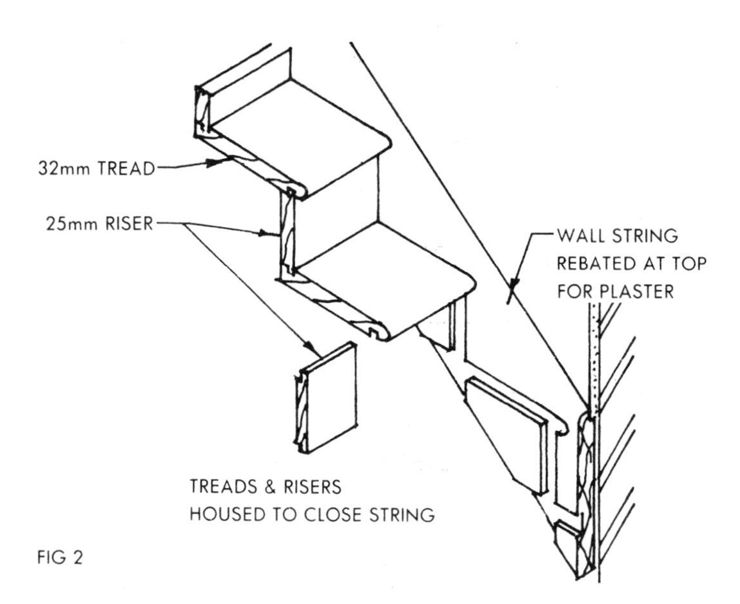

FIG 2

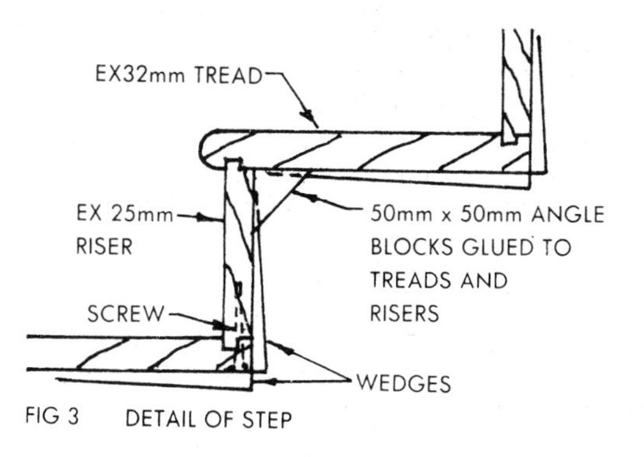

FIG 3 DETAIL OF STEP

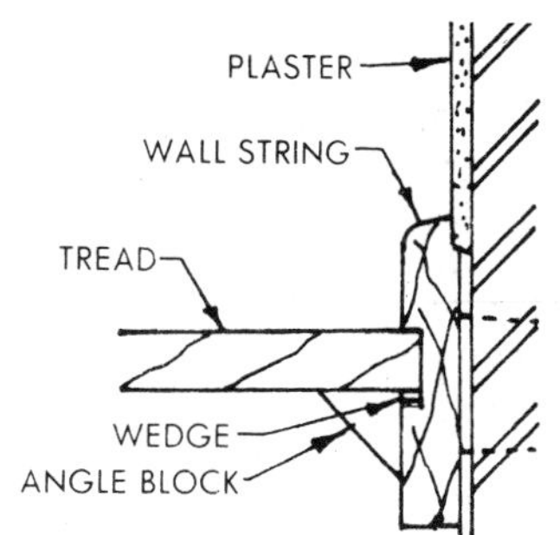

FIG 4 SECTION THRO' WALL STRING

Risers and treads shall be glue-blocked with angle blocks (Fig 3) not less than 75mm long and 38mm wide, glued into position. Two blocks per tread for stairs up to 900mm wide, three blocks for stairs exceeding 900mm and up to 990mm and four blocks for stairs exceeding 990mm and up to 1220mm wide.

The lower edges of risers shall be fixed to treads at centres not exceeding 230mm with No 10 gauge screws of length at least equal to twice the thickness of the material through which they are fixed, and in any case not less than 32mm (Fig 3). Treads and risers shall be housed, not less than 12mm deep, into tapered housings in strings and securely wedged and glued (Figs 1, 2, 3, & 4).

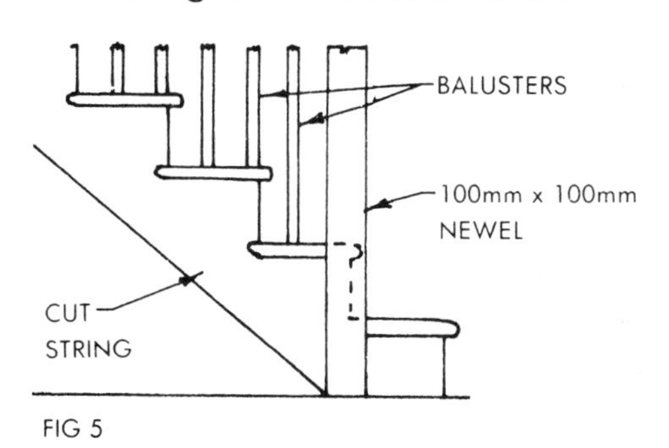

FIG 5

Strings :- Should not be less than 225mm nominal depth. They should be tenoned and prepared for pinning to newels where these occur. The tenons should be not less than 12mm thick and 50mm long.

STAIRS 5

Newels :- A newel post is placed at every change of direction of the flights. Newel posts are usually cut from 100mm x 100mm timber, although BS 585: 1972 states that outer newels shall have a cross-sectional area of not less than 5625mm² (eg 75mm x 75mm) and wall newels if any not less than 2500mm².

The newel posts are notched and bolted to the landing trimmer or floor joist

Newels are morticed and draw bored where required to receive the strings and morticed for handrails Figs 1 & 2. Treads, risers, winders and shaped end steps should be housed into newels to a depth of not less than 12mm.

Handrails & balustrades :- Must be designed and constructed as to provide a proper degree of safety and rigidity in use. BS 585 states:- The space between the handrail and string shall be filled in with balusters or other fillings as may be required.

Carriage :- Where the width of the stair is 900mm or more, it is usual to provide an intermediate support in the form of a 100mm x 50mm or

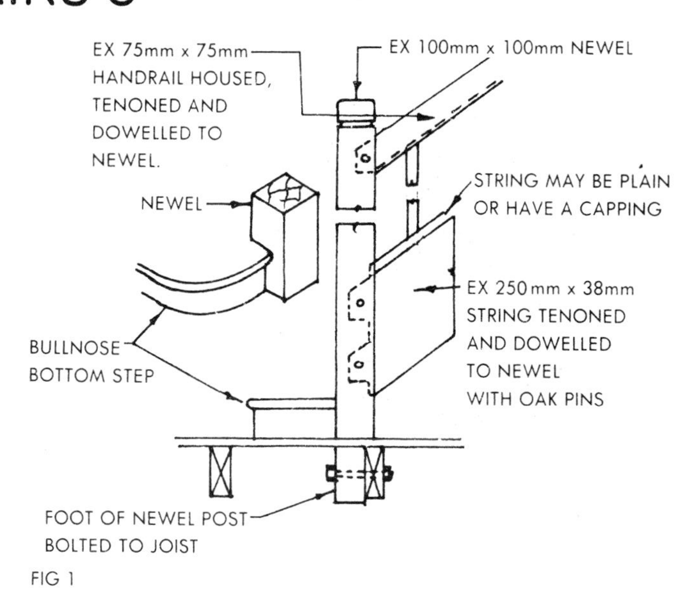

FIG 1

75mm, sawn softwood carriage, fixed below a flight. Rough brackets as shown are nailed to the carriage to provide additional support to the treads.

The bottom of the carriage is birdsmouthed over a plate, fillet or pitching piece nailed to the floor (Fig 3). The top of the carriage is fixed to the landing trimmer or to a pitching piece (Fig 4).

If the soffit is to be plastered then bearers similar to the central carriage will be required at the outer edges (Fig 3).

Nosings :- BS 585 requires that nosings shall have a rounded profile. Some authorities consider that the method shown Fig 5 may result in broken nosings and Figs 6 and 7 show alternative methods.

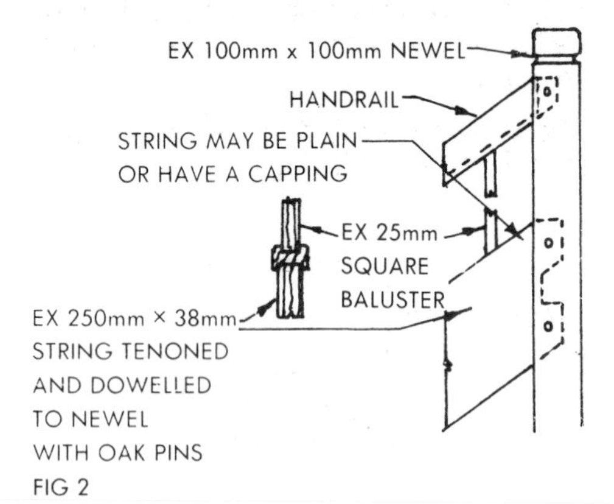

FIG 2

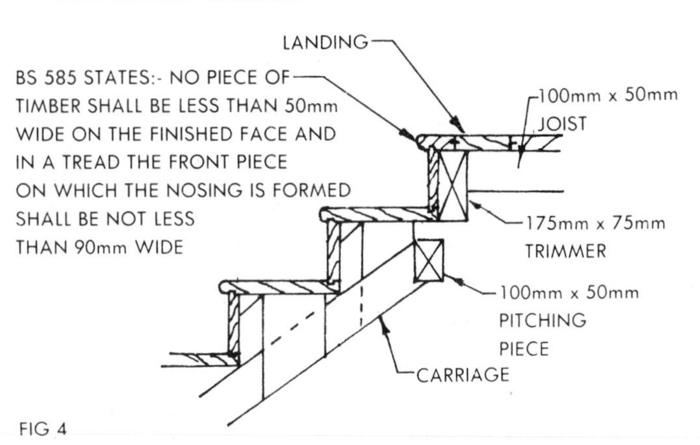

FIG 4

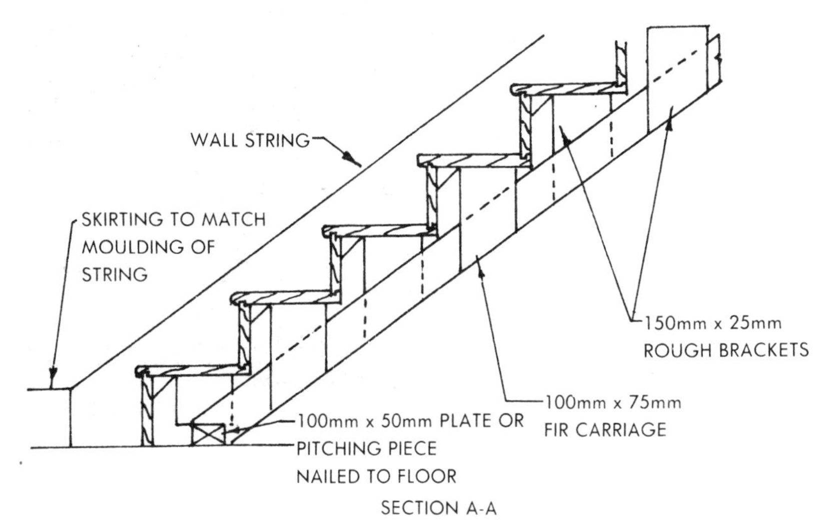

SECTION A-A

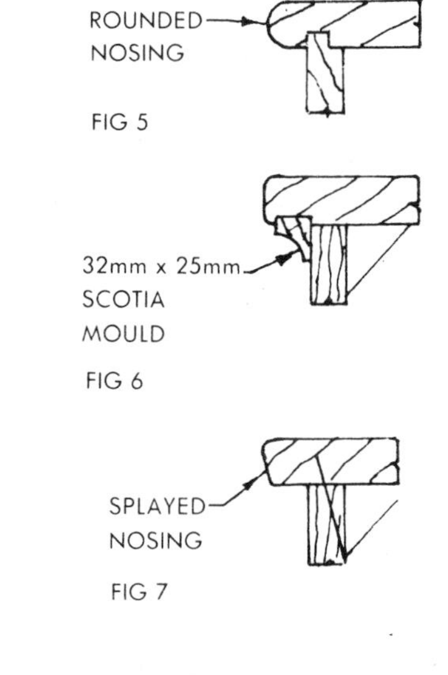

FIG 5

FIG 6

FIG 7

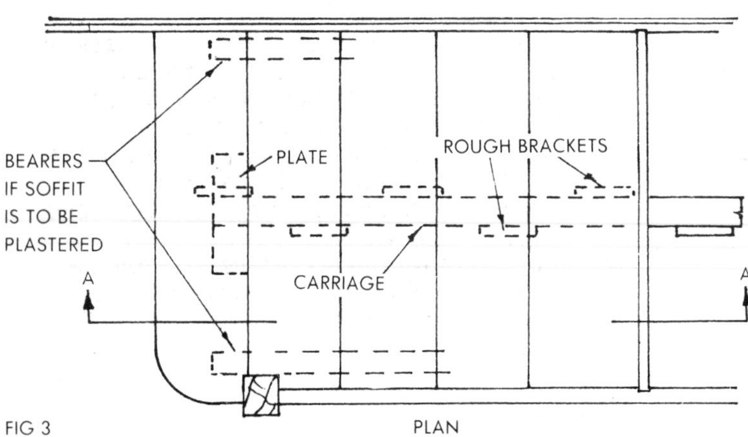

PLAN

FIG 3

STAIRS 6

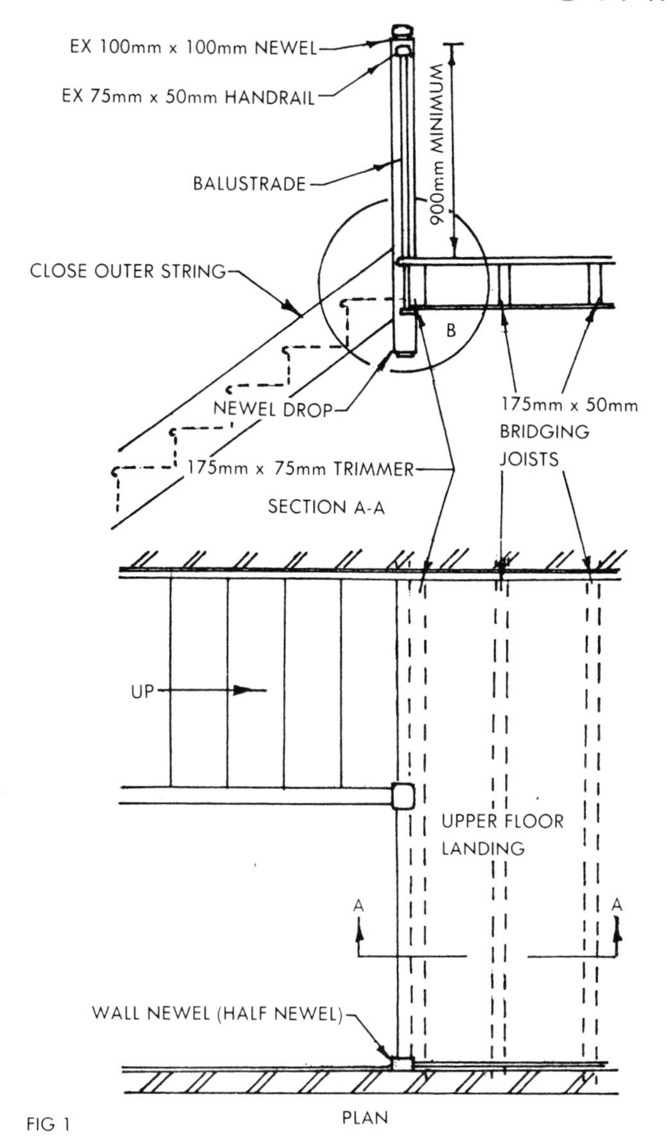

FIG 1

LANDINGS
Fig I of this sheet gives details of an upper floor landing. The newel is notched and bolted to the trimmer of the stair well and the balustrade of the landing runs from the upper stair newel to the half newel at the wall N.B. the landing balustrade must be at least 900mm high

Fig 2 shows an enlarged detail of the apron lining and trimmer at 'B' Fig I.

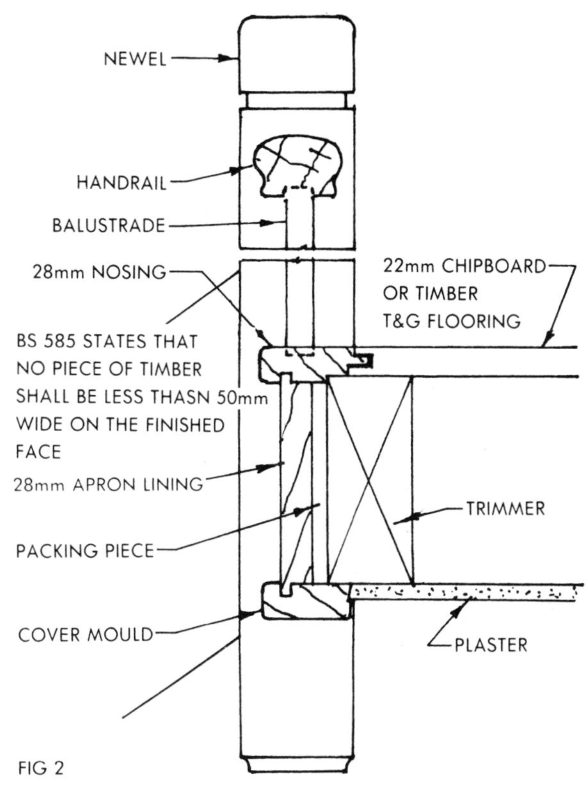

FIG 2

There are many ways of finishing balustrades. Ordinary balusters, balusters placed in groups of two or three with wider spaces between groups, balusters inclined at an angle to each other. Horizontal timbers with spaces between (Fig 3) are also sometimes used.

Metal balusters or wrought iron infill grills to balustrades may also be used. BS 585 states:-The space between the handrail and string shall be filled in with balusters or other filling as may be required. Approved Document K of the Building Regulations 1985 states that no opening in any balustrade shall be such as to permit the passage of the sphere of more than 100mm diameter.

Enclosed or panelled balustrades are often used and Fig 4 shows a section through a panelled balustrade. Alternatively the balustrade may be of timber studding cladded with plasterboard or plywood in this case a hardwood capping is necesary.

Spandrel:- The triangular surface between the outer string and the floor. This space may be panelled or built up and plastered as Fig 5.

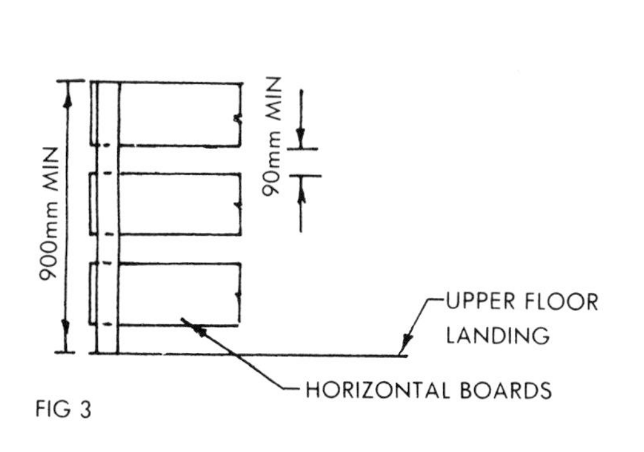

FIG 3

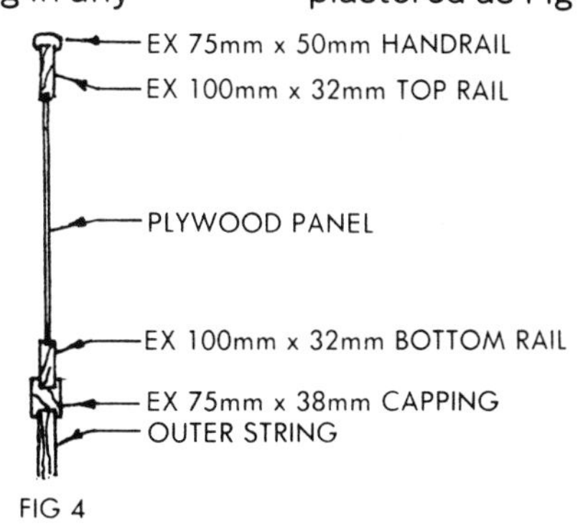

FIG 4

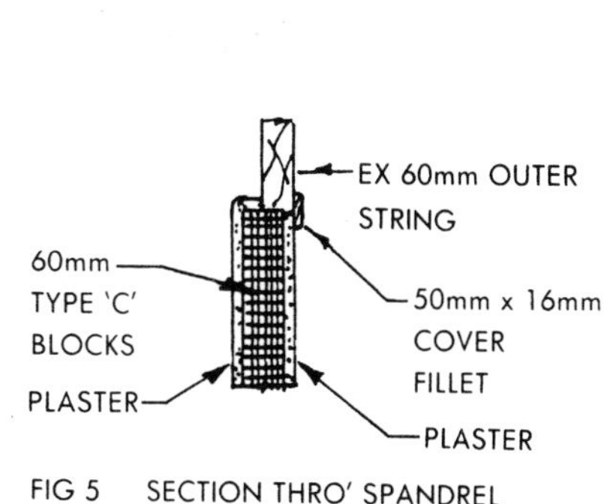

FIG 5 SECTION THRO' SPANDREL

SHAPED BOTTOM STEPS
A bullnose step is constructed by reducing the thickness of the riser (ie to form a veneer) at its end to enable it to be bent round a wood former built up of blocks Fig I.
Curtail step :- This type of step Fig 2, is formed in a manner somewhat similar to the bullnose step. As the step projects out from the newel post it can cause an obstruction and should not be used if the space at the side of the stairway is restricted.
Splayed step :- A splayed bottom step is usually formed as shown Fig 3.
The riser sections forming the splayed end are mitred and secured by ploughed and tongued joints (ie a hardwood tongue is inserted and glued into grooves as shown). The tread and nosing follows the shape of the step.

CUT STRING
Cut strings are not common since the construction is more costly. Where cut strings are used the balusters are dovetail housed to the treads. Each riser is mitred to the string, each tread is cut and mitred at the outer string and finished with a planted nosing, slot screwed to the end of the tread. Fig 4.

Since the treads and risers cannot be housed to the string, bearer brackets may be screwed to the treads, risers and string. Constructional details are shown Fig 4. (The glued blocks which would be placed in the usual manner have been omitted for clarity).

OPEN TREAD STAIRS
Steps may have open risers, but the treads should then overlap each other by at least I5mm. Steps with open risers in a flight which is
a) part of a private or common stair, or
b) is an institutional building and is likely to be used by children under 5, or
c) is in any other residential building
should be constructed so that a I00mm diameter sphere cannot pass through the open risers.

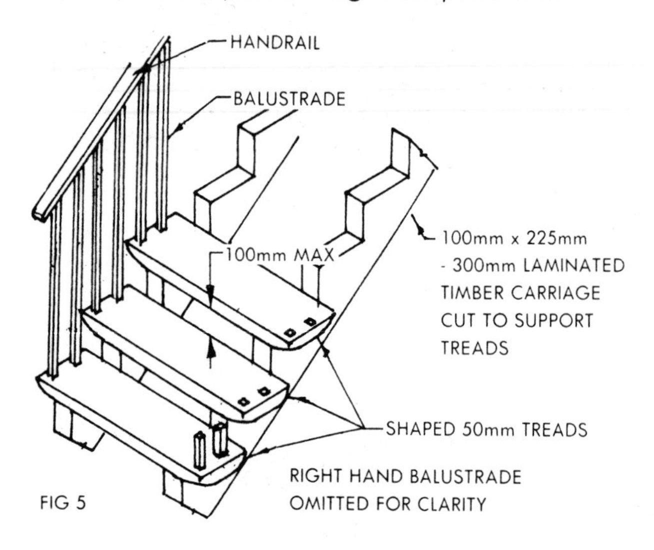

FIG 5

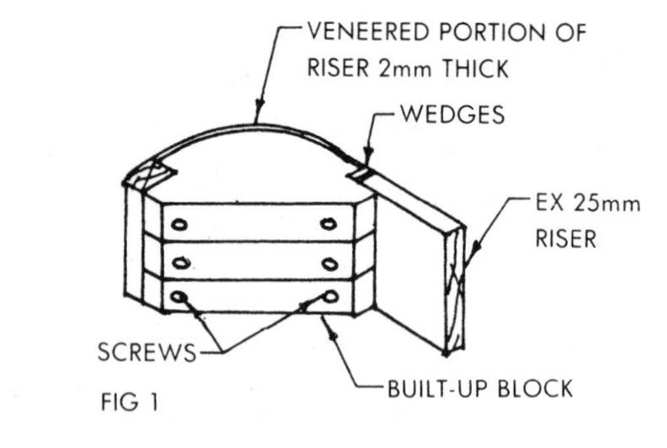

FIG 1

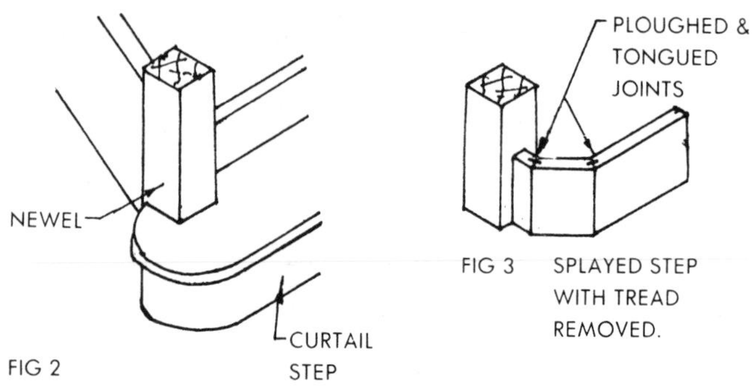

FIG 2

FIG 3 SPLAYED STEP WITH TREAD REMOVED.

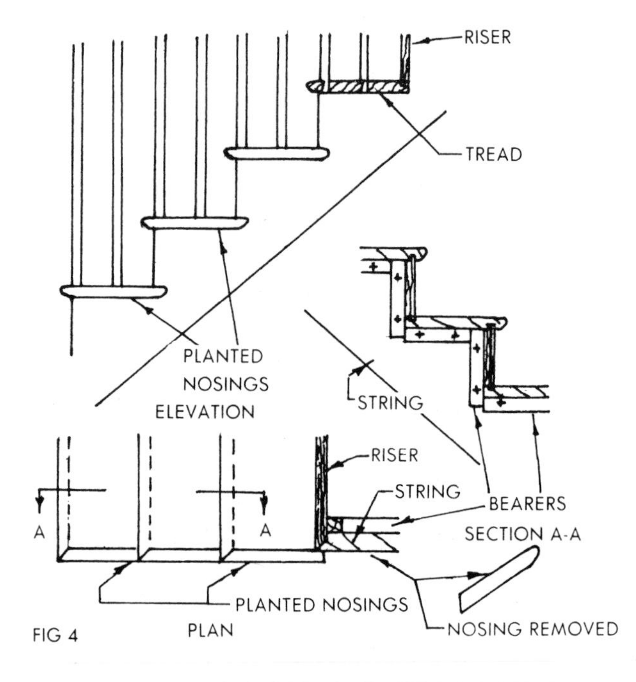

FIG 4

STAIRS 8

CONCRETE STAIRS

While timber stairs are satisfactory for normal domestic construction, they do not meet the requirements of the Regulations in respect of fire resistance for many buildings. Approved Document B of the Building Regulations 1985 deals with the spread of fire, and gives the following provisions:-

Item 1.44 stairways in buildings recommends that stairways (including landings) should be constructed with materials of limited combustibility if they are

a) within any storey of a house having more than four storeys (basement not counted), or
b) external and connecting the ground floor with a floor or a flat roof at a height of more than 6m above the ground.

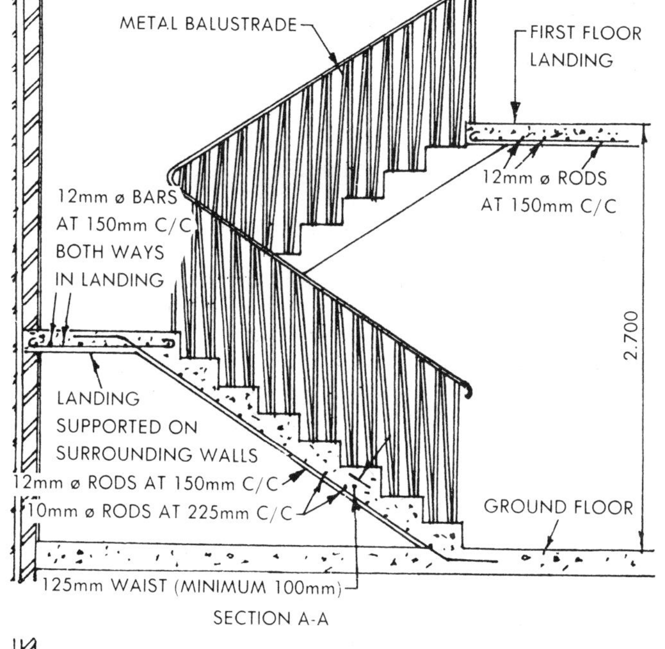

SECTION A-A

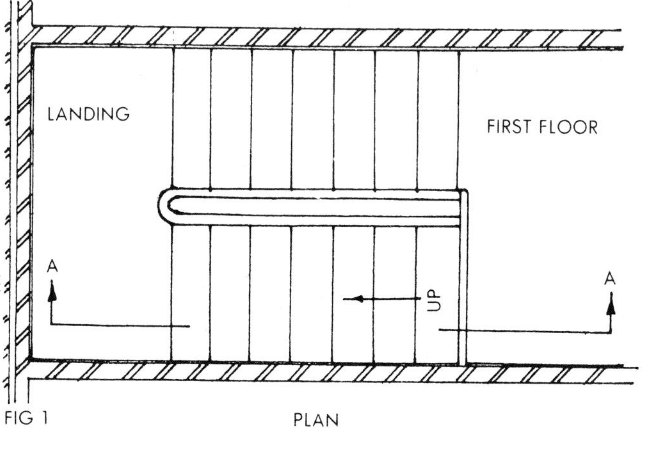

FIG 1 PLAN

Item 1.45 states that combustible materials may be added to the upper surface of these stairways and landings.

Item 1.46 indicates that an internal stairway in a house which has more than two storeys (basement not counted) may need to be enclosed and protected to meet the provisions of Section 1 of the document 'Mandatory rules for means of escape in case of fire' (from H.M.S.O.) in support of the requirements of paragraph B1 of Schedule 1 to the Building Regulations 1985.

Separate provisions are described for different building uses.

Reinforced concrete stairs, which have a higher fire resistance are commonly used where timber stairs would not satisfy the regulations.

The general principles of design in respect of rise, going, angle of ascent etc apply.

Reinforcement:-A simple concrete stairway may be designed as an inclined slab stair spanning between floor and landing or between landings. The size and spacing of the reinforcement will have to be calculated according to the required conditions, loading and span. An example of a 'cast-in-situ' reinforced concrete half-turn stair is shown Fig 1.

The thickness of the concrete may vary according to conditions, but the waist thickness (Fig 1) should not be less than 100mm. Reinforced concrete stairs may be constructed with open or closed strings, spanning between floor and landing trimmer (An example is shown Fig 2).

Design of string beams vary and an upstand string as Fig 3 may be used or a downstand string as Fig 4. The former has the advantage that

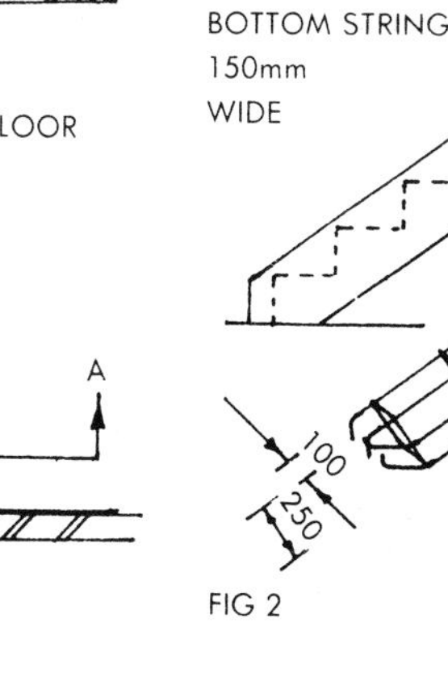

FIG 2

STRING BEAM

FIG 3

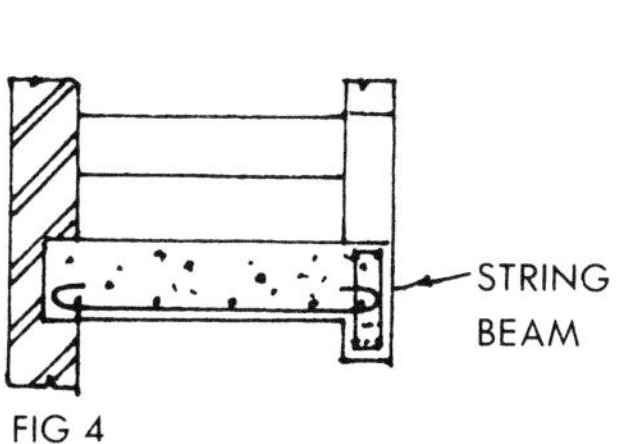

FIG 4

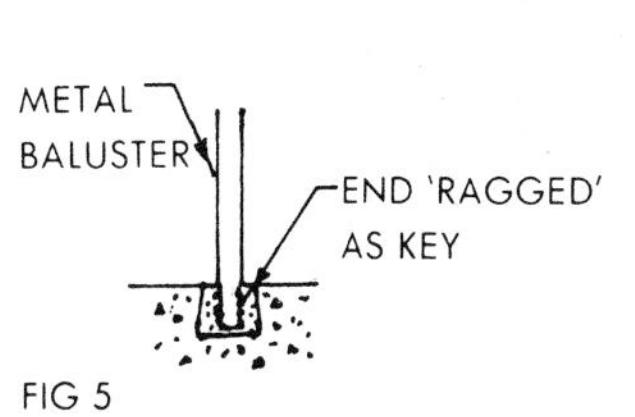

FIG 5

anything dropped on the stair does not roll over the edge.

Normal balustrading would be required for stairs shown in Figs 2, 3 and 4, but this has been omitted for clarity.

The balustrade is fixed by setting the balusters into holes cast in the concrete and either caulking with lead or grouting with cement. Fig 5.

STAIRS 9

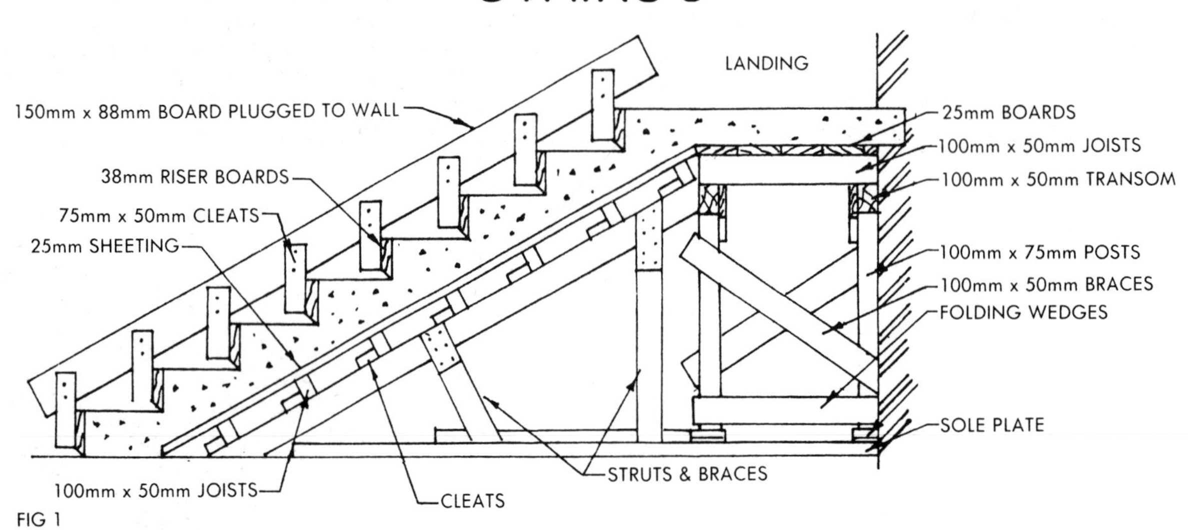

FIG 1

CAST-IN-SITU CONCRETE STAIRCASE

An example of formwork suitable for a cast-in-situ concrete staircase is given in Fig I.

It is important that the shuttering is adequately braced and supported to ensure rigidity. The posts should be supported on folding wedges to allow for final adjustment, levelling, and for easy dismantling when striking the formwork. Alternatively adjustable steel props may be used.

The steps are formed by fixing a board to the wall clear of the nosing line of the treads, and spiking vertical 75mm x 50mm cleats to it, which support the riser boards as shown. The bottom edges of the riser boards should be chamfered as shown to enable the tread of each step to be trowelled level for the full width of the step.

A fairly strong concrete mix should be used, say $1:1\frac{1}{2}:3$ and the formwork treated with mould oil, grease or soft soap before pouring the concrete to facilitate striking the timber when the concrete has matured.

The sketch Fig 2 shows an arrangement suitable for use where a cut outer string is required, and Fig 3 shows a method of constructing the formwork when a close string is required.

A method of supporting and restraining the formwork for a closed string is shown Fig 4.

The upper flight from the landing to the floor above is constructed in a similar manner to the example shown in Fig I. The supporting posts being long enough to be carried by the floor below.

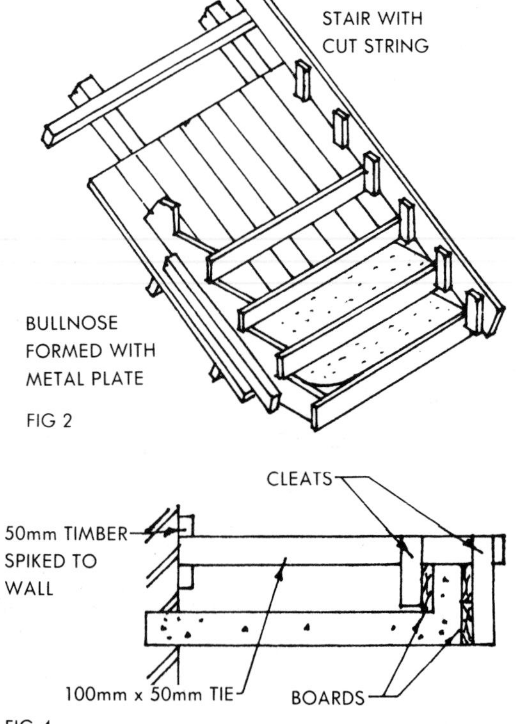

FIG 2

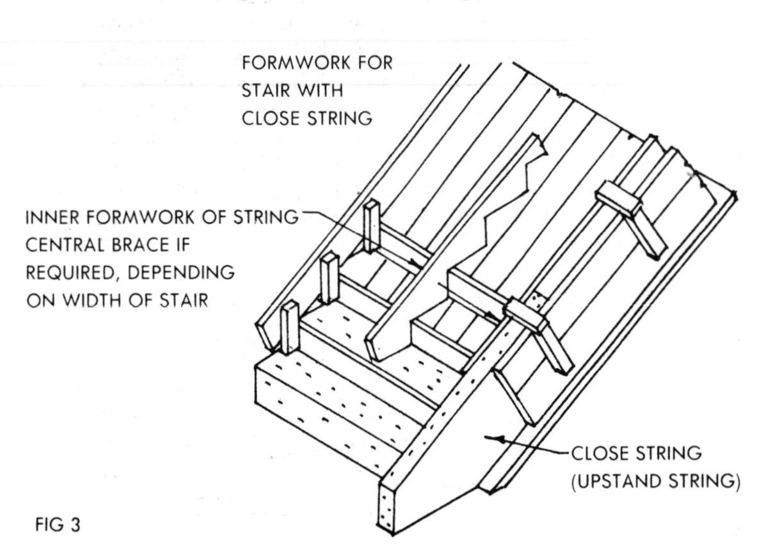

FIG 3

FIG 4

STAIRS 10

BUILT-UP STAIRS
STONE AND CONCRETE

Stone stairs are rarely used today because of the cost. Precast concrete steps however are used, and these and stone steps can be very effective, as an entrance feature to a building.

These steps may be supported at each side (Fig 1) or cantilevered (Fig 7). The steps may be rectangular, producing a stepped soffit, or chamfered on the underside to provide a straight soffit below a flight and to reduce weight (Fig 2).

Variations of these steps include skeleton steps (Fig 3) and built-up steps (Fig 4). A rebated form of solid steps is sometimes used (Fig 6).

Stairs may be cast in complete flights, sometimes with landings and hoisted in position by crane. Fig 5 shows an example of a precast flight with a stepped soffit. This gives a considerable reduction in weight.

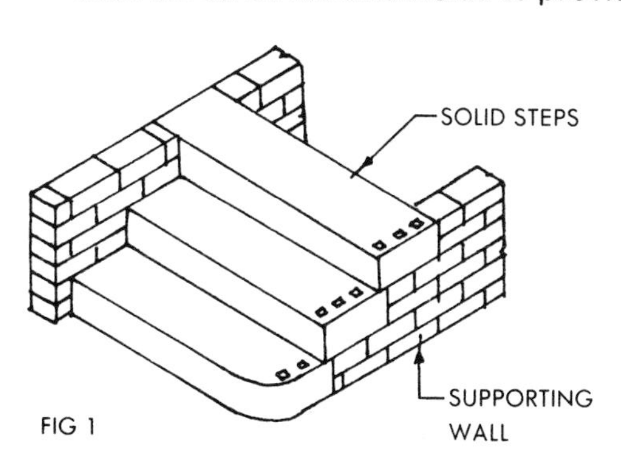

FIG 1

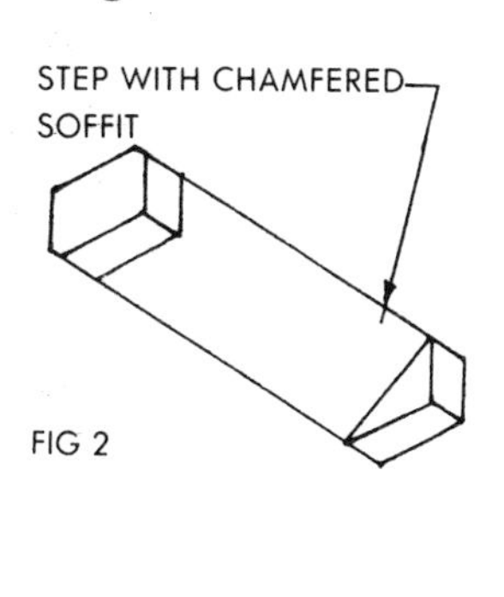

FIG 2

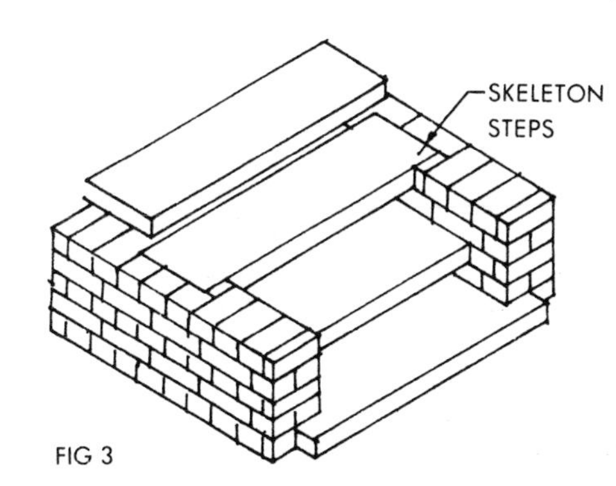

FIG 3

SPANDREL CANTILEVER STEPS

These are usually reinforced concrete steps of the type illustrated in Figs 7, 8 and 9. These steps are built into the wall at one end, and the triangular section provides greater headroom than steps of rectangular section, are lighter and have a plain soffit which is usually considered to improve the appearance. The step may be constructed with a nosing as Fig 8.

The steps are usually built-in as the work proceeds, or may be built-in after the walling is completed. The former method is considered to be easier.

A temporary support will be required at the free end of the flight during construction, Fig 10, and a 'storey' rod and 'going' rod will be required to check the accuracy of alignment.

If the steps must be built-in at a later stage, sand courses are built into the supporting wall, care being exercised in locating these. Before setting the steps, the sand courses are removed, the indents brushed clear of all loose sand and the wall dampened. Great care is needed in pinning-up the upper surface of each step at the wall.

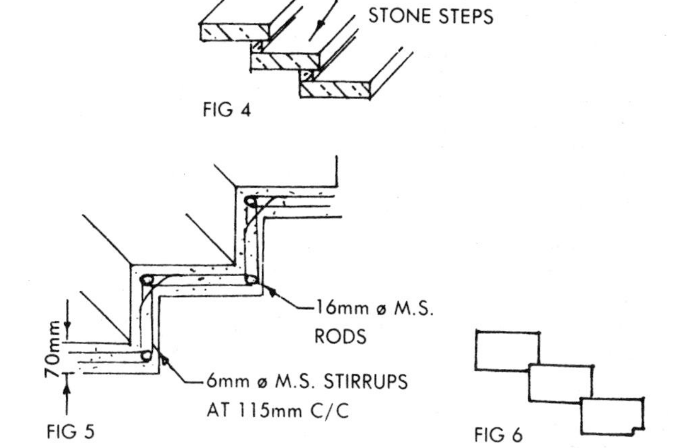

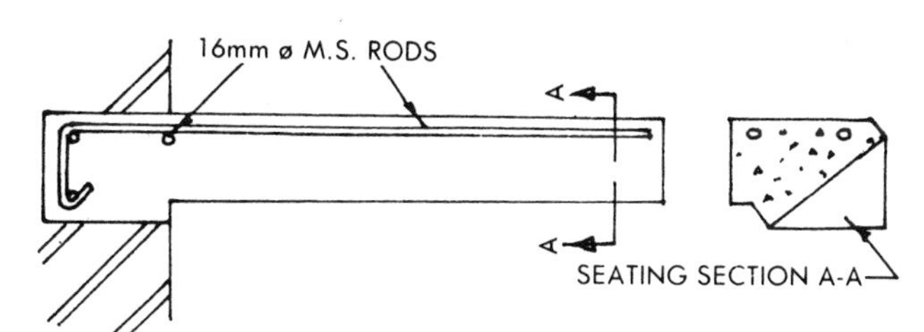

FIG 7 CANTILEVER STEP

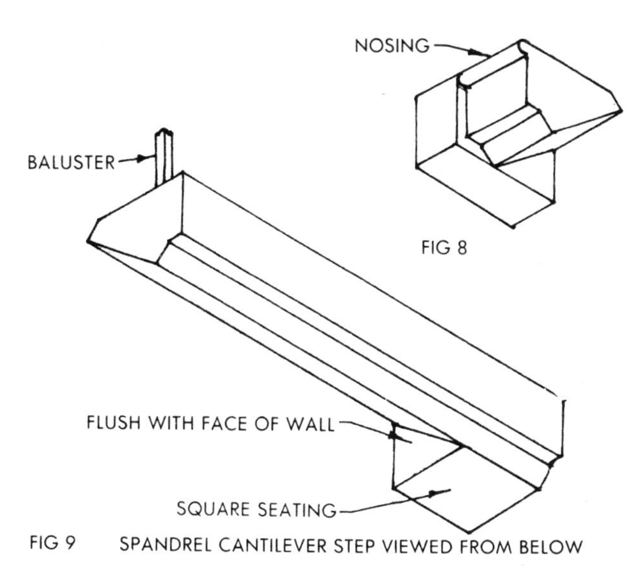

FIG 9 SPANDREL CANTILEVER STEP VIEWED FROM BELOW

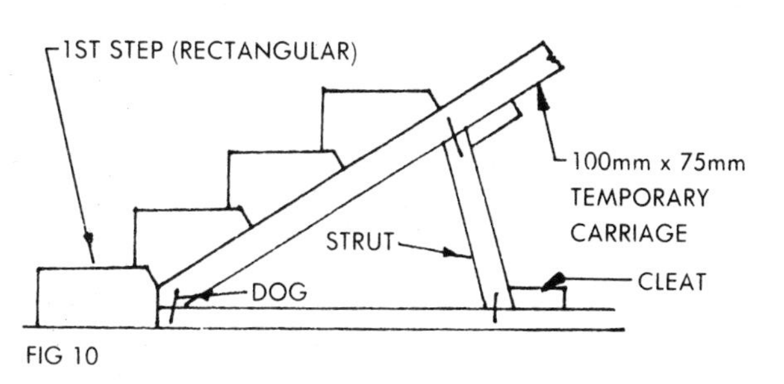

FIG 10

STAIRS II

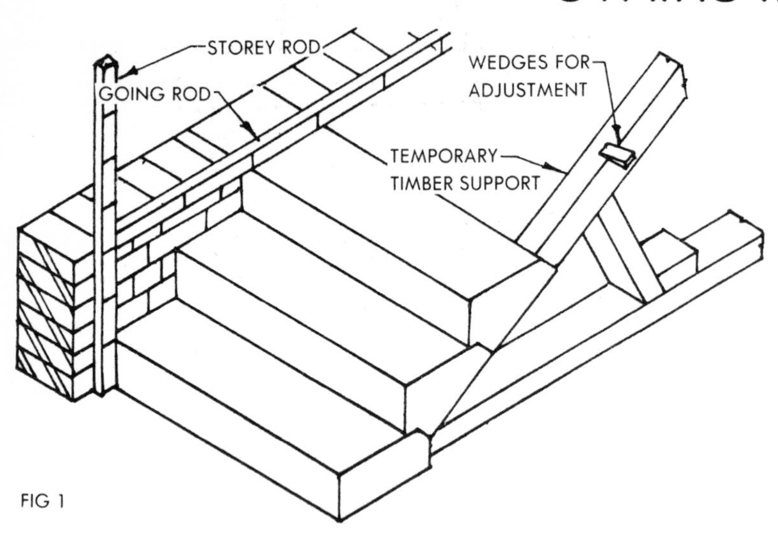

FIG 1

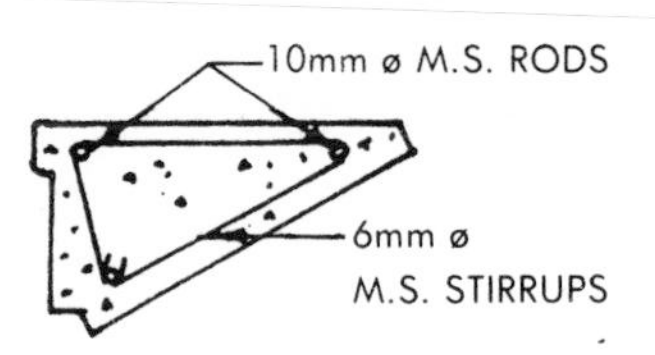

FIG 2 DETAIL OF SPANDREL STEP

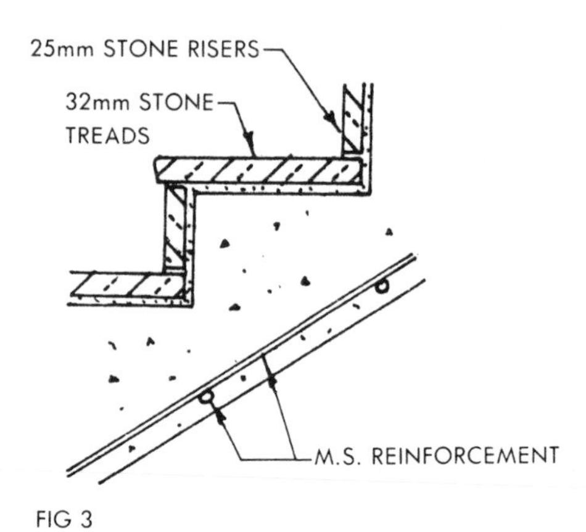

FIG 3

SPANDREL CANTILEVER STEPS

Spandrel steps require a temporary support at the free end during construction (Fig I) and a storey rod and going rod as shown will be found useful. A supply of small sharp timber wedges (Fig I) will be found useful for fine adjustment at the support timber as required.

The steps may be cast in a high grade granite concrete $(1:\frac{1}{2}:3)$. Fig 2 shows an example.

Concrete steps tend to become slippery with wear, and the addition of 0.15kg of carborundum per m^2 sprinkled over the tread area and 'trowelled in', will provide a non-slip surface and reduce the rate of wear.

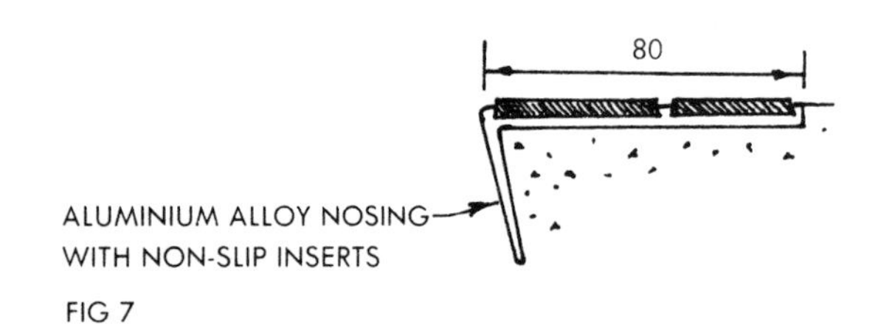

FIG 7

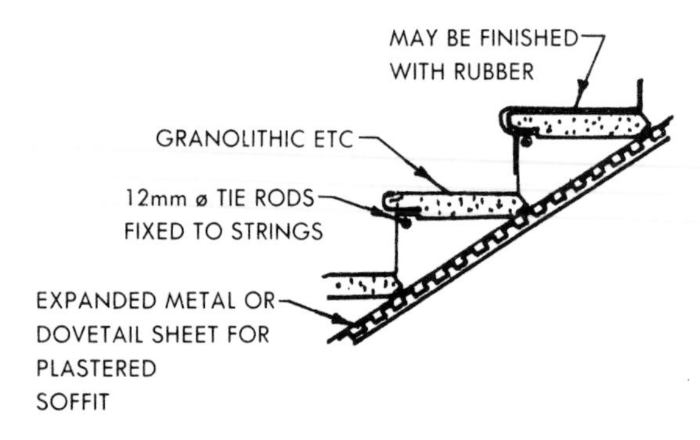

FIG 8

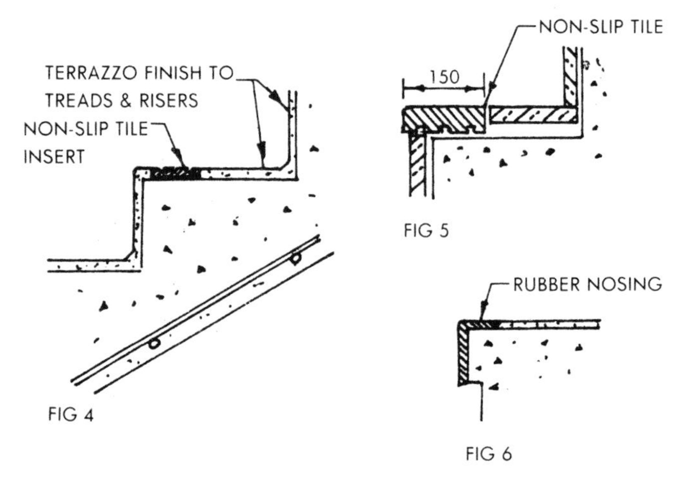

FIG 4

FIG 5

FIG 6

The surface of treads and risers may be finished in a variety of ways. Natural stone Fig 3, terrazzo (sometimes incorporating non-slip inserts of carborundum or rubber strip Fig 4). A variety of patent nosings in tile Fig 5, rubber Fig 6, and metal Fig 7 are available and may be used in conjunction with a wide range of floor coverings. Alternatively the treads and risers may be finished with one of the jointless composition floor finishes.

PRESSED STEEL STAIRS

Steel stairs are available of light pressed metal. Each tread and riser is of pressed steel and fixed at each side either to a pressed steel closed string or to a boxed channel section. The design of the treads allows them to be filled with a cement mix, granolithic etc. Alternatively the treads may be of timber or marble or terrazzo. The soffit can be plastered on expanded metal, fixed by clips.

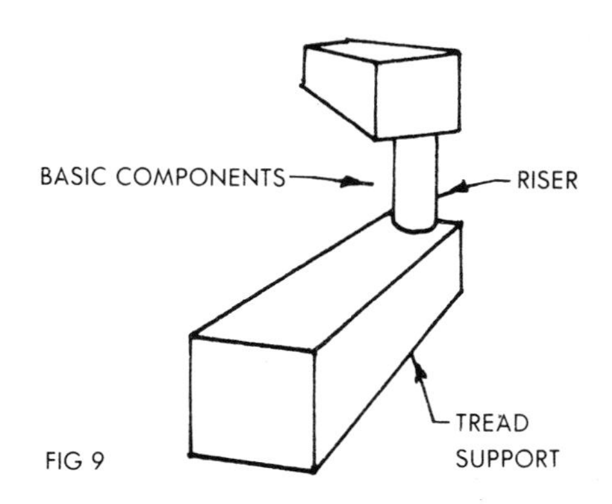

FIG 9

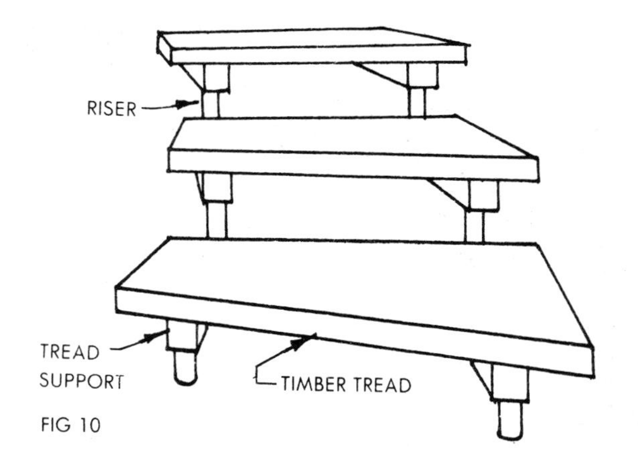

FIG 10

NUWAY STAIRS
A useful and very adaptable stair is the Nuway universal system. A wide variety of shaped staircases can be made from two basic components. A 60mm square tube tread support and a 40mm solid bar riser Fig 9 by varying the length of tread support and the riser it is possible to make a stair to suit any site conditions Fig 10.

ROOF COVERINGS I

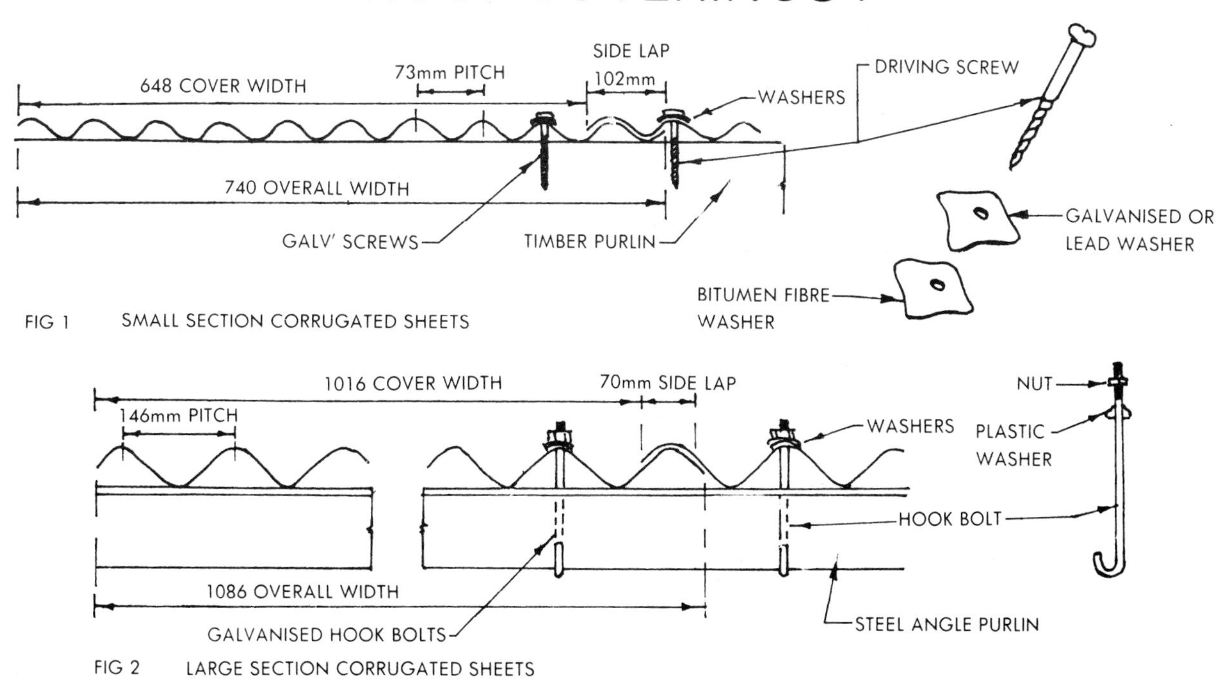

FIG 1 SMALL SECTION CORRUGATED SHEETS

FIG 2 LARGE SECTION CORRUGATED SHEETS

Flat roofs covered with felt, asphalt, lead, copper, zinc and aluminium were considered in earlier Detail Sheets.

ASBESTOS-CEMENT ROOFING
Asbestos-cement is manufactured from asbestos and portland cement. The material is reasonably strong, durable, noncombustible and light in weight. It is resistant to alkalis and most acids, although a higher degree of acid pollution in the atmosphere may affect it. If properly sealed or primed the material may be painted and this will assist preservation in unfavourable situations. (Sheets are available coloured by factory applied processes, which have a high resistance to fading and wear).

Sheet lengths available start at 1225mm increasing by increments of 150mm to 2425mm, then from 2450mm by 150mm increments up to 3050mm. One or two types are available up to 3650.

The pitch of the roof should normally be kept at a minimum of 22½ deg and end laps of sheets to a minimum of 150mm. However on comparatively small span roofs where single sheets can be used and horizontal laps avoided, pitches down to 5 deg are possible. In any case where laps are sealed with mastic extrusions as recommended by manufacturers, pitches may be considerably reduced.

Spacing of purlins will vary according to the type of sheet used, (see manufacturers' data), and as far as possible purlin spacing should be arranged to accommodate standard size sheets to avoid wasteful cutting. In general the maximum purlin centres for the small section sheets (Fig I) is 925mm, and 1375mm for the large section sheets (Fig 2).

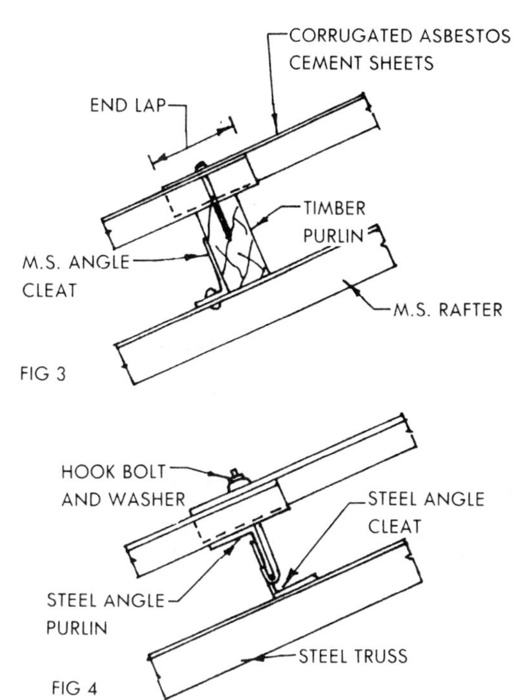

FIG 3

FIG 4

All sheets should be supported as near to their ends as practicable, but no hole should be nearer than 38mm to any edge of a sheet. The maximum overhang of any type of sheet should be 300mm. The top end of sheets should preferably extend 75mm beyond the centre line of timber purlins or 50mm beyond the back of metal or concrete purlins. Timber purlins may be used (Figs I & 3) but steel angle or tabular section steel purlins are more common (Figs 2&4). Fixings to concrete may be with drive screws and fibre plugs inserted in the purlins.

N.B. Sheets should be holed by drilling, not punched. Fixings should not be too rigid, allowance should be made for slight movement.

ROOF COVERINGS 2

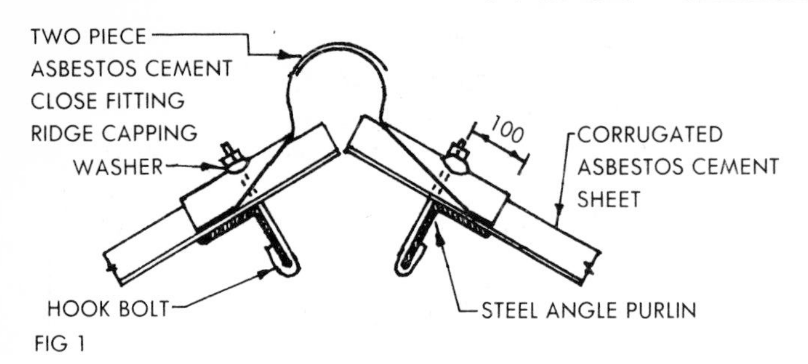

FIG 1

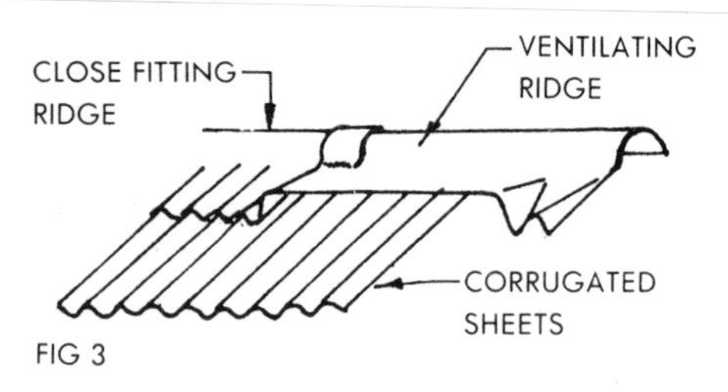

FIG 3

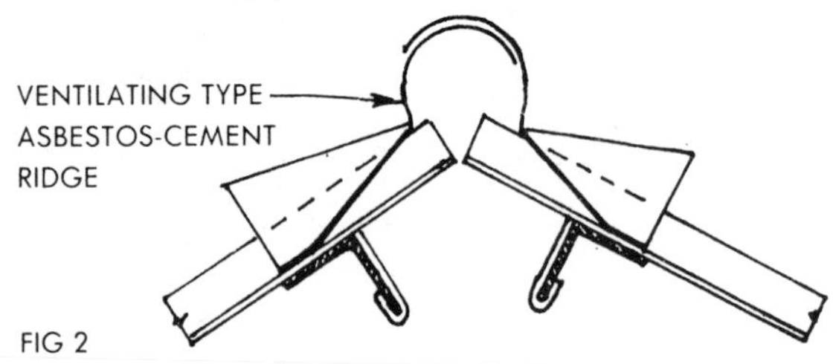

FIG 2

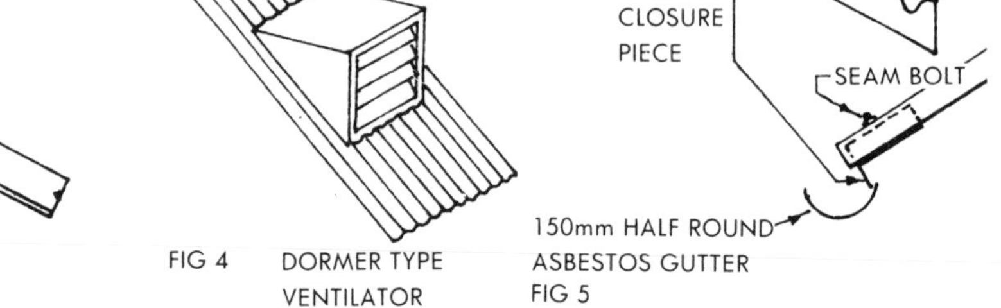

FIG 4 DORMER TYPE VENTILATOR

FIG 5

ASBESTOS-CEMENT ROOFING

Fixing corrugated asbestos-cement sheets on timber purlins – the sheets are fixed by using galvanised drive screws which are driven through holes (previously drilled through the crown at the corrugations) and then tightened.

A galvanised steel and a bitumen fibre washer being used to ensure a watertight job

Plastic washers may be used and these are available for both 6mm and 8mm bolts. In this case no separate bituminous washer is required. One type of washer has a dome shaped cap, which is clipped over the head of the fixing.

When fixing to steel angle purlins Fig I a galvanised hook bolt is used, and it is important to see that the sheets are carefully drilled in the correct positions to accommodate these fixings.

A wide range of accessories are available to meet a variety of situations. Fig I shows a section through a two-piece ridge fitment. Ventilating type ridge cappings are also available if required Figs 2 & 3. Dormer type ventilators Fig 4 are also available.

Eaves closure and filler pieces Figs 5 and 6 are also standard accessories. The latter may be used for finishing vertical asbestos cement sheeting at the eaves, where this is used (Fig 6). A wide range of fitments includes ridge finials, valley and abutment gutters, barge boards, flange sheets for flue pipes, expansion joint fitments etc.

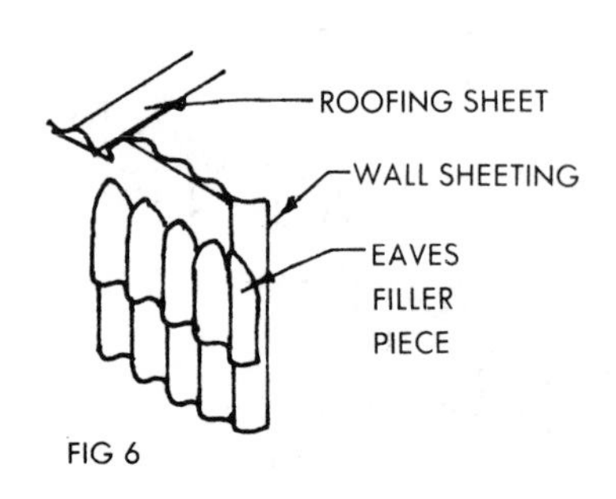

FIG 6

In addition to the usual corrugated sheets there is a variety of sheets of large section available. Fig 7 shows a section through one type of asbestos cement sheeting. In this case purlins may be at I675mm maximum centres. Other types of sheet are available which are similar in appearance but of thicker material (9.52mm compared with 6.52mm). When this is used, purlins may be spaced at I975mm centres.

Sheets are also available with corrugated ribs at a much wider spacing Fig 8 shows a section of a sheet where the appearance is an alternative to the usual corrugated sheet, and maximum purlin spacing is I375mm.

End laps to sheets are normally I50mm and the amount of side lap will depend upon the type of sheet used. Side laps should be arranged to face away from the prevailing wind.

Lap sealing materials:- A variety of plastic seals may be used. They may be hand or gun applied mastics, preformed dry strips, or bitumen impregnated plastic strips. Sealing of side and end laps will be necessary where a reasonably dust-tight covering is required.

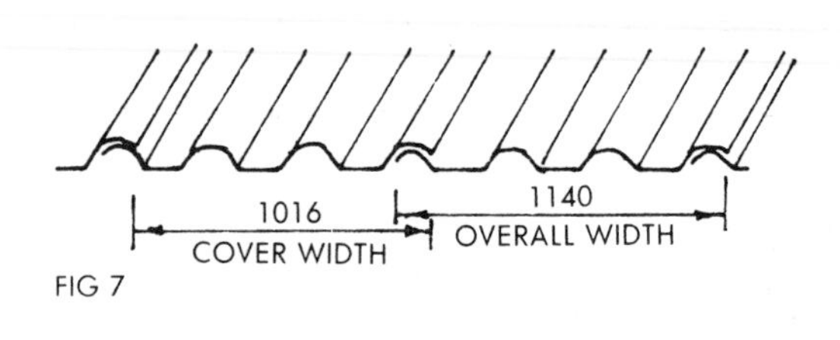

FIG 7

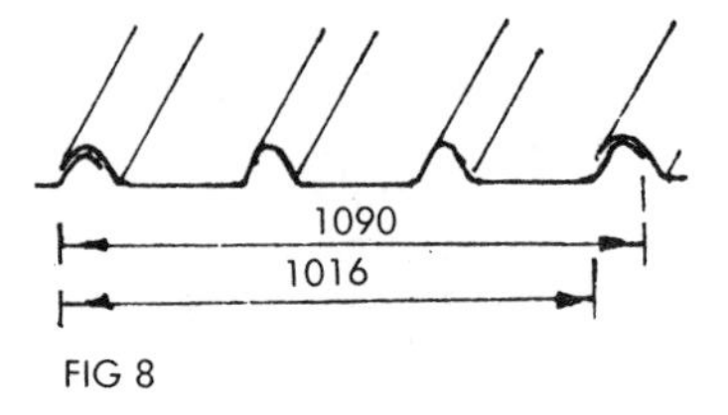

FIG 8

ROOF COVERINGS 3

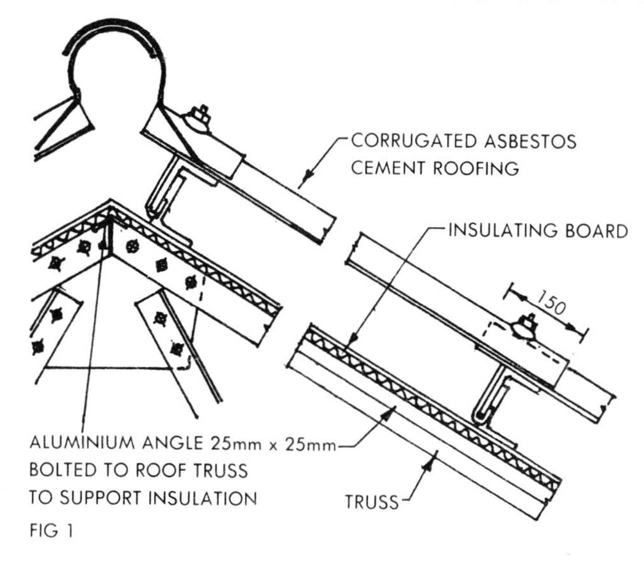

FIG 1

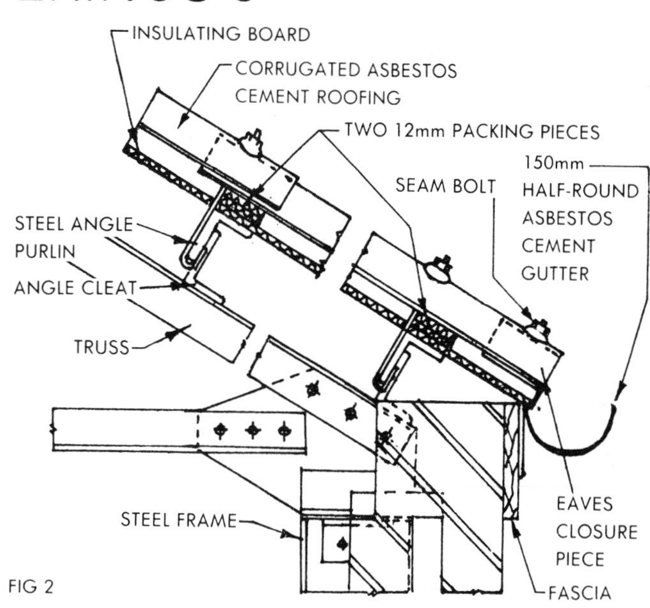

FIG 2

ASBESTOS-CEMENT ROOFING

Insulation :- Asbestos-cement sheets alone have little thermal insulation value. One method of insulating the roof is to provide a lining of insulation board, wood wool slabs etc beneath the roof covering either under or over the purlins. Fig I shows the insulation fixed below the purlins, in this case being supported on small aluminium angles bolted to the roof truss.

Fig 2 shows the insulation board laid over the purlins. In this case packing pieces of the insulating board are laid over the purlins, thus providing a 24mm air space between the insulation board and the corrugated asbestos-cement roofing sheet, which will further improve the insulation.

In addition to the aluminium angle supports to the insulation shown Fig I, sections as Fig 3 are fixed between the purlins, supported by steel hangers to give further support to the insulation board.

When over purlin insulation is used care must be taken to see that the insulation board is not exposed to rain, as this will lead to deterioration and distortion of the board. One disadvantage of under purlin insulation is that access to purlins and part of the truss is prevented so that painting cannot be effectively carried out.

A most effective method for thermal insulation is to use sandwich construction, ie an underlining sheet of asbestos cement, then a layer of glass fibre or similar insulating material and finally the capping sheet Fig 4.

Regulation L3 (2)(b) states subject to sub-paragraphs 4 and 5 the calculated rate of heat loss through the solid parts of the exposed elements shall be no greater than it would be if those parts had a 'U' value of

a) in the case of a residential building, shop, office or assembly building, of 0.6 W/m²K and
b) in the case of an industrial building or any other building of 0.7. W/m²K Fig 5.

Normal fixing bolts are used, penetrating right through the construction Fig 5.

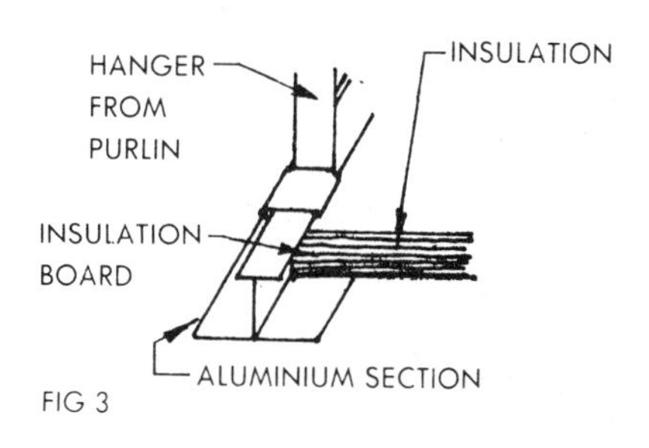

FIG 3

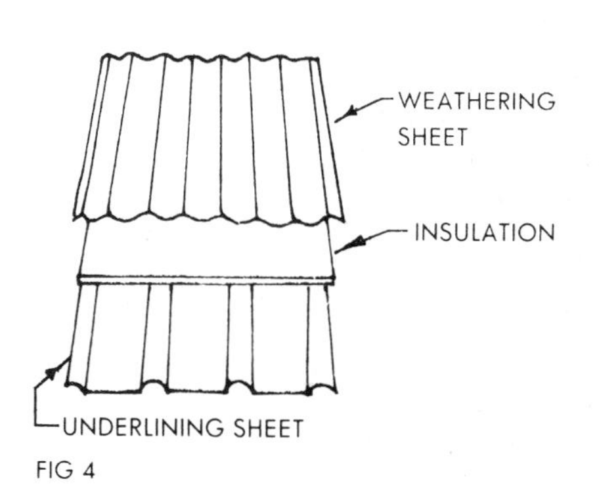

FIG 4

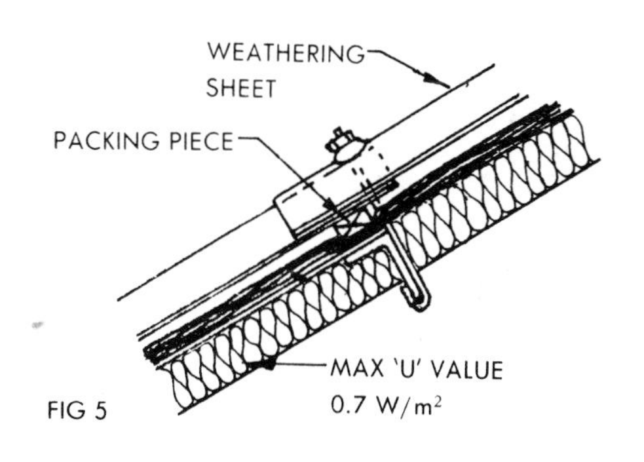

FIG 5

The insulating material must not be compressed and must be kept dry if it is to be fully effective. Wood packing pieces (which should be treated with a preservative are laid in position over the purlins Fig 5, to ensure that the insulating material is not compressed.

One proprietary system includes an asbestos channel spacer strip Fig 6. This is probably the most effective system of ensuring that the insulation is laid uncompressed, and so retains its full thermal insulation value.

WEATHERING SHEET
INSULATION
ASBESTOS CEMENT SPACER

FIG 6

ROOF COVERINGS 4

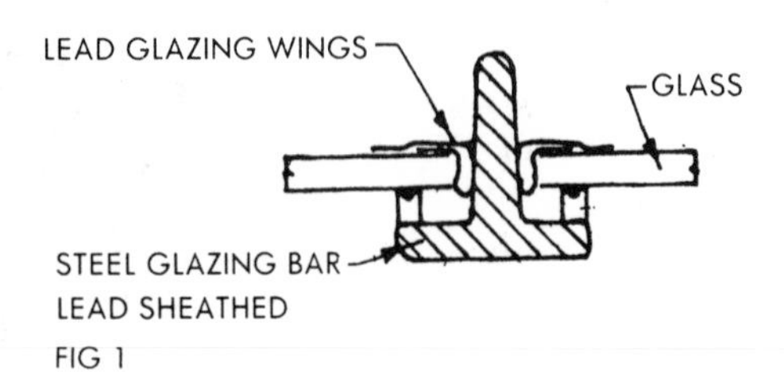

FIG 1

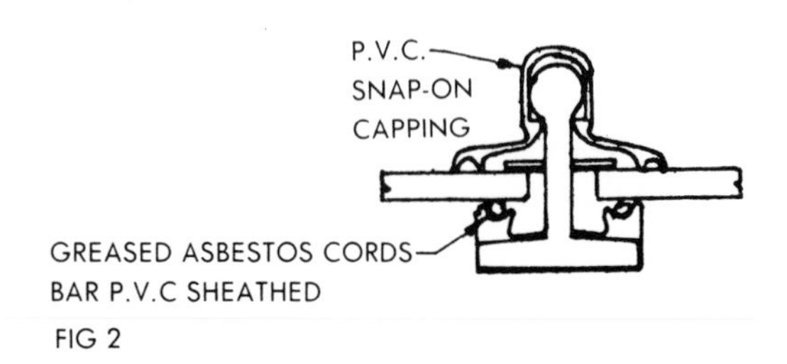

FIG 2

ROOF LIGHTS

In much factory and industrial building, roof lighting providing a good degree of natural daylight is widely used. The most common method of providing roof lights is to support sheets of glass by patent glazing bars which are secured to the roof purlins.

Steel glazing bars :- These are of three types

a) completely covered in an extruded jointless lead sheath, with which the lead glazing wings are integral, the sheath being sealed at both ends. Fig I shows an example of this type of bar.

b) complete covered with an extruded pvc sheath with sealed ends, and fitted with a plastic capping. Fig 2 shows an example of this type. In this case the glass is retained by an independent pvc snap-on capping, available in a range of colours. The glass is bedded on greased asbestos cords and held in position by a P.V.C. covered steel glass stop at the foot of the bar, similar to that shown Fig 5.

c) protected by hot dip galvanising and fitted with non-ferrous metal capping. It is more common today to treat steel bars with anti-rust composition and sheath in lead or pvc.

Aluminium alloy glazing bars :- These are either supplied with extruded lead wings inserted as an integral part of the bar, or with aluminium wings or capping. Fig 3 shows an example of the former type, having a glazing bar of an alloy of the silicon manganese range, with lead wings, and Fig 4 shows a bar of the latter type. In this system the extruded aluminium bar incorporates double drainage channels and cups to receive a butyl strip for bedding the glass which is secured with an aluminium spring cover strip.

Fig 6 shows a system where the glazing bar extruded from aluminium alloy is available in a range of depths, permitting an economical section to be selected for any span or situation. The design incorporates both drainage and condensation channels.

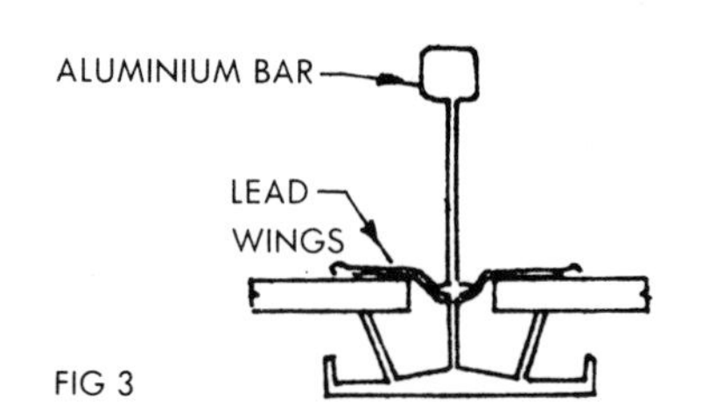

FIG 3

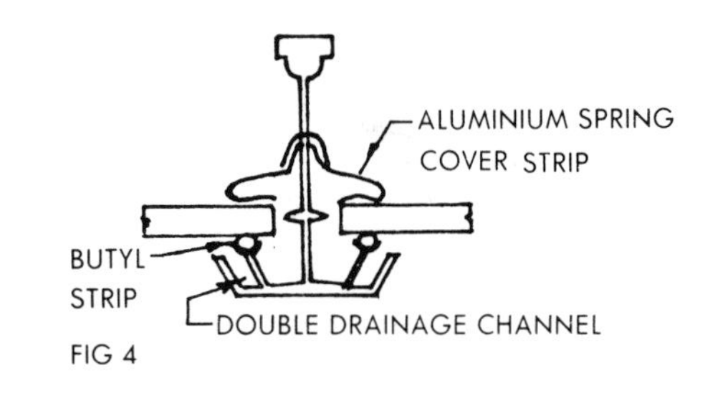

FIG 4

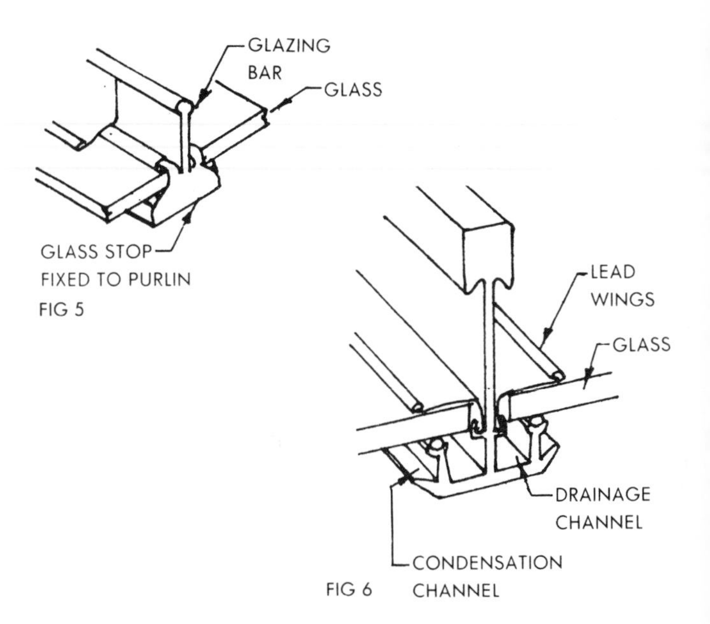

FIG 5

FIG 6

Weather-proofing is achieved either by aluminium or lead wings crimped into the bar, and subsequently dressed down onto the glass. Aluminium saddle clips bearing on the lead wings provide additional security against wind suction.

An aluminium glass stop, which clips into a shoe bolted to the purlin is shown Fig 7. The stop ensures that the glass is secure and will not slip down the slope of the roof.

Reinforced concrete glazing bars are available as Fig 8 shows. These are made of dense granolithic concrete for spans up to 10ft (3.048m). The glazing (usually 6mm rough cast or wired glass) rests on asbestos cord to provide a dust-tight level seating and is held by continuous rolled corrugated copper capping as shown.

Burglar preventative bars with the capping fixed from below the roof surface are used and Fig 9 shows this type.

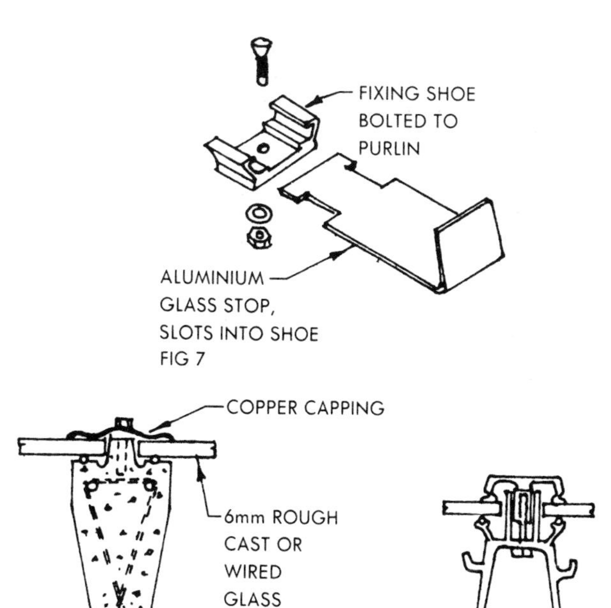

FIG 7

FIG 8 REINFORCED CONCRETE GLAZING BAR

FIG 9 SECURITY GLAZING BAR

ROOF COVERINGS 5

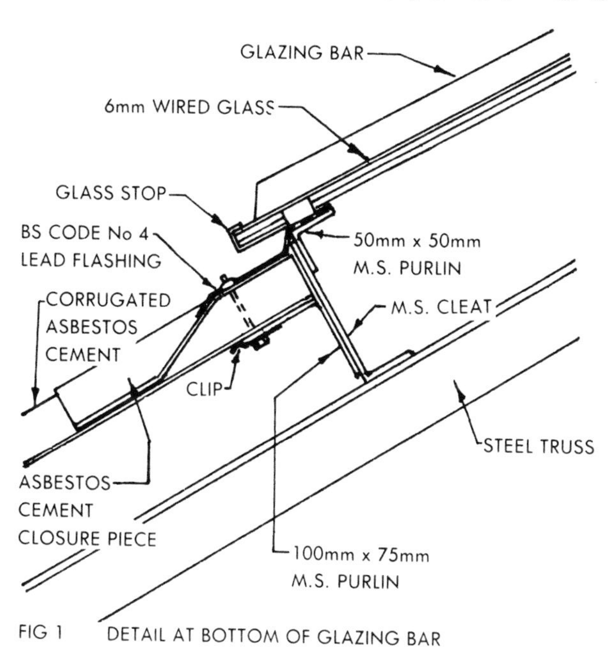

FIG 1 DETAIL AT BOTTOM OF GLAZING BAR

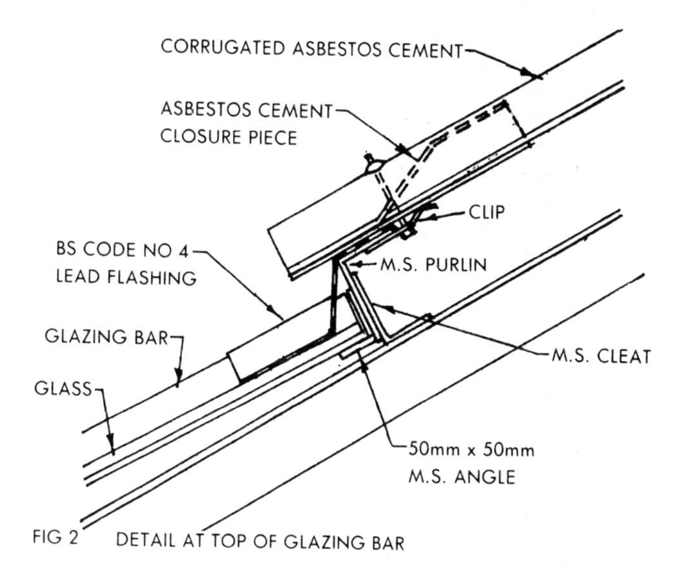

FIG 2 DETAIL AT TOP OF GLAZING BAR

ROOF GLAZING

Patent glazing bars are designed to accommodate glass sheets and provide adequate weatherproofing. Only dry materials are used to secure the glass and ensure weatherproofing, thus facilitating speedy erection.

The pitch of the glazing bars should not be less than 20 deg, and care is needed at the head and foot of the glazing to ensure that the structure is weathertight. This is usually achieved by means of lead flashings as shown (Figs 1 & 2) using BS Code No 4 lead.

The glazing is usually 6mm wire glass from 1m to 3m in length and 610mm in width, with glazing bars spaced at 620mm intervals. Wired glass is used for safety, so that in the event of breakage, fragments of glass do not fall onto people below.

When fixing proprietary sandwich insulation construction, the under lining sheets are laid first, the glass fibre or other insulation and then the corrugated sheets. Normal fixing bolts are used, penetrating through the sandwich construction Fig 3. The supporting sheets are rarely fixed separately. The lead flashing is sometimes dressed over the fixing (Fig 3), but if the bolts are taken through the flashing, and secured as shown (Fig 1). The flashing is secured against wind lift.

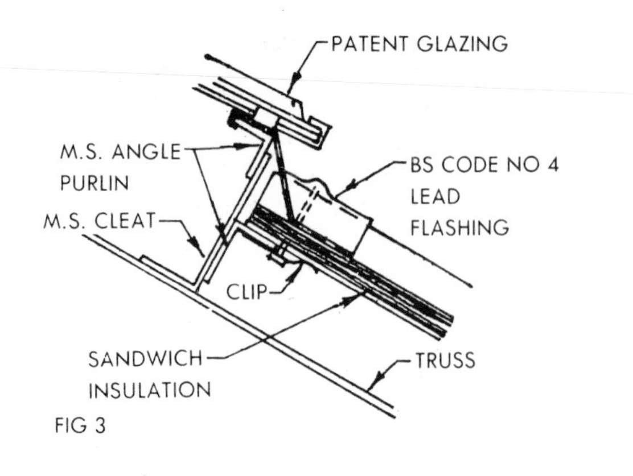

FIG 3

DOUBLE GLAZING

Although in many cases single glazing may be satisfactory, heat loss through the glass is considerable. Where thermal insulation is important double glazing is used. An air space 12mm to 18mm deep is created between two panes of glass, and approximately halves the heat loss, compared to single glazing. The risk of condensation is also reduced, although light transmission will be reduced (by some 20 per cent or so). Considerations of safety and handling limit the size of glass sheets and 600mm x 3300mm is usually considered maximum size. Wired glass is used for safety.

Fig 4 shows a double glazing aluminium system. Fig 5 double glazing with lead clothed steel, and Fig 6 a system which incorporates neoprene rubber as a spacing and sealing medium. Alternatively preformed double glazing units can be used.

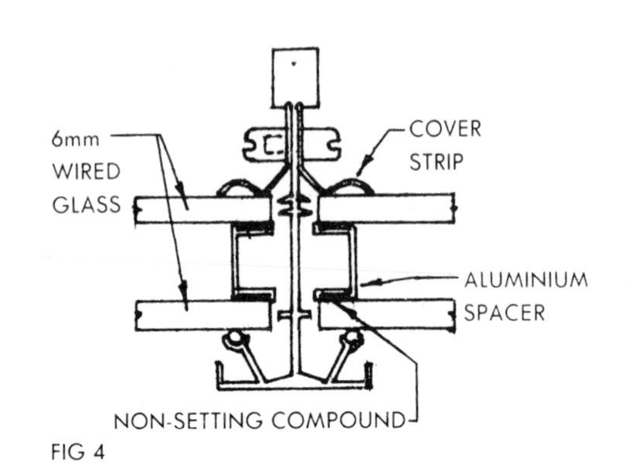

FIG 4

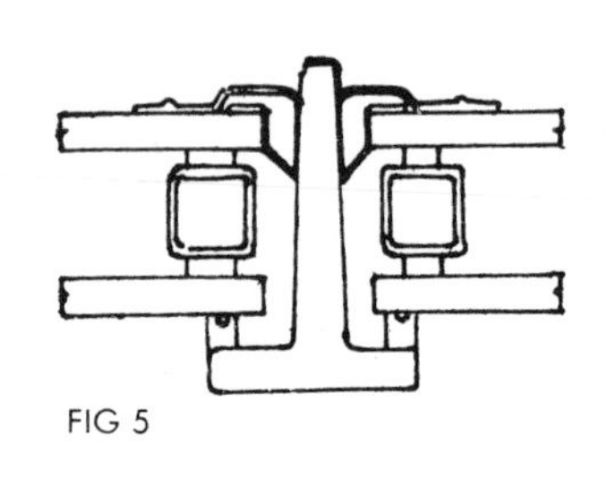

FIG 5

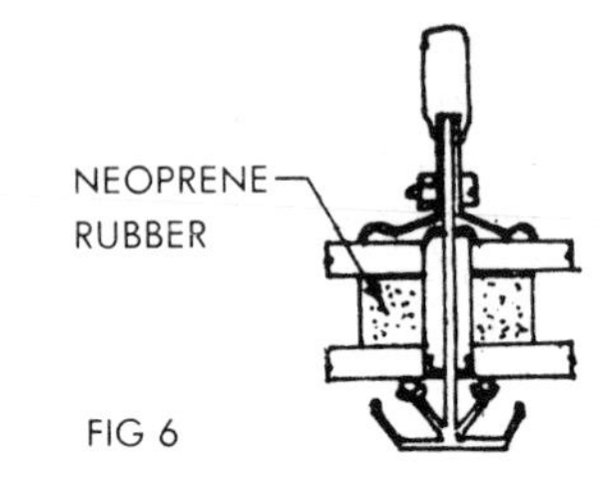

FIG 6

ROOF COVERINGS 6

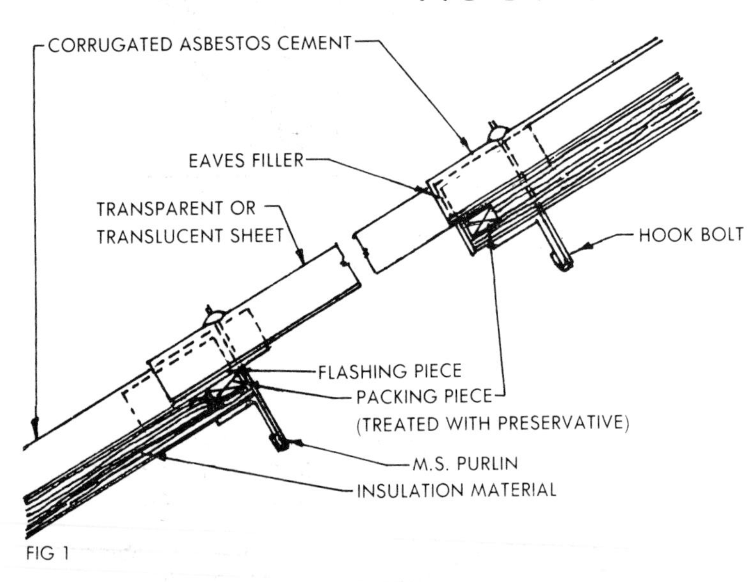

FIG 1

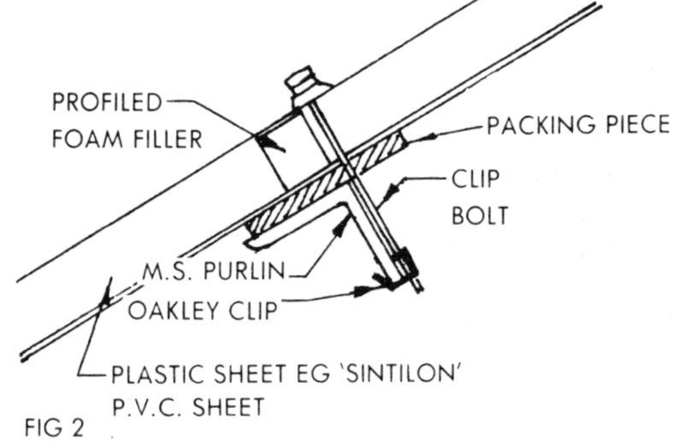

FIG 2

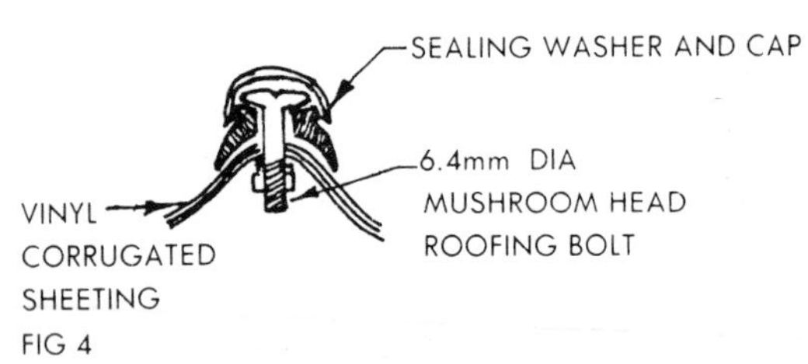

FIG 4

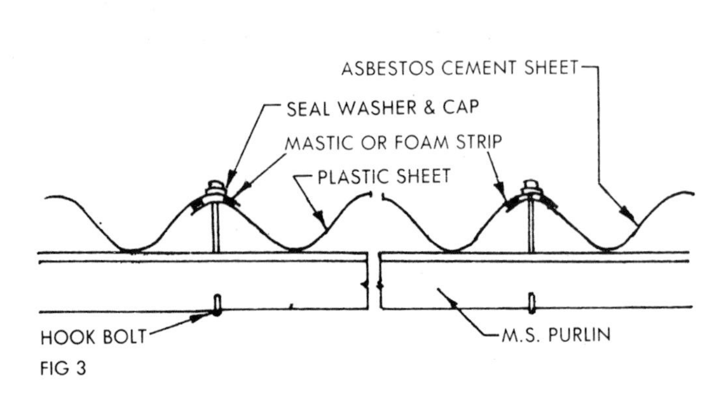

FIG 3

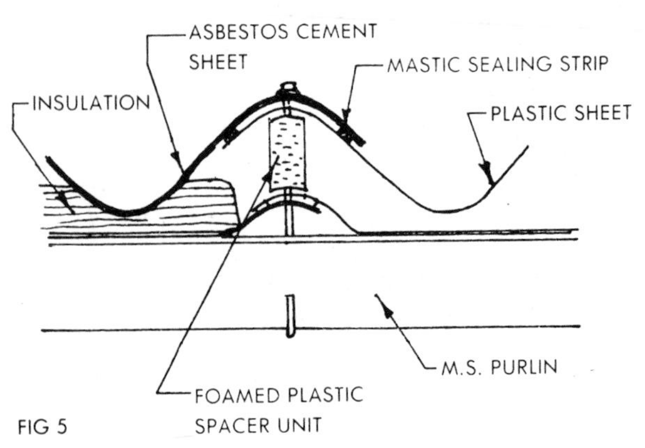

FIG 5

TRANSPARENT AND TRANSLUCENT SHEETS
Polyester/glass-fibre sheeting and pvc sheets having a range of corrugated profiles to match principal asbestos cement and metal profiles, are available either for use as dead lights in corrugated roofing or as a complete roofing medium for certain works.*

For roof lighting single sheets may be used or a continuous bay may be formed. Fig 6 shows the constructon at the head and foot of a section of transparent or translucent sheet used to introduce natural light into a corrugated asbestos cement roof. Fig 2 shows a detail at an intermediate purlin, and Fig 3 a cross section.

Standard hook bolts (Figs 1&3) may be used for fixing the sheets or a straight roofing bolt and Oakley clip, as shown Fig 2.

Side laps may be secured by means of stitch bolts Fig 4.

For fixing 3in (76mm) standard profile use $\frac{1}{4}$in (6.3mm) bolts or clip bolts with stitch bolts at 300mm C/C. For 6in (152mm) standard use $\frac{5}{16}$in (7.9mm) bolts or clip bolts with stitched bolts at 450mm C/C. Side laps should as far as possible be arranged to face away from the prevailing wind.

P.V.C. sheet is available in a wide range of profiles, to match standard 3in, Big 6 and other asbestos cement sheets profiles to match 'A7' and 'Box' corrugated aluminium sheets are available Fig 6. Single skin construction may be used or double skin as Figs 7 and 8. Foam spacer strips as shown are used for ordinary double skin construction, and where roof lighting is not required, foam profiled strips are available as shown Fig 9.

*Plastic sheeting may be used for roofing for industrial buildings, covered walkways, canopies, car ports etc.

N.B. Holes should be drilled never punched and should be 3mm larger than the diameter of the shank of the fixing to allow for thermal movement.

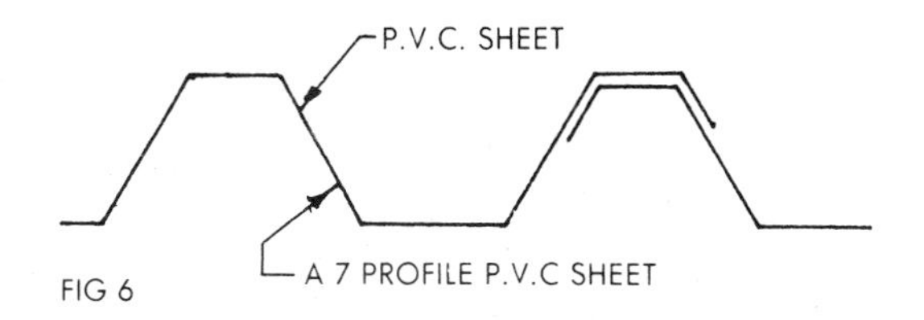

FIG 6

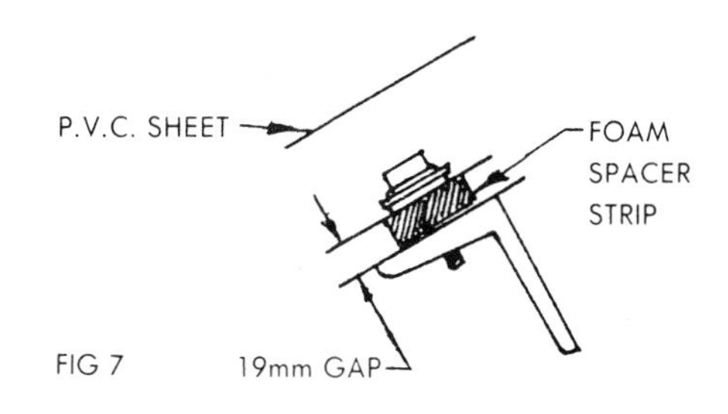

FIG 7

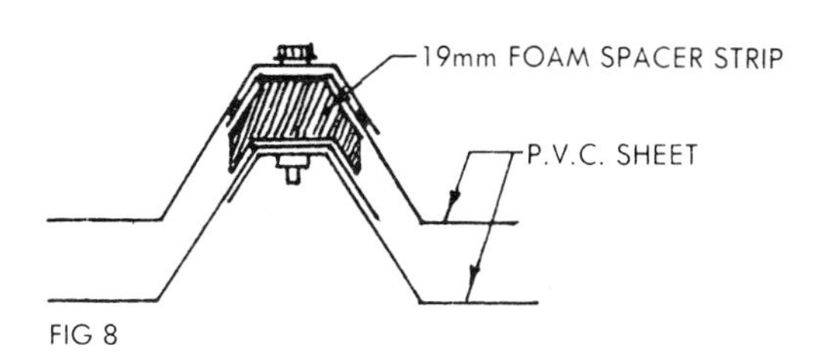

FIG 8

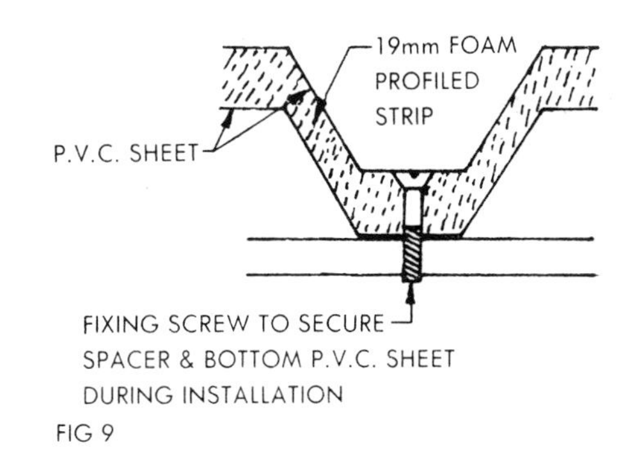

FIG 9

ROOF COVERINGS 7

TRANSPARENT SHEET
Solar heat gain:- Solar radiation passing through the glazing material (glass or plastic) causes walls, floors, furniture and fittings to heat up, the resulting general rise in temperature inside the building is known as 'solar heat gain'.. Where there are large areas of glazing adequate ventilation is most important. Both glass and plastic sheet are available in some profiles and hues designed to reduce the effects of solar heat.
Condensation :- With single skin construction some condensation is inevitable. It is likely to occur in cold, frosty weather and in buildings where processes used involve warm or high temperatures or humid conditions. Condensation may be reduced by adequate ventilation and the use of double skin construction.

Corrugated wired glass :- Annealed cast glass 6mm thick, reinforced with 25mm square mesh and corrugated to match 3in Big 6 etc is now available.

Certain corrugated glass has prismatic ribbing on the underside designed to eliminate dull areas and harsh light. The corrugated surface ensures the longest possible period of light transmission during daylight, longer than that provided by flat glass.

ALUMINIUM ROOFING
Fully supported aluminium using flat sheets was briefly mentioned . As only a brief reference was possible, it is proposed to deal with the construction in greater detail.

Boarding should be 25mm nominal thickness, and for roofs of less than 25 deg pitch laid diagonally or in the direction of the fall. The heads of nails used for fixing boards should be punched below the surface and all screws for securing batten rolls should be counter-sunk below the surface of the timber.

Underlays :- The purpose of an underlay is to provide a sympathetic surface to receive the aluminium, to insulate the roof against the drumming of hail or heavy rain, and to isolate certain types of decking from contact with the aluminium.

On timber structures type 4A(ii) brown sheathing felt No I, inodorous, complying with the requirements of BS 747 should be used. On concrete and screeded bases, 2000 gauge clear polyethylene sheeting should be used.

Nails :- Clout nails for securing aluminium clips and felt underlays should comply with the requirements of BS I202 Part 3. Galvanised clout nails may be used, but aluminium is to be preferred. For fastening clips, nails at least 25mm long and 3.5mm diameter should be used. Clout nails for fixing felt underlays should be at least 20mm long and 3mm diameter.

Copper, brass or uncoated nails should not be used.

On concrete, aluminium clips or batten rolls should be secured by screws or nails driven into 75mm x 48mm wooden dovetail battens or plugs set into the concrete substructure. Such timbers should be treated with creosote or other preservative not containing copper salts.

The traditional methods of fixing aluminium are either using batten rolls as Fig I, or standing seams as Fig 2. An alternative method is to use extruded aluminium alloy rolls as Fig 3.

A recent development is the use of the long system.

Sheet sizes :- Stock sheets I000mm x 2000mm and I250mm x 2500mm are available and other sizes can be supplied to individual requirements. For roofing however it is convenient to use coiled strip for general roofing. Widths of 450mm, 500mm and 600mm are available, while I50mm and 300mm widths are useful for flashings, aprons etc.

N.B. Tools that have been used for working copper or brass must be thoroughly cleaned, as a small amount of copper transfer can cause aluminium to be attacked.

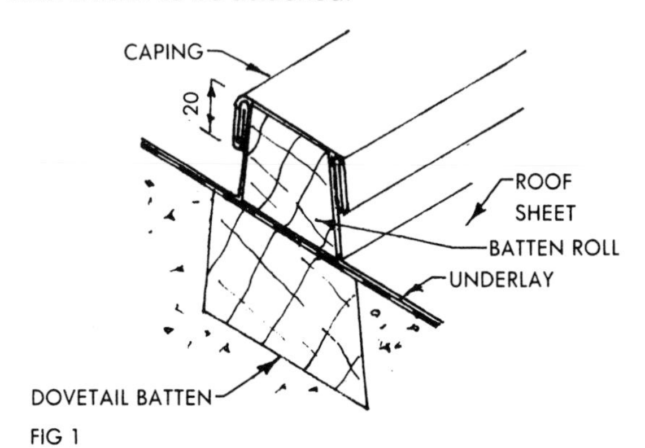

FIG 1

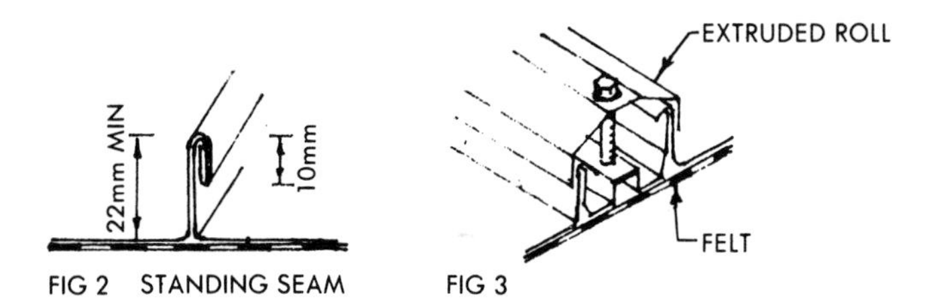

FIG 2 STANDING SEAM FIG 3

ROOF COVERINGS 8

The batten roll method of fixing aluminium strip is illustrated in Fig I. A detail at the junction of the batten and ridge rolls, and Fig 2 a section through a drip a (layout and forming of rolls is similar to that for copper

Standing seams are used for the long strip system, and where a drip occurs in this system it should be formed to allow for expansion. Fig 3.

The traditional alternative to the batten roll method is to use standing seams. Transverse joints, ie across the fall of the roof are made with single or double welts (Figs 4 & 5). Single welts should only be used for roofs having a pitch greater than 40 deg and for aprons, cappings etc. Welted transverse joints should be fixed with two clips evenly spaced in each bay on roofs where the pitch is 20 deg or less. The double lock cross welts should be sealed by coating the edges with boiled linseed oil before forming the welt. Seams and cross welts are secured by aluminium clips (Fig 4).

The long strip system :- Based on aluminium strip used in long lengths 0.7mm and 0.9mm strip may be used. The following table from CP I43 indicates recommended maximum width of strips. Lengths up to 7m can be laid. This system has advantages particularly for low pitched roofs as cross welts can be avoided. Expansion clips are used and Fig 7 shows two types.

If the roof is to take occasional traffic, the roll cap system is preferable. Standing seams are cheaper and are used for roofs where access is unlikely to be required.

Where cross welts are used they should be staggered in adjacent bays. Fig 6.

Eaves :- A traditional type of eaves construction is shown Fig 8. Two clips per bay are secured to the timber eaves fillet with 25mm aluminium alloy nails. An eaves detail for the long strip system, to allow for expansion is shown in Fig 9. The lining plate is secured by aluminium alloy nails.

Thickness mm	Width of strip mm	Approx' ctrs of standings seams mm
0.7	500	410
0.9	600	500

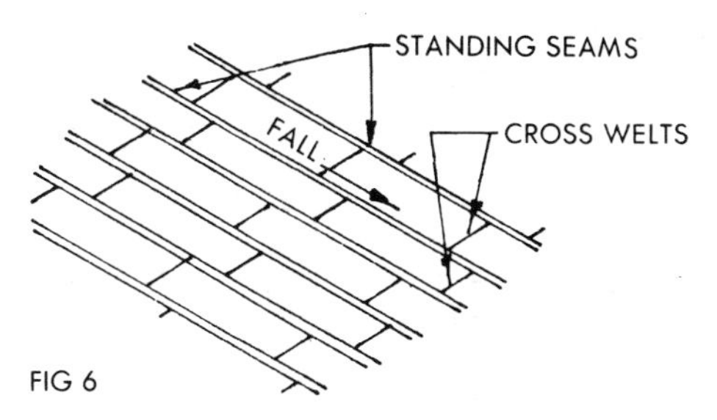

FIG 6

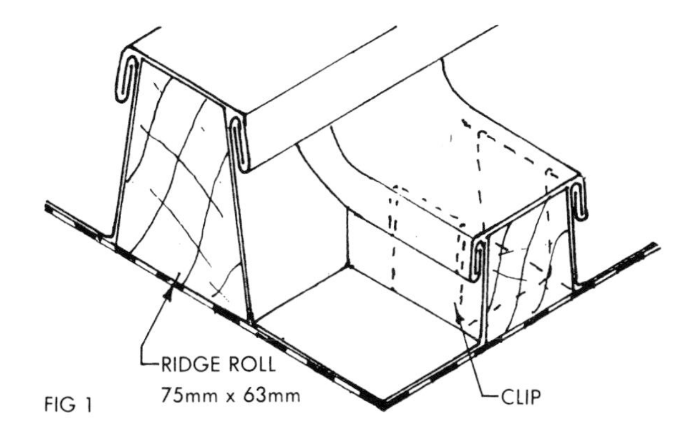

FIG 1

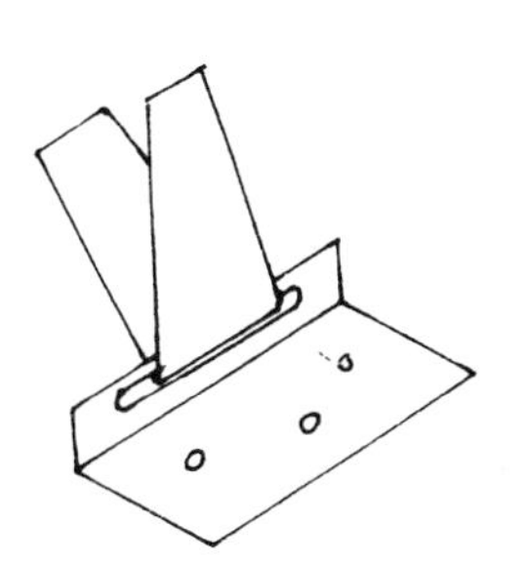
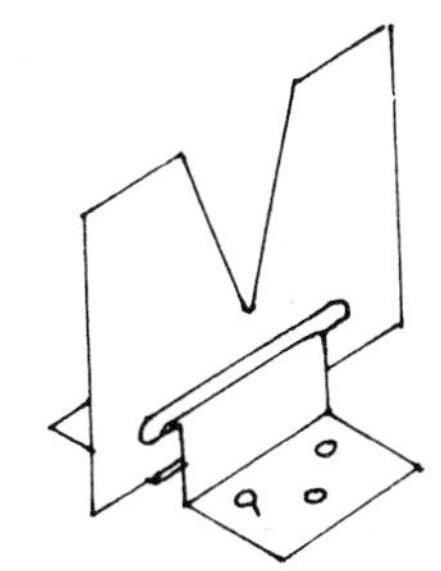
FIG 7 EXPANSION CLIPS

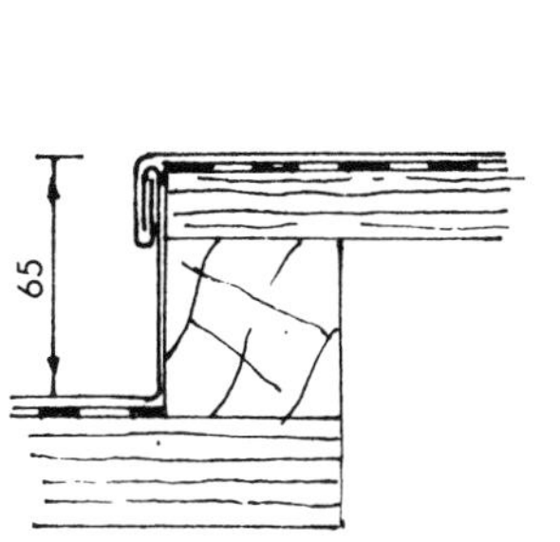

FIG 2

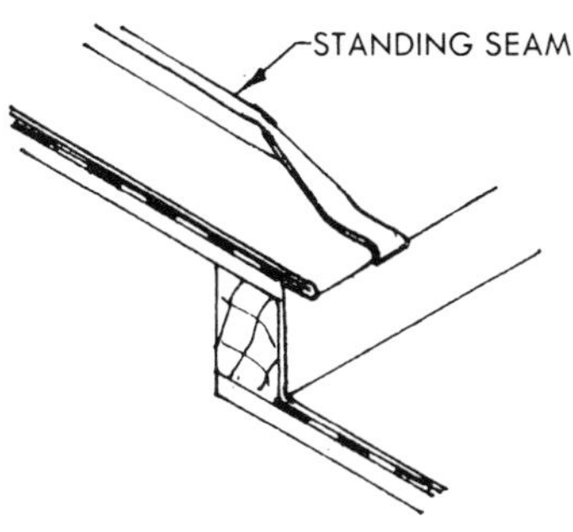
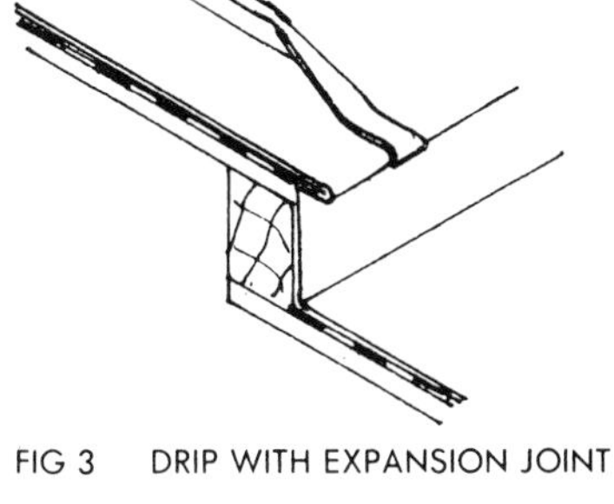

FIG 3 DRIP WITH EXPANSION JOINT

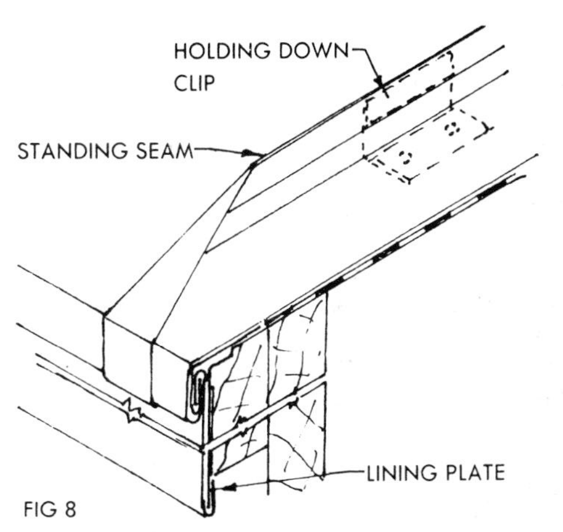

FIG 8

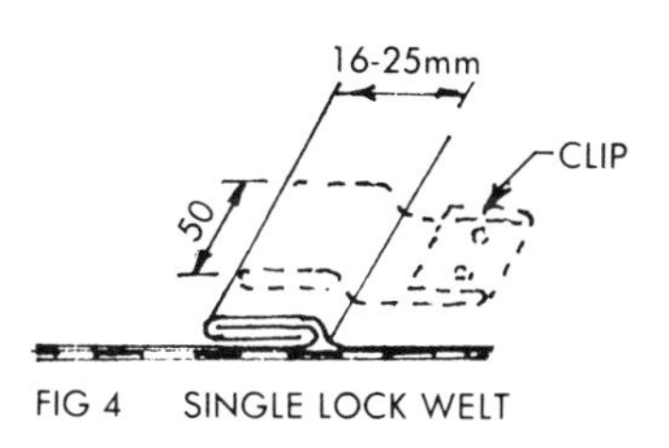

FIG 4 SINGLE LOCK WELT

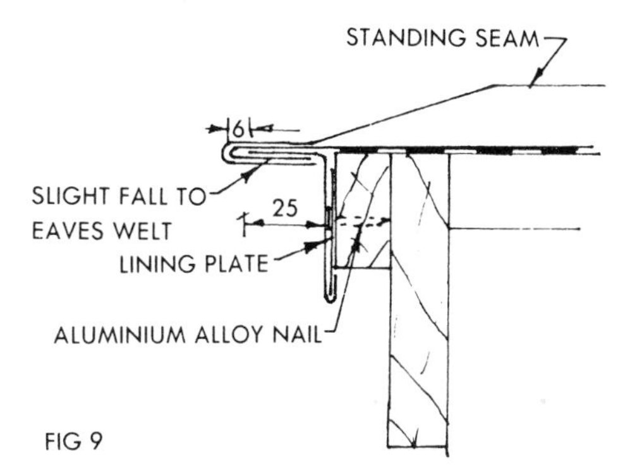

FIG 9

FIG 5 DOUBLE LOCK CROSS WELT

ROOF COVERINGS 9

ALUMINIUM ROOFING

A detail showing construction at a roof verge, suitable for both traditional and long strip roofing systems is shown Fig I.

Precautions :-

i) Care should be taken in manipulating the metal to avoid work hardening by repeated dressing. A few firm blows are better than many light ones.
N.B. Machine forming of the joints may be used, particularly in the long-strip system. This allows strips of greater hardness to be used and speeds up installation.

ii) It is recommended that aluminium be given a coat of bituminous paint where in contact with cement, concrete or lime mortar. Mixes containing lime are more likely to affect the metal than Portland cement.

iii) Some wood preservatives, particularly those containing zinc chloride, mercuric salts or copper sulphate are unsuitable for use with aluminium. Copper derivatives should generally not be used.

iv) Water draining from copper onto aluminium

can cause corrosion, and water should not be allowed to drain onto aluminium surfaces from copper roofs, lightning conductors, expansion pipes or overflows. Water draining in the opposite direction is harmless, providing there is no direct contact.

v) Where aluminium is supported on cellular, foamed or aerated concrete, anchorages for securing battens should be by bolts passing through the full thickness of the concrete layer.

vi) The fall of the roof should be not less than 1 in 60, but in order to avoid puddles a minimum fall of 1 in 40 is normally recommended.

vii) Standing seams are secured by holding down clips These should be at least 50mm wide and spaced at 300mm intervals.

Aprons :- These are made of separate strips of metal not more than 2m long, jointed by single-lock welts. They are secured by lining plates Fig 1.

Cover flashings :- Let into a wall joint a minimum of 25mm, with a turn-back acting as a drip Fig 2. Adjacent lengths are either joined with a single welt, or lapped 75mm. The flashing is secured by folded aluminium wedges at 400mm c/c before pointing. If the metal is painted as recommended, this should be done after folding.

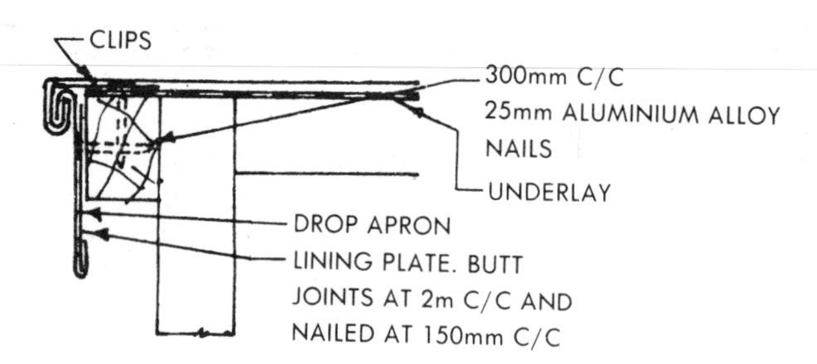

FIG 1 VERGE DETAIL SUITABLE FOR BOTH TRADITIONAL AND LONG STRIP SYSTEMS

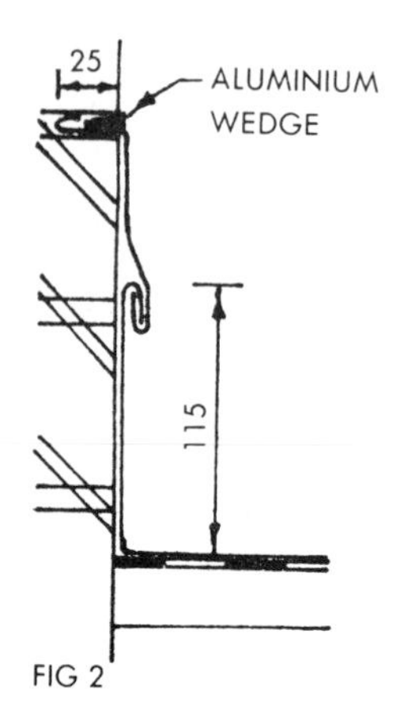

FIG 2

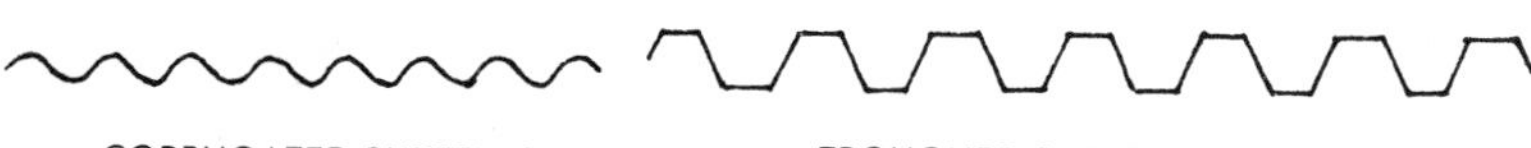

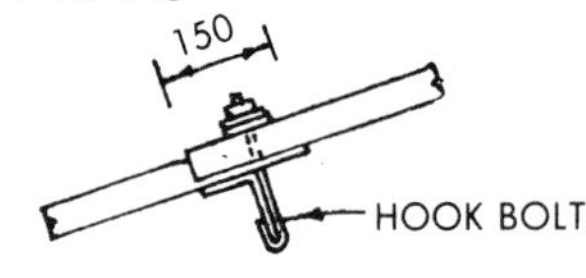

FIG 3

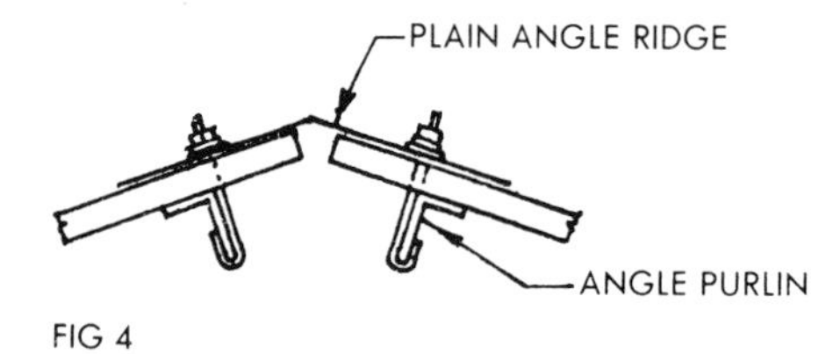

FIG 4

ALUMINIUM CORRUGATED AND TROUGHED SHEETING

An aluminium manganese alloy is normally used for these types of sheeting. They are produced in a range of gauges and trough depths (Fig 3) and a minimum slope of 15 deg is recommended, but lower slopes are possible if provision is made to avoid penetration at eaves and end laps.

Fixing is usually by hook bolts (Fig 3). Plastic foam filler strips are available and are fixed by the hook bolts and by self tapping screws (Fig 5). Plain ridges (Fig 4) and roll top ridges (Fig 5) are secured as shown. Side wall flashing strips (Fig 6) are also available.

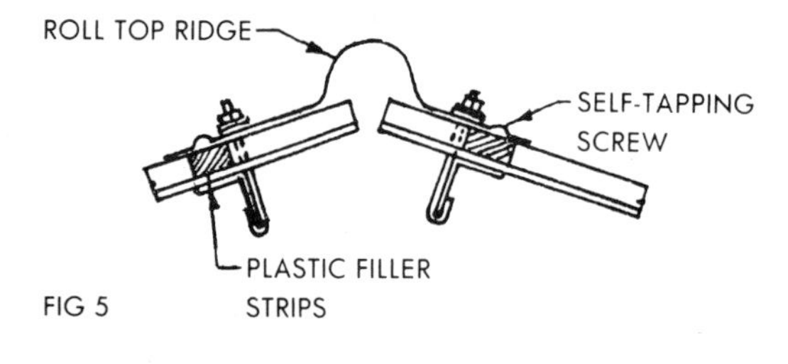

FIG 5

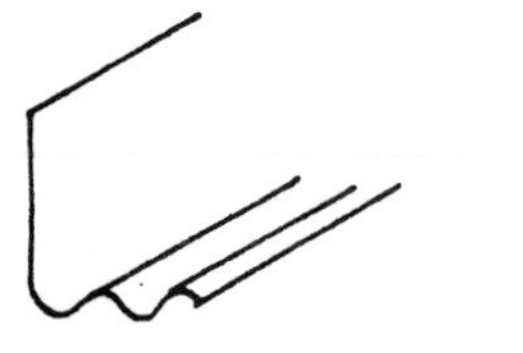

FIG 6 FLASHING STRIP

ROOF DECKINGS

ROOF DECKING

Consists of a supporting deck of troughed galvanised steel, aluminium or asbestos cement units with a built-up finish of an insulating layer covered with felt. Roof decks may be used for flat roofs, and also for pitched roofs where the decking either spans up the slope or between roof frames. Woodwool cement slabs may also be used, either over intermediate supports or as reinforced decking slabs.

Steel and aluminium decking:- Metal trough decking is widely used and many proprietary systems include both steel and aluminium decks. Manufacturers often provide a complete roof including insulation and weatherproofing and should be consulted on suitability and structural details.

Fig I shows a roof which comprises a metal deck, vapour barrier, thermal insulation and built-up roofing. The built-up roofing normally consists of two layers on a pitched roof and three layers on flat roofs, up to a fall of 5 deg.

Fig 2 shows a roof deck comprising a steel sheet decking with dovetail shaped ribs in the sheet. Side laps are formed by an overlap fitting into a rib of the adjoining sheet and leaving the top surface flush. The working width of the sheets is 610mm. An insulation layer may be bonded to the surface of the decking, or if the bonding coat is not required as a vapour barrier, the insulation may be fixed with clips fitting into the dovetail ribs.

Where galvanised steel deck is exposed to external atmosphere e.g. in cantilevered slabs etc, it should be painted at regular intervals to prevent rusting. Where damp, humid conditions exist it is preferable to use aluminium decking. Deck units range from 600mm to 900mm in width, and are normally limited to 10m in length.

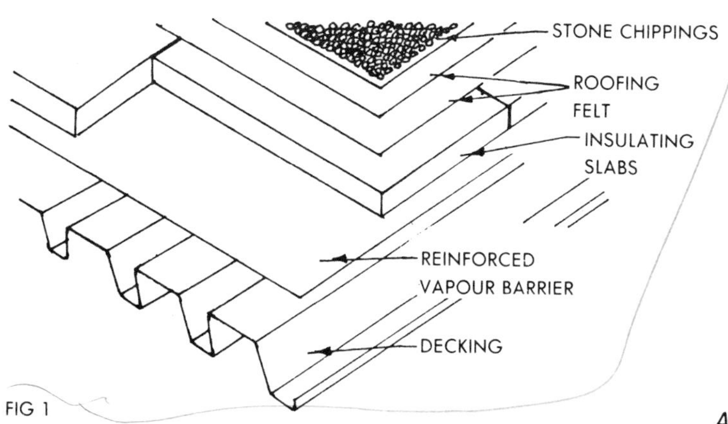

FIG 1

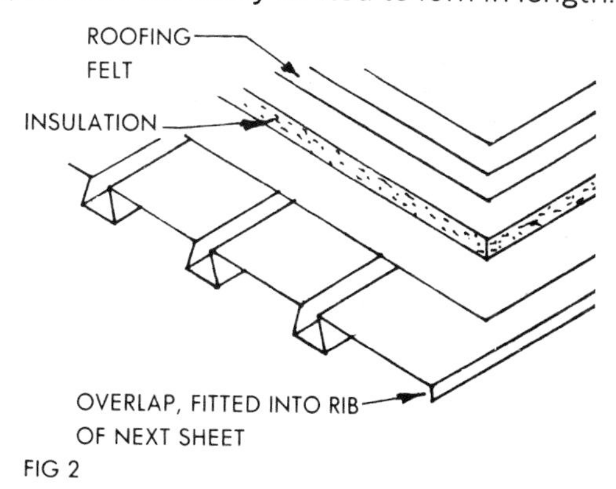

FIG 2

Asbestos cement decking:- One type of decking is shown Fig 3. Insulation boards may be bonded to the upper surface and the roof finished with two or three layer felt. The depths of the troughs are varied to suit the span required (Maximum span 3m approx')

A number of asbestos cement roof decks are available to suit requirements and manufacturers should be consulted for particular circumstances.

Wood wool slabs:- Woodwool/cement slabs are available in thicknesses of 50mm, 75mm and 100mm and may be obtained in eight lengths up to 3600mm. Fig 4 shows woodwool/cement slabs which are primarily as insulation over intermediate supports. (Support is normally required every 600mm). This type of slab is sometimes used as permanent formwork and insulating lining. Fig 5 shows a decking, reinforced on the edges with interlocking steel channels. 100mm decking has a maximum span of 4m. Fig 6 shows reinforced decking. Spans up to 6m are possible.

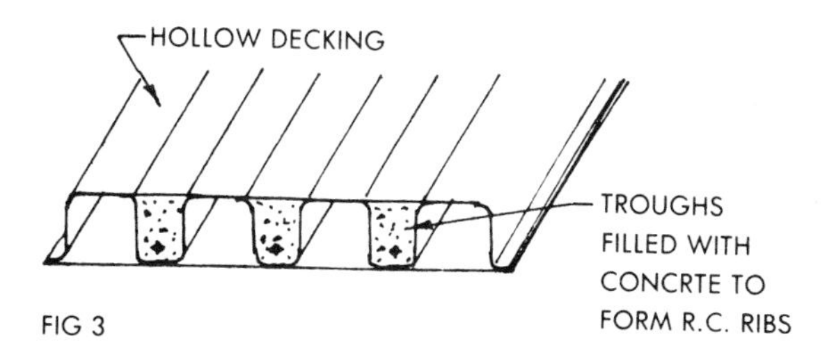

FIG 3

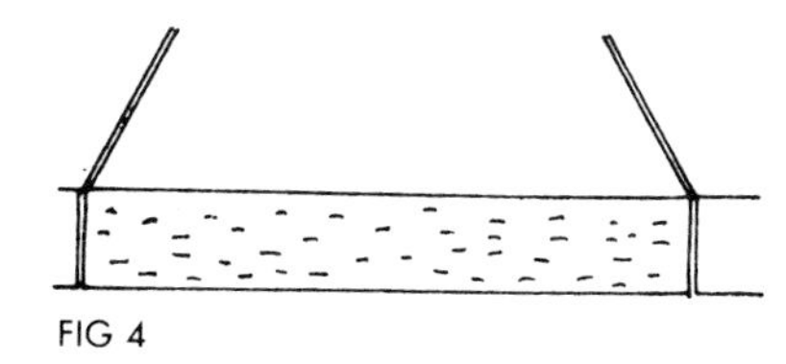

FIG 4

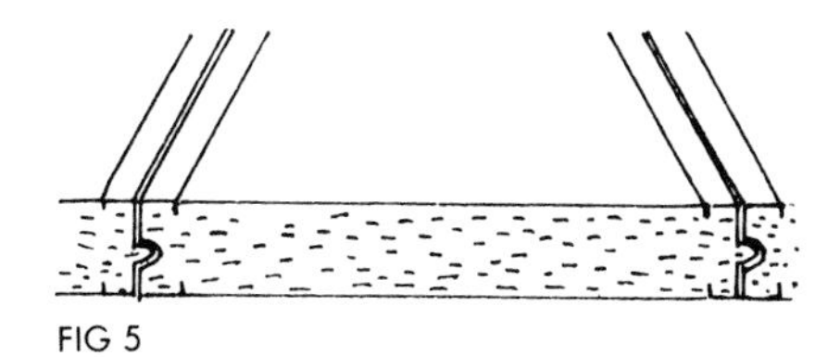

FIG 5

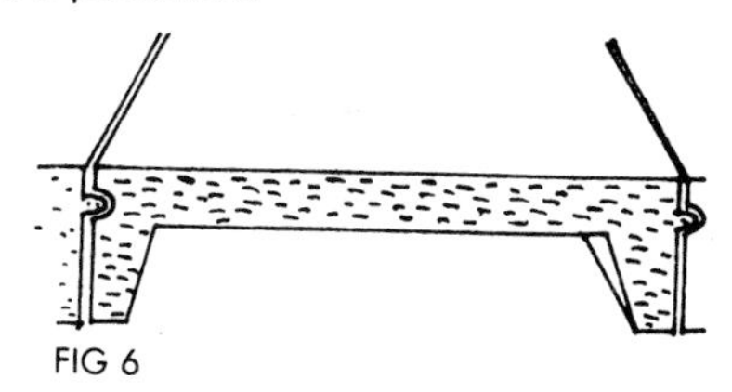

FIG 6

STEEL ROOF TRUSSES

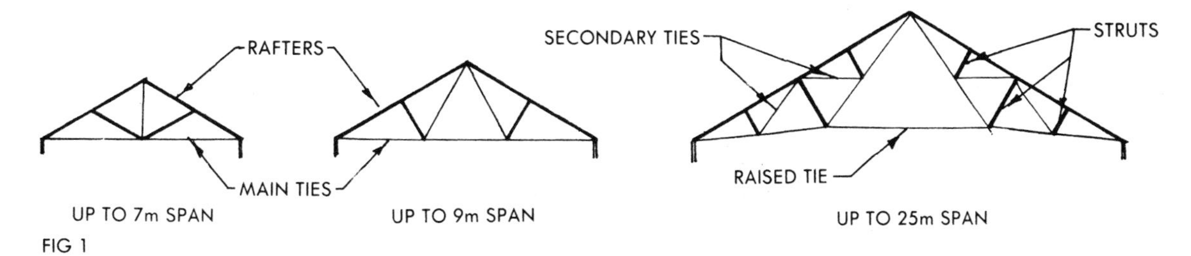

FIG 1

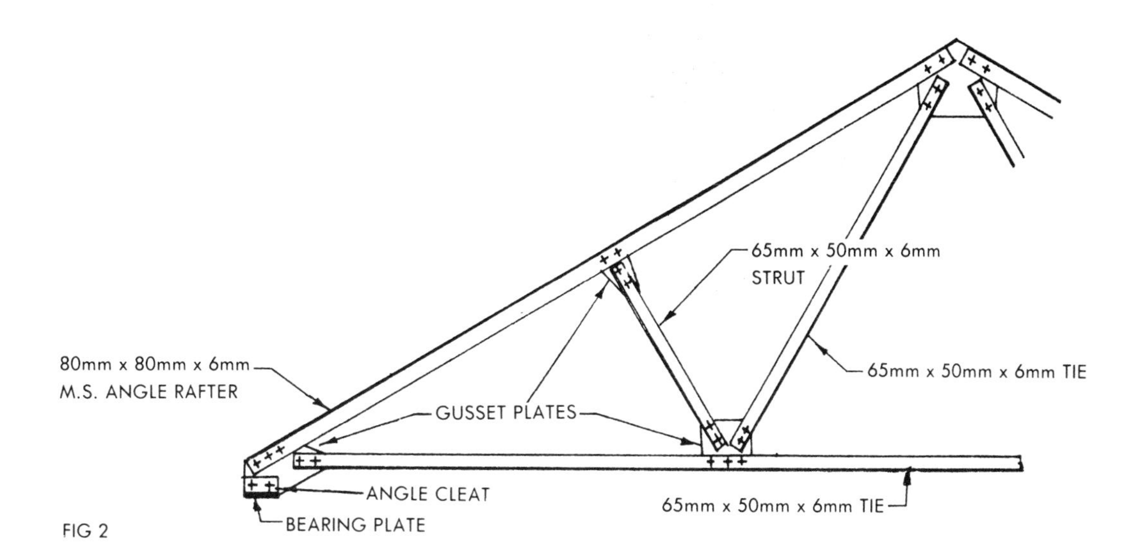

FIG 2

Timber trusses were covered in earlier Builder's Detail Sheets. Steel trusses are fabricated of light steel sections welded, bolted or rivetted together. They are fairly simple to prefabricate, light in weight and are economical for medium and large span roofs.

Typical truss arrangements for spans from 7m to 25m are shown Fig I. Principal rafters and struts are in compression and main and secondary ties in tension. Truss diagrams usually show members in compression by thick lines and those in tension by thin lines and this convention has been adopted in Fig I.

For longer spans the main tie may be raised as shown. The connections for a typical truss to span 7m is shown Fig 2. The rafters and main tie may consist of single angles or of two angles placed back to back with space between them for connecting Gusset plates Fig 4. Single angles are used for the remaining struts and ties. The members making up the truss are arranged so that their centre lines intersect at the connecting points Fig 3. Sizes of steel sections will vary according to the span and loading.

The pitch of the roof is governed by the roof covering used, and for asbestos cement sheet may be 22½ deg to 30 deg. The trusses are spaced from 3.0m to 7.5m apart and may be supported on stone templates on brick walls as Fig 4 or on steel columns as Fig 5.

The roof covering of asbestos cement sheets, plastic sheets or metal sheets are supported on steel angle purlins. Spacing of purlins will depend on the roof covering used, position of roof lights, any services or gear to be supported etc. The purlins are bolted to M.S. angle cleats welded or otherwise fixed to the truss rafters.

Tubular steel trusses :- Lattice trusses of welded tubular construction are available for spans up to 24m. They are lighter in weight, and present less area to paint. Fig 7 shows a truss for a span of 24m.

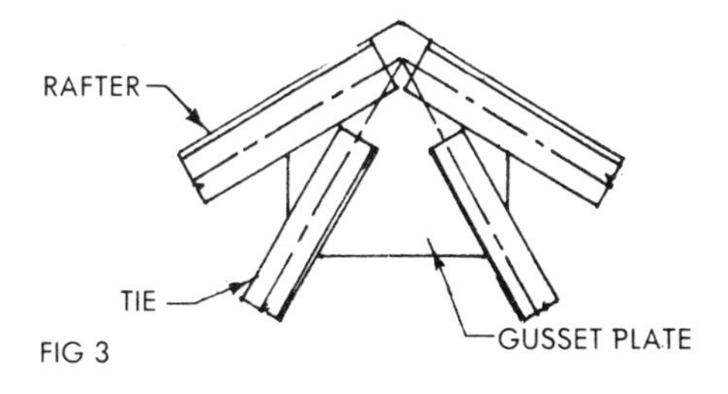

FIG 3

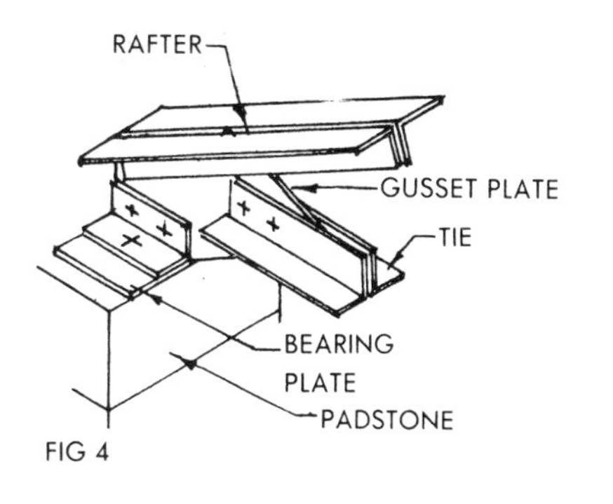

FIG 4

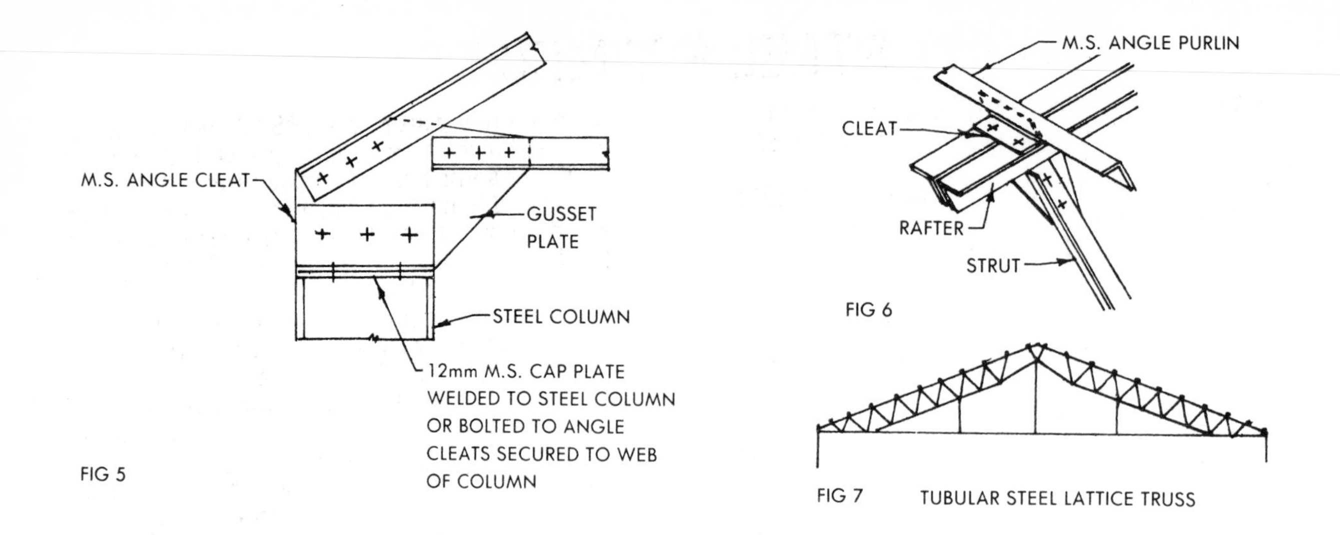

FIG 5

FIG 6

FIG 7 TUBULAR STEEL LATTICE TRUSS

STEEL FRAMES I

BS 449:1969 'The use of structural steel' is the relevant BS for the design and construction of structural steelwork.

Structual steel now conforms to BS 4360: 1972 'Weldable structural steel', which is based on I.O.S. (International Organisation for Standardisation) recommendation R360 – Structural steels. Weathering steel is now included.

Weathering steel:-This has been developed for external use unprotected, and provides a maintenance free finish. The steel has high strength and possesses good welding properties. The exposed surface builds up a stable oxide coating which protects the metal beneath, and can present an attractive purplish or dark brown hue. Care must be taken in the design and in handling and storage. For successful results and to ensure a uniform appearance a simple design avoiding crevices is required and care must be taken to avoid staining by cement mortar, chalk marks etc.

The physical dimensions of steel sections are governed by BS 4: 1980. This presents data in the form of metric equivalents of the original imperial values.

BS 4848 is also applicable. Part 2 deals with Hollow sections; Part 4 with Equal and unequal angles; and Part 5 with Bulb flats.

Formerly rolled steel joist (RSJs) sections were produced having tapered flanges while being suitable for some beams and columns. Under certain conditions it was necessary to strengthen the section by the addition of steel plates rivetted to the flange. (A range of joists with 5 deg taper flanges is still specified).

Universal beams (or broad flange beams) having either a slightly tapered flange (up to 2 deg 52 min) or a parallel flange are now produced. In these sections the inner profile remains constant, heavier and stronger sections having thicker flanges and webs. Fig I.

Designation:- BS 4 : Part I : 1980 contains a series of tables giving ranges of sizes and properties for beams, columns, bearing piles, jcists with 5 deg taper flanges, T-bars, channels, angles, bulb angles, and bulb flats. These are grouped under a series of 'designations' which state an approximate size – the serial size. This may cover a group of up to seven different sizes. Thus the serial size does not indicate the actual size of a particular section, but this is indicated by the weight of the section per metre run. So that a beam 920.5mm x 420.5mm at 388 kg should be designated 914 x 419 x 388 beam. A beam 911.4 x 418.4 at 343 kg per metre, comes within the same serial size and would be designated 914 x 419 x 343 beam. Fig 2 indicates the range of various steel sections.

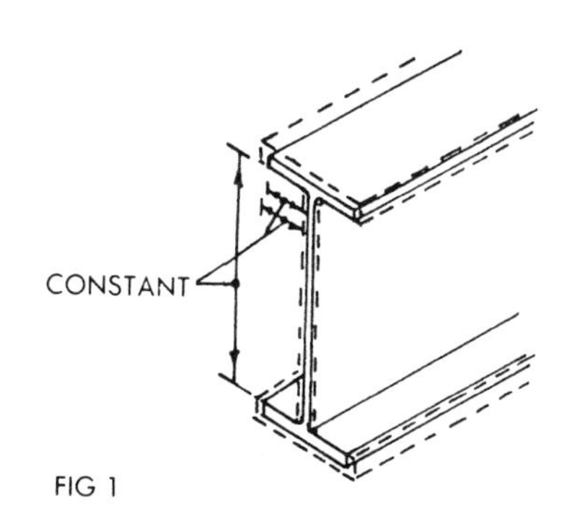

FIG 1

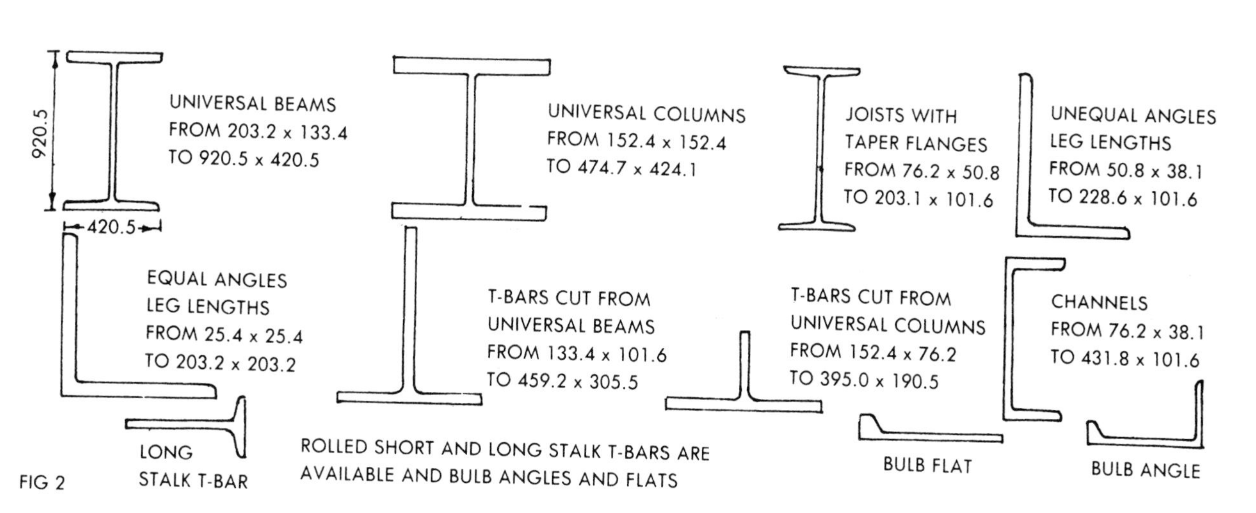

FIG 2

STEEL FRAMES 2

Castellated beams:- Where a number of services, water, gas, cables etc are required to pass through beams, a castellated beam may be used. The web of a standard joist section is cut along a castellated line Fig l. The two halves are then placed one above the other as Fig 2 and welded at the connecting points. The beam thus made is one-and-a-half times the depth of the original beam from which it was cut, and so has less deflection under load.

Steelwork connections:- Steel sections may be supported and secured by steel angles connected to the flange or web of the beam or stanchion. Fig 3 shows typical connections of beams to a stanchion and the method of connecting a further section of stanchion. If the upper section is smaller, then a bearing plate, tapered splice plates and packing pieces will be required. Fig 4.

Rivetted connections:- Rarely used today, as connecting cleats and plates are rivetted in the shop and the final assembly on site secured by bolts.

Bolts:-Are of three kinds (i) black bolts used mainly for site connections. These bolts do not fit tightly into the holes, and are secured by tightening a hexagonal nut. They are weaker than rivets or turned bolts, and are only used for end connections, where a supporting angle bracket will resist the effects of shear.

(ii) Turned and fitted bolts, are turned to fit tightly into their holes. The bolt is driven tightly and secured by a nut. Have about the same strength as rivets, but are more expensive. Where connections to a bevelled surface are made, eg a flange connection, a tapered washer is used.

(iii)High strength friction grip bolts. These are also known as torque or torque controlled bolts. They are made of steel having a greater yield point than mild steel, and the combined effect of the extra strength and the friction induced by the tight clampling together of the plates being secured, enables these bolts to carry much greater loads than ordinary bolts or rivets. The nut is tightened by using a torque wrench which measures the tightness of the connection.

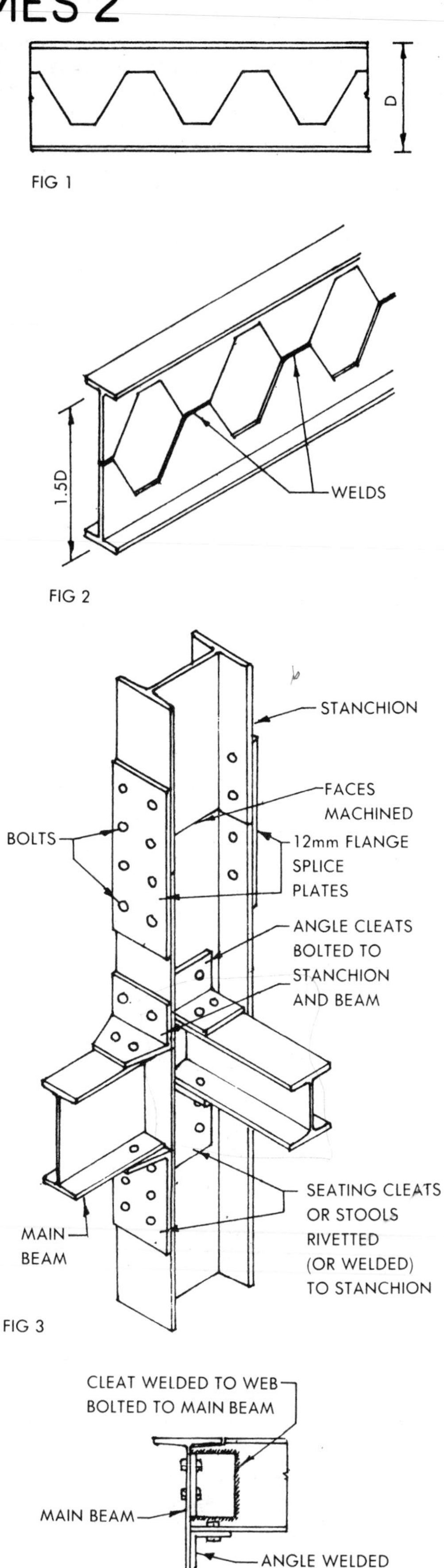

FIG 1

FIG 2

FIG 3

FIG 5

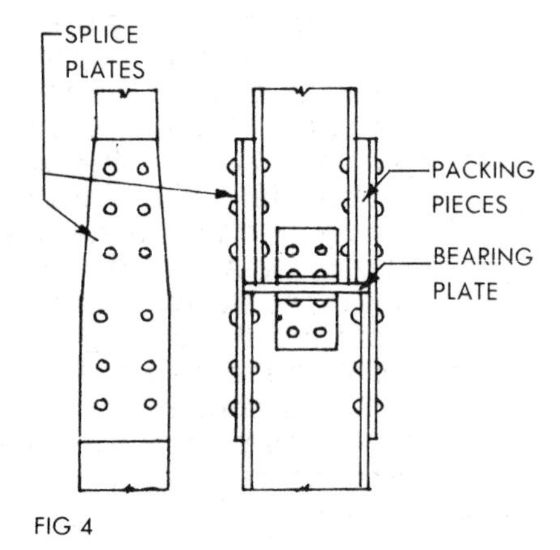

FIG 4

Welding:- Has almost completely replaced rivetting for shop connections. Site welding may be used at times, but presents certain difficulties and is expensive. Butt welds are tending to be used more than fillet welds. the former giving a more direct transference of load between members without abrupt changes of section and having a higher resistance to repeated loadings. Butt welds can more easily be examined by X-rays. Fig 6. Shop welded and site bolted connections are widely used. Fig 5 shows an example of the connection of a secondary beam to a main beam.

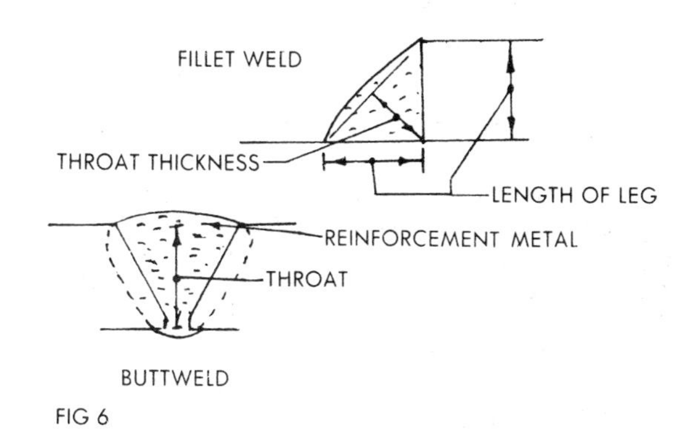

FIG 6

STEEL FRAMES 3

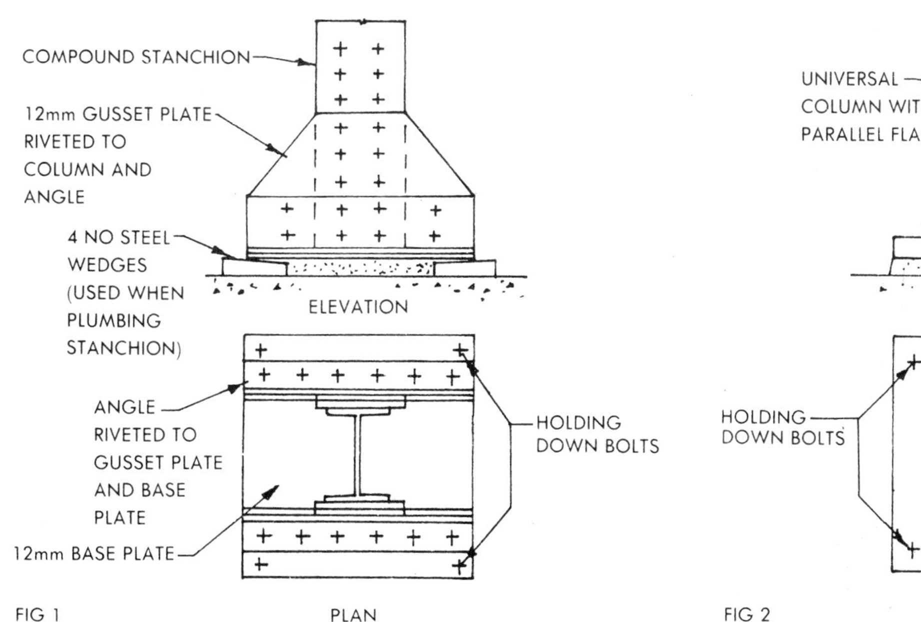

FIG 1

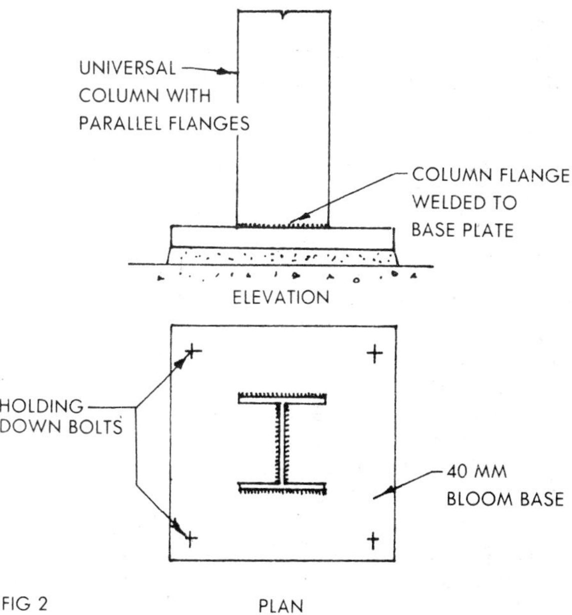

FIG 2

WOODEN BOSS NAILED TO UNDERSIDE OF TEMPLATE

TIMBER TEMPLATE

HOLDING DOWN BOLT

EXPAMET BOLT BOX

ANCHOR PLATE

FIG 3

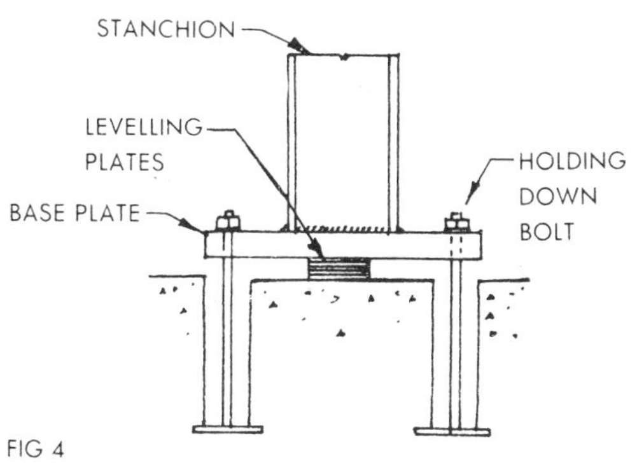

FIG 4

Column or stanchion bases:- Gusset bases as Fig I are rarely used today, and have been replaced by slab or 'bloom' bases as Fig 2. The size of the bloom will depend upon the load carried and the allowable pressure on the concrete base. The size of the concrete base will depend on the total load supported, and the safe bearing pressure permitted for the particular subsoil. The calculations are undertaken by the responsible engineer.

Alignment of stanchions:- The base plate is secured to the concrete foundation by means of special anchor bolts. These are held in position by means of a wooden template during the concrete pour. Fig 3. Bolt boxes or collars of expanded metal as shown are placed round the holding down bolts, and the bolts grouted in position when the stanchion has been plumbed and levelled. A gap of about 40mm is left between the underside of the baseplate and the surface of the base concrete, steel wedges being driven in to facilitate plumbing of the stanchion (Fig I) and the final adjustment completed using thin steel levelling plates Fig 4. The space between the base plate and the foundation concrete used to be filled with cement mortar poured in as a liquid grout, but modern practice is to ram in a cement mortar (l:l sand/cement) in a fairly dry condition.

This ensures stronger consolidation and largely obviates shrinkage.

The work is usually carried out by the steelwork specialist responsible for the erection of all the steelwork.

It is common practice for the main contractor to obtain a certificate from the steel erectors, certifying that the steelwork is lined, levelled and plumbed, upon handing over each phase of erection and prior to encasing.

The underside of the column shaft, and the area of the surface of the bloom base where in contact, are machined dead square, so that the stanchion bears truly on the steel base plate.

Solid round steel columns are sometimes used and also hollow circular, square or rectangular steel sections. Fig 5. Connections may be made by bolting through the sections, but a more common method is to use connecting cleats welded to the sections.

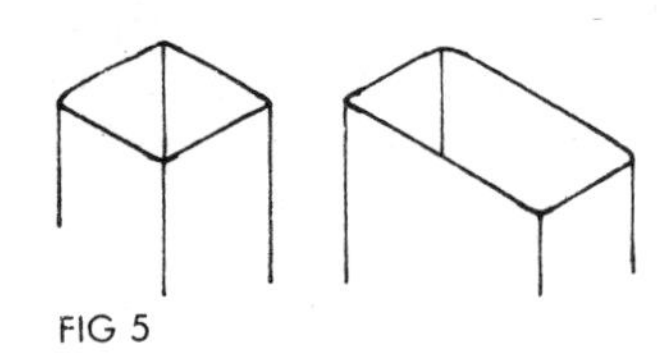
FIG 5

PLASTERING I

EXTERNAL RENDERING

Purpose :- To provide a weatherproof finish to a wall, which is durable, pleasing to look at and reasonably economical. Types of surface finishes normally used are, float, textured, rough cast, pebble dash and machine applied finishes.

The ideal rendering should prevent penetration of water, be well keyed to the wall, free from cracks and flaws and require little or no maintenance. In practice this is not easy to achieve and the choice of a mix and the type of finish will be governed by (i) appearance, (ii) the background to which the rendering is to be applied, (iii) the degree of exposure of the building, (iv) cost.

Backgrounds :- A satisfactory 'key' must be provided to ensure sound adhesion. Concrete may have to be hacked or bush hammered to provide a mechanical key, or a bonding coat of spatterdash applied. On dense concrete, effective hacking removes laitance and is recommended. Bonding 'agents' (eg P.V.A. emulsions) are sometimes used, but it is essential to see that manufacturers' instructions are strictly observed, and that any preliminary cleaning of the surface required is effective. Any dust, algae or fungi must be removed. Failure due to dampness or weak backgrounds may occur, and a good mechanical key is to be preferred.

Smooth backgrounds with little suction should be hacked, and the joints of brickwork and blockwork well raked out and brushed down to provide a key. Alternatively a spatterdash coat may be applied. Keyed bricks, e.g. grooved flettons or keyed hollow clay blocks provide a satisfactory background.

Spatterdash :- This is a wet mix of Portland cement and sand 1:1½ to 2 (by volume) thrown forcibly on to the surface and allowed to harden before the renderinig undercoat is applied. N.B. The spatterdash should not be allowed to dry too quickly, and if necessary should be sprayed with water for two or three days, and shielded from the effects of sun and wind.

In difficult cases it may be necessary to staple a suitable steel mesh or metal lathing to the wall surface. Timber battens (treated with a preservative) may be plugged to the wall first. When rendering on metal lathing, it should be back-plastered (plastered on both sides). This cannot be done when the metal lathing is on battens and in this case self-firring nails or distance pieces should be used, to allow the pricking-up coat to completely encase the lathing.

No-fines concrete contains large voids and provides a good key. Most bricks and clinker blocks provide a moderately strong and porous background. Well raked out joints will normally suffice. If suction is too high or is irregular, a spatterdash coat (p.c./sand 1:2 or3) may be applied. Lightweight concrete blocks or bricks of low strength provide a moderately weak and porous background. The rendering should be weaker than the background. N.B. Strong mixes tend to shrink and this imposes stress on the backing which may result in cracking and failure.

In severe conditions e.g. exposure to driving rain, the first coat of the rendering should be fairly impervious. No coat should be richer than the preceding coat or stronger than the background.

Where very severe conditions exist and there is a likelihood of the wall being exposed to rain and hard frost smooth finishes should not be used. In these circumstances, rough cast or pebble dash are recommended, as these finishes tend to shed the water and are not prone to crack.

FINISHES

Smooth renderings :- Tend to show surface crazing or cracking. Very strong mixes using fine sands and finished with a steel trowel are most vulnerable. Over trowelling particularly, must be avoided, as this brings a layer of rich cementitious material to the surface resulting in excessive shrinkage., A weaker mix tends to spread the effects of contraction and not develop large cracks as may happen with a stronger mix.

A wood float or a felt-faced float should be used for finishing rather than a steel trowel and the rendering should not be allowed to dry too quickly.

Textured finishes :- The final coat is treated in various ways to give a variety of surface effects. Examples are, English Cottage finish, an uneven effect obtained by applying small quantities of material on to the first coat with short upward strokes of a laying trowel. Stipple finish, obtained using a stipple brush on the final coat. Scraped surfaces using a hacksaw blade in a wood frame can be quite effective. A scraped finish is much less likely to craze.

PLASTERING 2

EXTERNAL RENDERING

Pebble dash (SPA dashes etc) :- A clean aggregate of pebbles or crushed stone (6mm to 13mm) thrown on to a fresh coat of mortar. The finish is sometimes lightly pressed into the mortar.

Rough cast :- A set mixture of sand, crushed stone or gravel (6mm to 13mm) and Portland cement or cement and lime, thrown on to a prepared backing coat and left rough. Mixes vary according to the texture and colour required. Two typical mixes are 1PC:1 coarse aggregate: 2 sand and 1 PC:1 lime:1 sand: 4 coarse aggregate.

A pebble dash or rough cast finish on an undercoat with a spatterdash coat beneath are probably the most durable of all finishes.

Mechanically applied finishing coats:- Vary in texture with materials used and type of machine. the Tyrolean machine Fig 1, is held in the hand, and when the handle is cranked blobs of the mix are flicked on to the surface. The application is usually in three stages, (i) at an angle of 45 deg from the right, (ii) at an angle of 45 deg from the left, (iii) the final coat applied perpendicular to the wall surface.

Proprietary materials, such as colours are supplied ready for mixing.

One such colour mix is composed of a cement and a special aggregate, supplied as a mixture ready for use, in a range of colours. If no special undercoating is supplied a 1:1:6 cement-lime-sand may be used. Machines may be hired.

N.B. It is not advisable to carry external rendering down to ground level as it may provide a path for rising damp. The rendering is best stopped above the D.P.C. Galvanised expanded metal sections and stop beads are available for finishing renderings at the bottom edge. Fig 2.

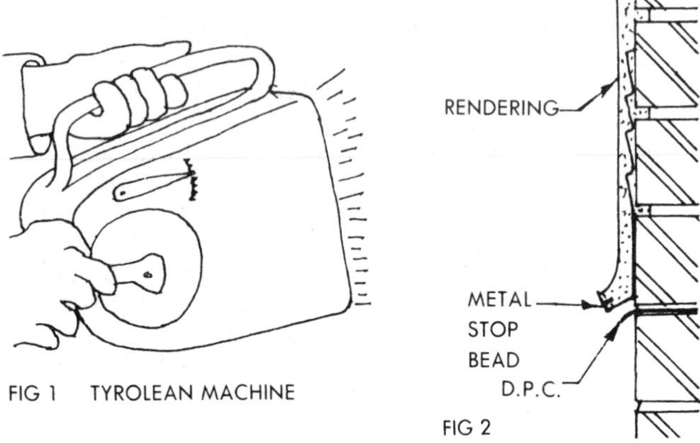

FIG 1 TYROLEAN MACHINE

FIG 2

Mixes :- In sheltered positions a 1:2:8 to 9 pc, lime, sand mix may be used. In exposed positions a 1:1:5 to 6 mix is preferable. In very severe conditions, eg a wall exposed to driving rain, the first coat of the rendering should be fairly impervious.

Lime/cement mixes tend to harden more slowly and may be vulnerable to frost action. Thus in cold weather a stronger mix is recommended.

Where abrasion-resistant rendering is required a 1:3 pc/sand mix is recommended.

As a general guide, use the same mix for the undercoat as for the finishing coat. If variations are adopted, then the undercoat should be stronger than the finishing coat.

Undercoats should be allowed to dry out before applying the next coat. (As a guide, allow at least two days in summer and at least a week in cold or wet weather).

Renderings should not be allowed to dry out too quickly. The surface may be sprayed with water at intervals and should be protected from the effects of wind and sun.

Number of coats :- Single coat work is rarely satisfactory, as the joints tend to 'grin' through. On high density backgrounds, (very dense bricks, concrete blocks with low suction and smooth surfaces and structural concrete), spatterdash coat 3mm to 6mm thick should be applied to provide a key. On very uneven backgrounds more than one undercoat may be needed to build up to an even surface.

Undercoats should be not less than 9mm thick nor more than 16mm thick. Each coat should be 'combed' to provide a key for the next coat.

In conditions of moderate exposure two-coat work is usual, but in conditions of severe exposure three-coat work is recommended.

Sand :- It is most important that clean, well-graded sand is used for rendering mixes. Dirty, poorly graded sands can cause serious defects through irregular setting, poor adhesion and shrinkage cracking of the finished rendering. Sand should be specified to comply with the requirements of BS 1199 sands for external renderings. If the sand is suspect (eg it stains the hands when rubbed through the fingers) a field settling test should be carried out and the amount of clay and silt present should preferably not exceed 5 per cent.

PLASTERING 3

INTERNAL PLASTERING

Purpose :- To provide a smooth, hard, level, hygienic, crack-free surface to walls and ceilings. It may further be used to increase fire resistance and to improve the acoustic and thermal insulation properties of a structure. CP 211: 1966 – Internal plastering, lays down general principles, deals with materials and backgrounds, and suggests specifications for various purposes.

Lime plaster :- At one time a mixture of lime and sand (1:3) was used for plastering. It is however weak, easily indented and tends to shrink on setting. Animal hair (5kg/m^3) was included in the mix to restrain the shrinkage. The addition of suitable Gypsum plaster to give 'gauged lime plaster' increases resistance to impact and gives a tighter finish. Both mixes tend to slow hardening, and each undercoat must be allowed to dry out before subsequent coats are applied.

Cement-lime-sand plasters :- Provide a relatively strong, hard surface and are suitable as undercoats on most backgrounds. Are useful in damp conditions and for textured finishes. A wood float (Fig 1) finish is usual, as a steel trowel (Fig 2) finish tends to produce laitence which results in crazing. A permeable decorative finish, which is not affected by alkalis, should be used at first. Common mixes used are 1:1:6 and 1:2:8 to 9 pc:lime:sand.

FIG 1 HAND OR SKIMMING FLOAT

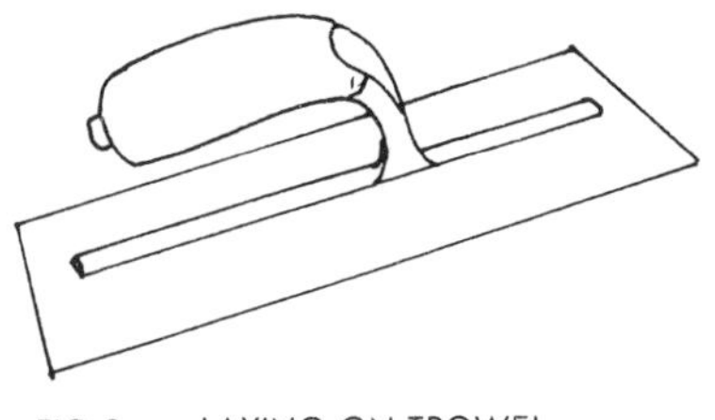

FIG 2 LAYING-ON TROWEL

GYPSUM PLASTERS

Class A (Plaster of Paris):- Manufactured from Gypsum, which is crushed and ground to fine powder. When heated to about 170 deg C, 75 per cent of the combined moisture is driven off, producing a hemihydrate, gypsum plaster ($CaSO_4 \frac{1}{2} H_2O$). When mixed with water sets quickly, and is used for gauging final lime coats, for patching and repair work and as casting plaster for fibrous plaster work. For gauging lime final coats a mix of ¼-1:1, plaster, lime with sand if appropriate. For fibrous plaster work used neat or with lime added.

Mixing:- Should be mixed in a clean pail with clean water. The plaster should be added to the water slowly enough to allow air bubbles to escape. As little water as possible should be used consistent with a satisfactory workable mix. The mix should be allowed to soak for two or three minutes, then stirred, making sure all lumps are broken up, before pouring. Excessive stirring should be avoided, as this introduces air bubbles and accelerates the set.

When casting plaster is poured into a mould, petroleum jelly diluted with white spirit, may be brushed on to the mould first, to facilitate removal after setting.

Drying time of casts will depend on thickness, average casts dry in about 7 days at room temperature.

Class B (Retarded Hemihydrate plaster):- The setting action of ordinary hemihydrate plaster is slowed down by the addition of retarders.

Thistle plaster is a retarded hemihydrate and is produced in four undercoat and two finishing grades.

Thistle browning:- An undercoat for most normal solid backgrounds.

Thistle slow setting browning:- Suitable for mechanical mixing.

Thistle fibred (containing Rayon fibre):- An undercoat for thistle baseboard, Gyproc lath and expanded polystyrene.

Thistle metal lathing (incorporates Rayon fibre and a rust inhibitor):- For use on expanded metal lathing, Newtonite lath and for wood wool slabs.

Thistle finish:- A finishing coat for the foregoing thistle undercoats.

Thistle board finish:- A finish coat for thistle baseboard, Gyproc lath and concrete.

Thistle ascoustic plaster:- A textured finishing-coat plaster with sound-absorbing qualities. A retarded hemihydrate plaster with a graded pumice aggregate. For use where sound absorption is required. Must be used with the correct grade of Carlite or Thistle undercoat, depending on the type of background. Textured surface will vary depending on whether finishing has been completed with a wood, cork or carpet float. Dries a light beige colour. Sound absorbent properties depend on surface texture. Surface is slightly friable, and should not be used where it will be subject to abrasion.

PLASTERING 4

INTERNAL PLASTERING

Dimensions and applications of thistle plasters

The following table is reproduced from the British Gypsum 'White Book' 1984.

Lime may be mixed with Class 'A' and Class 'B' finishing plaster, with or without sand. The addition of lime increases the workability of some plaster mixes, especially if quicklime is run to putty and allowed to mature before use. Hydrated lime should be soaked overnight. (A fat chalk lime gives best results). The amount of lime used with Class 'B' finishing plaster should not exceed 25 per cent by volume. The addition of lime to this plaster accelerates the set.

N.B. For one-coat work on plasterboard, neat plaster should be used. No lime.

Type of plaster Background	Recommended coat thickness mm	Approx' weight set & dry kg/m²	Approx' coverage m²/1000 kg	Type of coat (vol' plaster vol' sand)	Method of application
Thistle Browning Clay bricks Clay tiles	11	20.8	234	Floating (1 : 3)	Floating coat ruled to an even surface and lightly scratched to form a key for Thistle finish
Concrete bricks with raked joints No-fines concrete Clinker blocks. Foamed slag blocks Masonry			181	Floating (1 : 2)	
Engineering bricks			144	Floating (1 : 1½)	
Thistle fibred Thistle baseboard and Gyproc lath paramount dry partition	8	13.9	206	Floating (1 : 1½)	Floating coat ruled to an even surface and lightly scratched to form a key for Thistle finish
Expanded polystyrene soffits			181	Floating (1 : 1)	
Expanded polystyrene walls	11	19.5	131		
Thistle metal lathing Metal lathing Newtonite lath	11*	19.5	115	Rendering (1 : 1½)	Rendering coat deeply cross – scratched to form a key for floating coat.
Wood wool slabs			181	Floating (1:2)	Floating coat ruled to an even surface and lightly scratched to form a key for Thistle finish
Thistle finish Thistle plaster undercoats	2	2.3	390	Finishing	Used neat with not more than 1 part putty lime by volume to 4 parts plaster at architects discretion. Trowelled to a smooth (not polished) surface
Thistle Board Finish Thistle baseboard and Gyproc lath	5	6.8	165	Finishing (one or twp coats)	Trowelled to smooth finish putty lime should not be added to one-coat work.
Concrete					Applied as above. Bonding agent necessary on very smooth concrete

* From face of lath

PLASTERING 5

GYPSUM PLASTERS

Class 'C' (anhydrous Gypsum plaster) :- Gypsum rock is heated in a kiln until the combined moisture has been driven off. The rock is then crushed, ground to a fine powder, and accelerating chemicals added to control the setting time of the plaster. This plaster is bascially anhydrous calcium sulphate ($Ca\ SO_4$) and gives a harder surface than that of Class 'B' plasters. Class 'C' plasters are slow setting and thus are more easily brought to a good finish than Class 'B' plasters.

Sirapite :- A gradual setting anhydrous Gypsum plaster. Is used to provide a smooth finish or textured finishing coat (when mixed with sand). To sanded plaster undercoats, or over cement:lime:sand (1:1:6) undercoats. Used neat or with the addition of up to one-quarter part of lime putty. Recommended thickness 3mm.

When used over pc:sand or pc:lime:sand undercoats, these must be allowed to mature before applying the setting coat or shrinkage cracks from the undercoat may be transmitted to the surface and cause failure.

To obtain a shallow stippled finish use 1:1 clean, sharp, river or pit sand. The texture is achieved using a wood float, and will be governed by the type of sand used. (Coarse or fine). A stippled finish may be obtained by mixing the Sirapite to a fairly stiff consistency and then immediately after applying the finish coat, stipping with a hair or rubber brush.

Sirapite may show a double set and may be retempered if desired.

N.B. Plasters should not normally be retempered after 30min from mixing, should not be allowed to dry too quickly, should not be applied direct to boards.

Coverage :- At 3mm thick coverage is approximately 263m²/tonne.

Class 'D' (Keene's cement) :- Obtainable in three grades, standard Keenes, standard polar white and fine polar white. For final coats (applied neat) only, in two or three-coat work over strong cement:sand or strong Gypsum (Class 'B') sand undercoats.

N.B. Class 'D' plaster must not be mixed with lime.

The plaster should be applied evenly, then laid down with a hand float. When sufficiently stiff, plaster is scoured lightly to ensure a flat surface, closed in with a fresh mix and trowelled to a final polish. Provides a very hard, smooth surface, resistant to impact damage. Can be brought to a very true flat surface, suitable for use where very strong lighting would emphasise any uneveness on other plasters.

Class 'D' plaster must not be allowed to dry too quickly or delayed expansion may occur causing blistering and shelling.

Standard and fine polar white may be used in cold storage and refrigeration chambers. They may also be used to give a hygienic finish to laggings of heating installations. Used neat on cork, expanded polystyrene slabs or sanded undercoats.

Because of its strength and resistance to impact, Keenes is often specified for narrow reveals, splays and returns under 300mm.

Carlite premixed plaster :- A lightweight, retarded hemihydrate premixed Gypsum plaster containing additives to produce required properties. Requires only the addition of clean water for use. For general use, especially to increase fire resistance or reduce condensation five grades are available (i) Carlite finish, a finishing coat for the following undercoats. (ii) Carlite browning, an undercoat for solid backgrounds of normal suction. (iii) Carlite browing H.S.B., an undercoat for solid backgrounds of very high suction (iv) Carlite metal lathing, an undercoat for expanded metal lathing, Newtonite lath, wood wool slabs and expanded polystyrene, (v) Carlite bonding coat, an undercoat for low suction backgrounds, composite surfaces and plasterboard.

Carlite welterweight bonding coat plaster :- Heavier and stronger than carlite bonding coat, and more easily worked. May be used as an undecoat on the following:- Thistle baseboard, Gyproc lath, Paramount panel, stone, cork slabs, Lignacite blocks, low suction engineering bricks, composite ceilings with concrete beams, backgrounds treated with proprietary bonding fluids.

Carlite plasters are slightly resilient which reduces cracking caused by structural movement.

Precautions :- (i) don't mix different grades of plaster

(ii) suitable, *clean* sand is essential for sanded mixes

(iii) the mixing water must be clean, sea water, polluted or brackish water must not be used. Water should be potable.

(iv) Cement must not be used in a mix with Gypsum plaster.

(v) Gypsum plaster mixes should not be retempered after plaster has set.

PLASTERING 6

THISTLE PROJECTION PLASTER

Projection plastering is a development of mechanical plastering, requiring the use of special plaster, specialised equipment and techniques that are different from the normal methods.

Thistle projection plaster is a blend of Gypsum plasters which have been formulated for one-coat application by machine. Supplied pre-mixed and only requires the addition of clean water to prepare it for use.

The plaster is applied using a specially designed machine. the nozzle being held fairly close to the background and the plaster sprayed in ribbon form, at a consistency which allows the ribbons to run together. The recommended thickness for each application of plaster is 13mm, thinner layers can be used on backgrounds such as plasterboard, normal concrete and similar low or non-absorption backgrounds on low suction backgrounds, e.g. smooth dense concrete, ceretain types of block and smooth-faced or high density brick, the use of a reputable P.V.A. bonding medium, applied in accordance with manufacturers instructions should be considered. A grade of thistle projection plaster is available for high suction backgrounds.

When the background has been covered to the required thickness, a metal feather-edge should be used to spread the plaster to a reasonable surface. Any hollows left should be re-sprayed lightly. When the plaster is sufficiently stiff (initial set), a shorter feather-edge should be used to pare down any ridges or undulations. At this stage the plaster should be lightly open in texture. Water should be finely sprayed onto the surface, and a hand or power float used to prepare the plaster for trowelling.

Internal angles should be planed to remove undulations. Lightly floated and then trowelled using the angle trowel (Fig 1). The whole surface is then trowelled off with a two-handed trowell (Fig 2) and completed using a normal hand trowel.

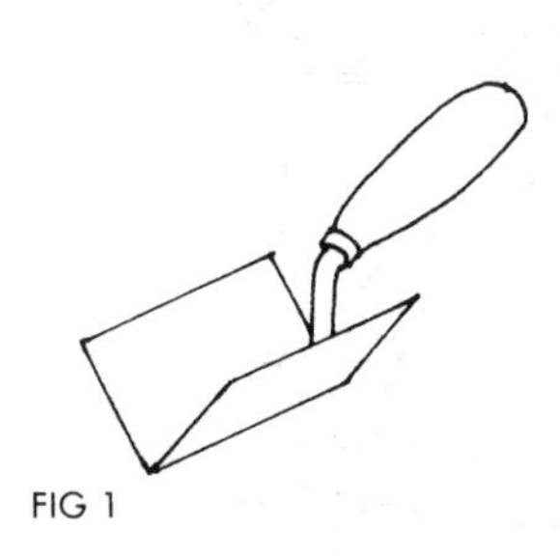

FIG 1

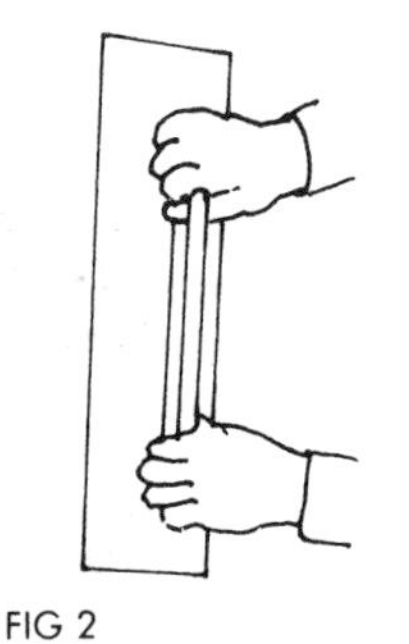

FIG 2

PREMIXED GYPSUM X-RAY PLASTER
A retarded hemihydrate premixed plaster, with a barium sulphate aggregate. Provides protection against X-rays. For full protection the plaster must be completely free from cracks, and considerable care is necessary in its application.

Three grades are available. (i) Rough, an undercoat grade for use on most solid walls. (ii) Fibred, an undercoat grade for use on metal lathing. (iii) Finish plaster, a finishing grade for use with either of the two undercoats.
Thickness of coats :- The coat thickness for a particular situation should be specified by the National Radiological Protection Board, or the local hospital physicist. (Undercoats may normally be 10mm, 16mm or 22mm thick). This plaster should be applied to an even thickness of 3mm as soon as the undercoat has set. Up to 16mm thick undercoat plaster should be applied in two coats, each being laid to an even thickness. The rendering coat should be deeply scratched to form a crossed undercut key for the floating coat, and the floating coat thoroughly scratched to form a key for the finishing coat. Over 16mm thick the undercoat plaster should be applied in three coats, each coat should be keyed as previously described.

PLASTERBOARDS
The use of plasterboards, including square edge, tapered edge, bevelled edge, Gypsum lath and Gypsum plank, were dealt with in earlier Detail Sheets.
Gyproc thermal board and vapour check thermal board :- This is a laminate consisting of gyproc wallboard bonded to a backing of self-extinguishing expanded polystyrene. It is also available with a special polythene membrane incorporated at the intersurface between the wallboard and the polystyrene. This membrane acts as a vapour check. Available with square or tapered edges Fig 3. The board is used to provide an insulated ceiling or wall lining, and is available with an ivory surface or direct decoration or a grey surface for plastering. Reduces risk of condensation and eliminates cold bridges.

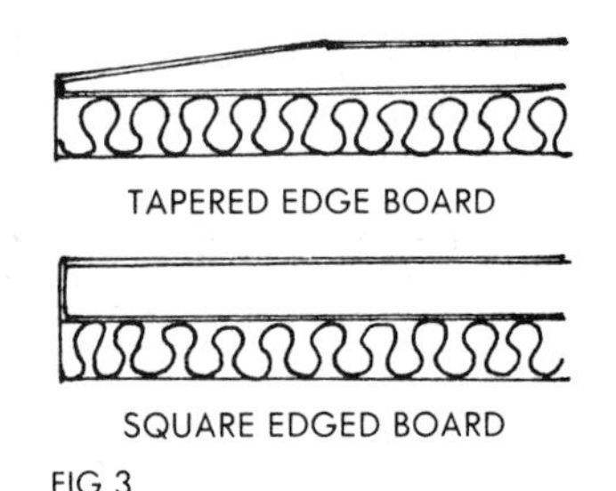

FIG 3

PLASTERING 7

PLASTERBOARDS
In addition to the plasterboards previously mentioned, the following boards are also available:-
Gyproc vapour check wallboard :- A wallboard backed with a blue tinted polythene film bonded to the back of the board. The use of the board as an inner lining of a building, provides a check to the passage of water vapour, and helps to prevent structural decay and loss of thermal insulation. The boards do not provide a complete vapour barrier because of gaps at the perimeter and at joints between boards. On timber frame vapour resistance can be improved by (i) applying self adhesive vapour resistant tape or (ii) a thick application of chlorinated rubber paint. These treatments must be applied to the timber framing to ensure complete and continuous contact with the polythene film at all edges of the boards.
N.B. (i) the boards are only suitable for fixing by mechanical means and cannot be fixed in systems employing adhesives.
(ii) Buildings must be properly dried out before the boards are fixed, or moisture may be sealed in, and impair the performance of the boards.

Industrial grade board :- Gypsum plasterboard faced with white P.V.C. film and backed with polished aluminium foil to improve thermal insulation, when used in conjunction with an air space. Does not require decoration and is easily cleaned. May be used for metal fixing systems or may be nailed to timber supports. Available in 12.7mm and 9.5mm thickness. For use in patent ceilings, the latter designated as M/G (modular grid).

PLASTERING ON BUILDING BOARDS

Joints between boards (except Gypsum lath) should be reinforced to prevent cracking Fig I. Jute scrim may be used as shown, or a galvanised wire scrim. If the latter is used, any cut ends should be painted to avoid rust stains. The joint is covered with the same mix as is used on the boards. The scrim is pressed into the wet mix and trowelled as flat as possible. The scrim should not overlap and joints should be allowed to set, but not to dry out before plastering. End joints between boards should be staggered. Fig 4.

Joints between walls and ceiling can be similarly treated, or alternatively, unsightly cracks may be avoided by making a straight cut with a thin blade before the plaster sets. Fig 2.

An alternative method of concealing the join between wall and ceiling is to use a cornice, either 'run' in the traditional manner, or cast as fibrous plaster, or to use a manufactured cornice, e.g. Gyproc cove. Fig 3. This is a Gypsum plaster core encased in Millboard, which may be either fixed with a suitable adhesive, or where the background is not suitable for adhesives, it may be nailed or screwed into timber grounds or plugs.

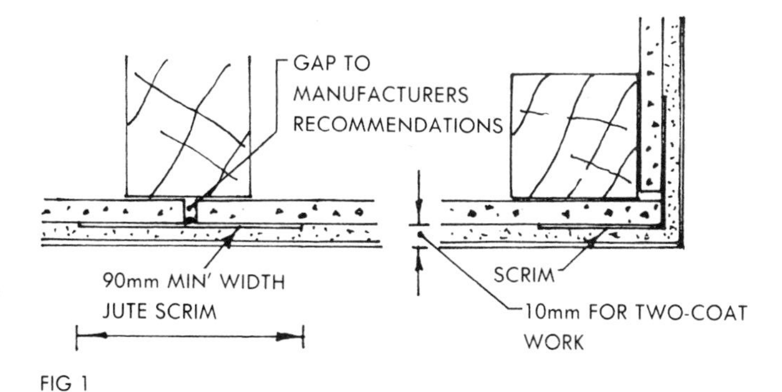

FIG 1

Mixes :- For single coat work on plasterboards and asbestos boards, use a neat gypsum board plaster. Any other plaster having a low setting-expansion may be used, provided it is recommended by the manufacturer for single-coat work. Single coat work may be used on plastics boards provided there is no risk of impact damage, but two-coat work is recommended. For two-coat work use an undercoat of sanded Class 'B' Gypsum plaster (I Plaster :I½ Type I sand (or : I Type 2 sand). Any neat Gypsum plaster or lime gauged with a suitable Gypsum plaster may be used for the final coat.

Premixed lightweight Gypsum plaster may be used on boards. Neat plaster used, adding only water. The final coat should be the same brand as the undercoat.

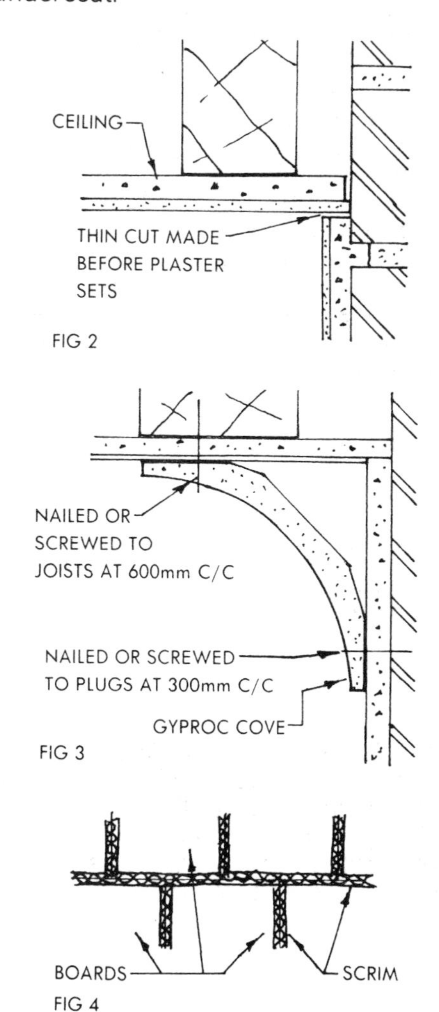

PLASTERING 8

DRY LININGS
Plasterboards used for dry linings have one ivory face, which is exposed for direct decoration. Wallboards are available with three different types of edge:- tapered for seamless jointing, square for cover strip jointing and bevelled for V-jointing. Fig 1.

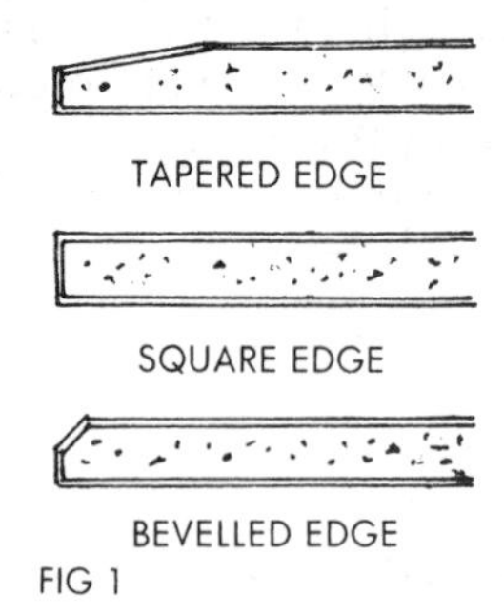

FIG 1

Fixing:- One method of fixing boards to a solid background is the 'dot and dab' technique. The dots consist of small bitumen-impregnated fibreboard pads which are fixed to the wall with dots of plaster and carefully plumbed and aligned. When the dots are set firm, dabs of plaster, (Carlite Welterweight or thistle board finish) about 75mm wide are applied to the wall. Plaster dabs should be the length of the laying trowel, thick enough to stand proud of the pads, and a gap of 50mm or so should be left between dabs. Fig 2. The dabs should be applied for one board at a time, making sure that those adjacent to joints between boards are approx' 25mm from the edge.
N.B. Carlite Welterweight plaster has a setting time of 2 hrs approx', and may be used for complete systems. For small areas or where timing is important, the alignment dots may be fixed with thistle board finish plaster which has a setting time of 1 to 1½ hrs.
Dots are fixed 230mm from the ceiling level

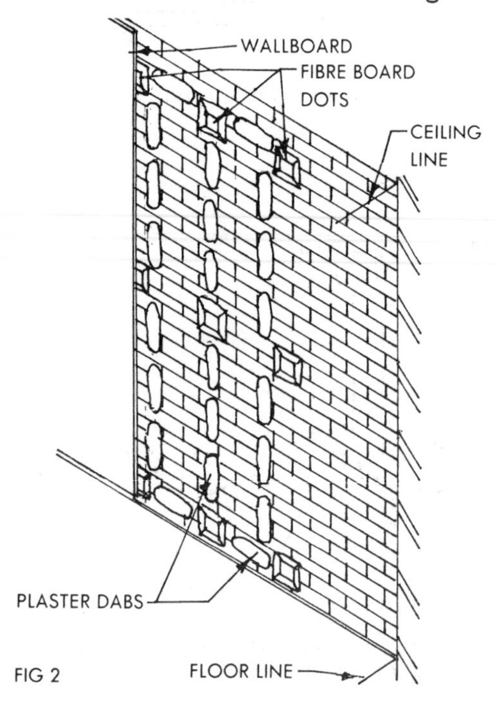

FIG 2

and 100mm up from floor level and carefully plumbed. Intermediate dots are fixed midway between and lined up with the previous dots. The dots should be 1070mm maximum centres vertically, thus a wall higher than 2440mm requires two intermediate dots. Horizontal spacing should be such as to give three rows for each board, and so arranged that when the boards are fixed, fibreboard dots bridge the junctions between them.
Fixing the boards:- If necessary the boards are cut (using a fine-toothed saw or by scoring and snapping) so that their length is 25mm or so less than the floor to ceiling height. To position the board, a foot-lift is used. Fig 3. The board is placed in position against the plaster dabs with the bottom end resting on the foot-lift. It should be tapped back firmly with a straight-edge until it is tight against the dots, and then lifted gently with the foot-lift until it is tight against the ceiling. the board is temporarily secured by nailing through the board into the fibreboard dots. Galvanised nails may be used, but double headed nails (Fig 4) are preferable. These can be removed after the dabs have hardened and re-used. The advantage of using these nails is that, when removed there is only a small hole to fill instead of a nail head to spot. Nailing is normally only required at the edges. An alternative method is the Gyproc metal furring (M/F) system. Metal furring channels Fig 5 are used to align the walls and secure the plasterboard. The channels are bonded to the wall with Gyproc drywall adhesive and the boards screwed to them. The system is suitable for 12.7mm wallboard in plain or insulating grades, vapour check wallboard and 25.4mm thermal board. The plasterboards are screwed to the channels with 22mm screws with countersunk Pozidriv heads using a powered screwdriver. For increased insulation a Drywall screw 42mm in length is available for use with 32mm thick Gyproc thermal board. Using brick outer sink and P.F.A. block inner skin lined in this way gives a 'U' value of 0.56 W/m^2 deg C.
Jointing:- With tapered edge boards a flush seamless surface is possible. A band of suitable filler is applied to the trough of the tapered-edged joints. A length of joint tape is then pressed into place with a filling knife and a final coat of filler applied and finished smooth with a jointing sponge.

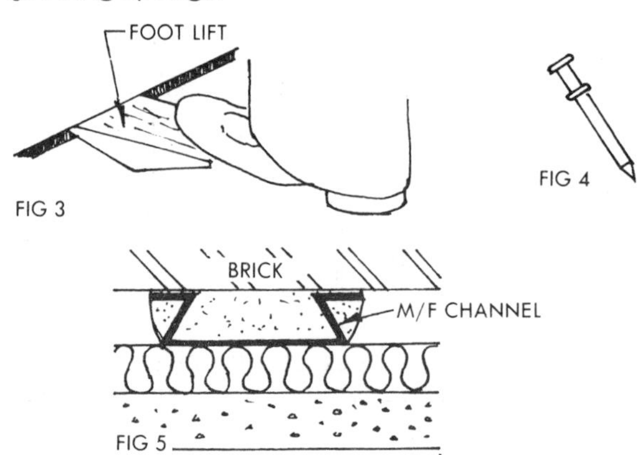

FIG 3
FIG 4
FIG 5

SCAFFOLDING I

Putlog scaffold :- Has a single row of standards as shown Fig I. For a five board scaffold set 1.245m from the wall face, allows 100mm clearance for plumbing. Scaffold is partly supported by wall.
Tubes :- Steel or aluminium alloy to BS 1139 Metal Scaffolding. Do not mix steel and aluminium as strengths differ. Keep aluminium clear of damp lime, cement, or salt water.
Standards :- Vertical members, for bricklayers 1.829m (6ft) to 2.438m (8ft) apart depending on load. Where joints are required they should be staggered Fig 2. With steel standards 2.44m apart loads should not exceed 146 kg/m² (ie 408 kg/span). This is approx' two men 25kg of mortar and 80 bricks with standards 1.83m apart load may be increased to 273 kg/m².
Ledgers :- Horizontal members. Spacing depends on height of lift, usually 1.35m. Fig 2. Joints should be staggered as shown. Fig 2.
Putlogs :- Cross members Fig I. Spacing depends on thickness and length of boards. (See table). To prevent tipping, the end of a board must not overlap its support by more than four times its thickness. Fig 3. Fig I shows a working platform conforming to the regulations.

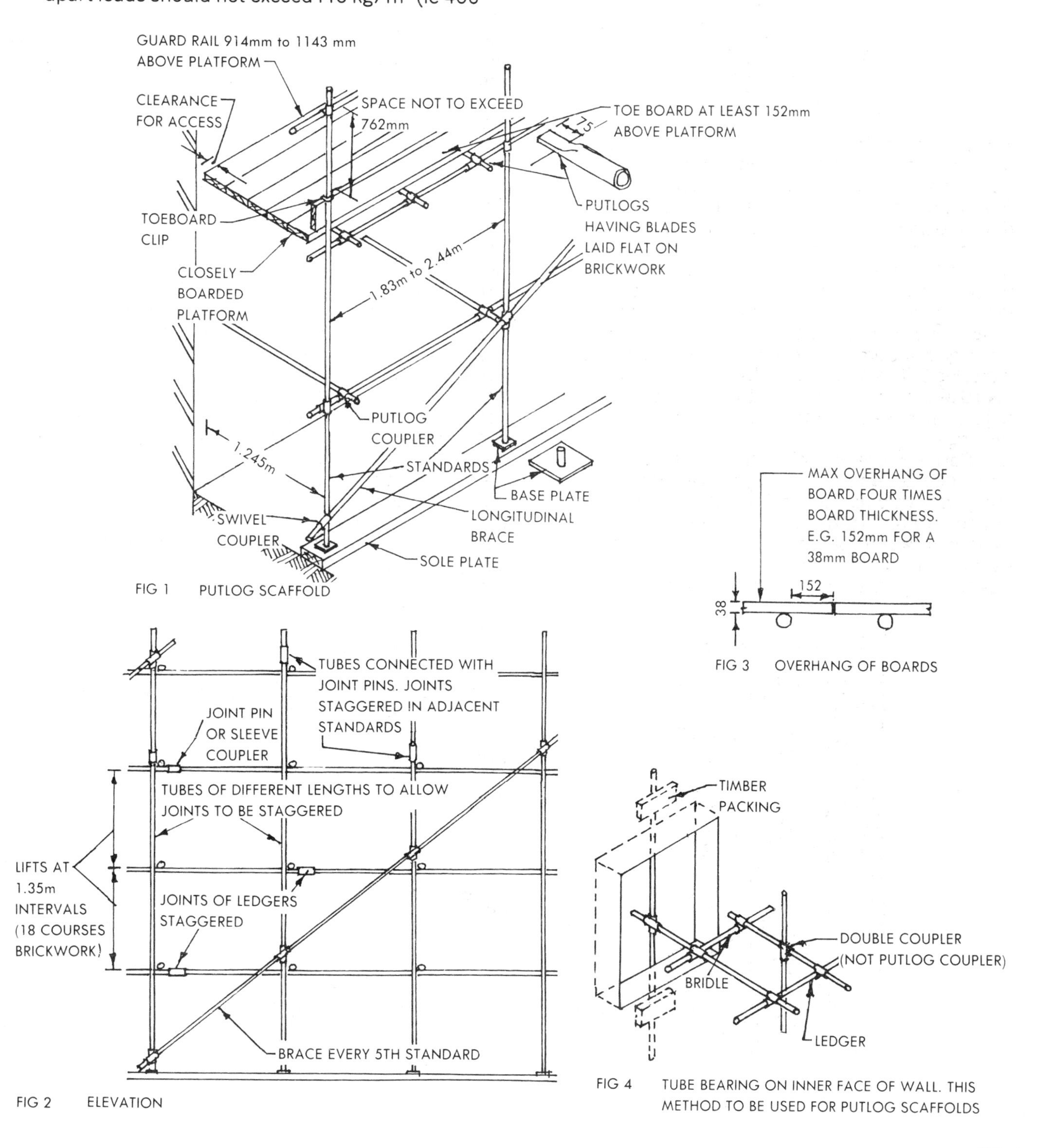

FIG 1 PUTLOG SCAFFOLD

FIG 3 OVERHANG OF BOARDS

FIG 2 ELEVATION

FIG 4 TUBE BEARING ON INNER FACE OF WALL. THIS METHOD TO BE USED FOR PUTLOG SCAFFOLDS

Tying-in :- Every scaffold should be securely tied to the building at every other lift. (2.7m approx' vertically) and every 6m horizontally. Figs 4 and 5. All couplers used to make these ties shall be right-angle couplers and each tie should be made at, or as close as possible to, the junction of a standard and a ledger.

Where there are insufficient openings in the wall, strut the scaffold by raking tubes inclined towards the building. Fig 6.

Spacing of putlogs	
Thickness of boards	Maximum spacing
32mm ($1\frac{1}{4}$")	991mm (3'-3")
38mm ($1\frac{1}{2}$")	1.524m (5'-0")
51mm (2")	2.591m (8'-6")

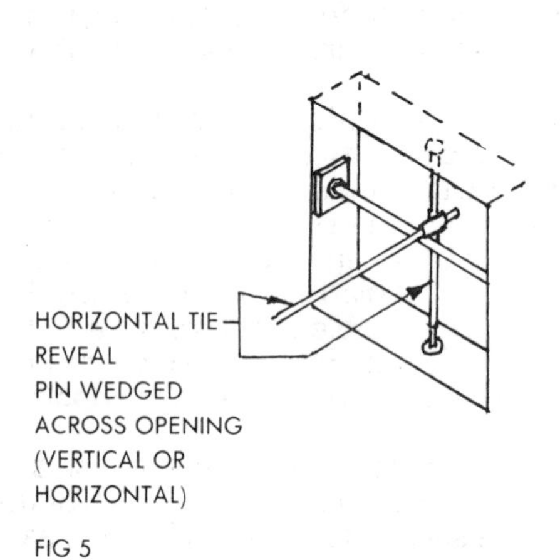

FIG 5

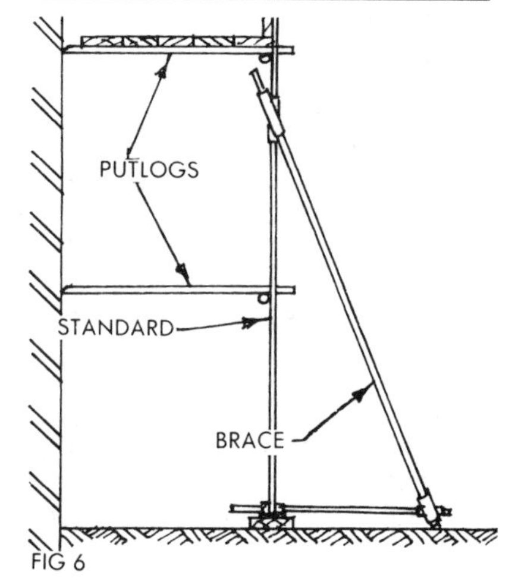

FIG 6

SCAFFOLDING 2

LADDERS

1. Timber ladders to conform to BS 1129: 1966.
2. Aluminium ladders to BS 2037: 1964
3. Ladders should be the proper length for job to be done. If used as a means of access or as a working place they should rise to a height of at least 1.066m (3'-6") above the working platform, or above the highest rung to be reached by the feet of the person working on them. Fig 2.
4. Ladders should be at a safe angle (four up to one out) and lashed at the top. Fig 2.

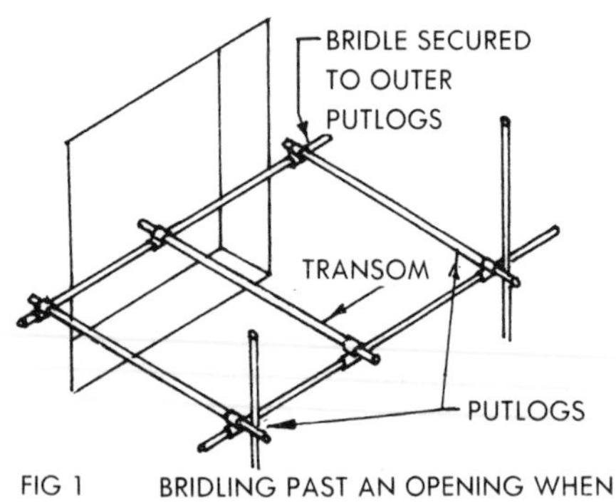

FIG 1 BRIDLING PAST AN OPENING WHEN THERE IS NO TIE IN

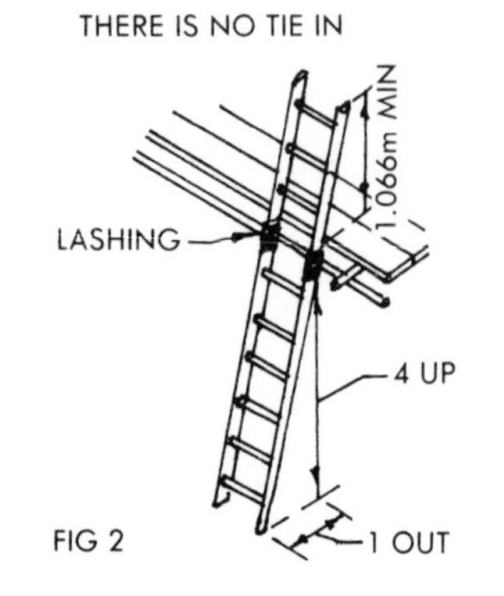

FIG 2

INDEPENDENT TIED SCAFFOLD

Has double row of standards about 1070mm apart. (Depending on the width of the working platform). For bricklayers the inner row of standards placed 330mm to 380mm from the wall. This allows one board to rest on the projection of the transoms. Fig 3. For plasterers set standards 530mm from the wall, to allow two boards to be placed in front. For decorating standards can be as close to the wall as possible.

The scaffold should be tied to the building in a similar manner to the putlog scaffold.

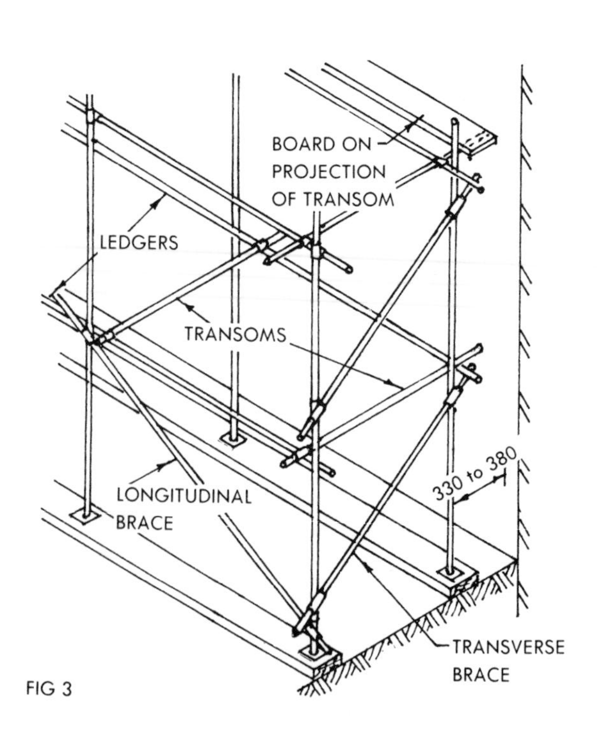

FIG 3

Working platforms should be close boarded and boards should have a minimum of three supports. For 38mm thick boards the maximum spacing for putlog or transoms is 1.52m.

Inspection of scaffolds The Construction (Working Places) Regulations require that,

1. All material for any scaffold shall be inspected by a competent person on each occasion before being taken into use.
2. At least once a week.
3. After adverse weather conditions.

Records of inspection must be kept in the prescribed register. (Form 91 Section A).

Checks to be made on inspection :-

1. Standards correctly aligned and properly supported at their bases.
2. No undue deflection in ledgers, putlogs or transoms.
3. No essential members have been removed.
4. All ties and braces are effective in stabilising the structure.
5. All couplers are properly tightened.
6. All boards are sound and properly supported.
7. All guardrails and toeboards are in place. All ladders are sound, properly supported and secured.

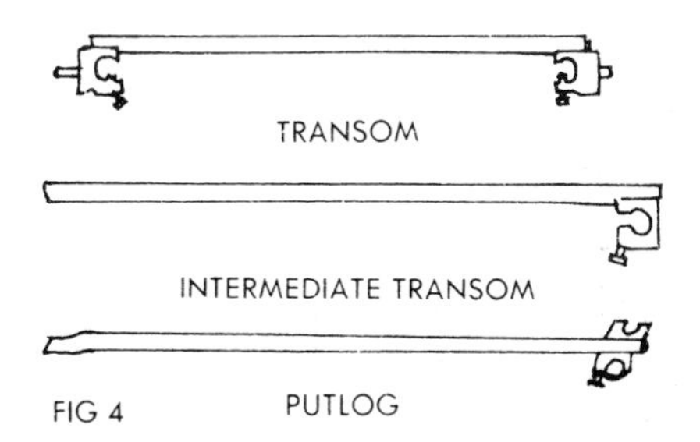

FIG 4

Partly dismantled scaffolds must either comply with regulations or else display prominent warning notices indicating that they must not be used.

Pre-coupled units :- These units (Fig 4) are being increasingly used. The units are made to predetermined length with dual couplings welded in position as shown. Advantages are speed of erection, elimination of 80 per cent of loose fittings with corresponding reduction of site losses, no need for ledger bracing in a properly tied structure and saving of labour costs.

Prefabricated scaffolds :- Usually consist of 'H' frames made up of two short standards to which one or more transoms are fixed, with couplers for ledgers welded in position.

Trestle scaffolds:- Fig 5. No trestle scaffold shall be used (a) if the platform is so situated that a person would be liable to fall from it a distance of more than 4.572m (15ft), or (b) if constructed of more than one tier where folding supports are used.

No trestle scaffold shall be erected on a scaffold platform unless (a) the width of the said platform is such as to leave sufficient clear space for the transport of materials along the platform and (b) the trestles or supports are firmly attached to the platform and adequately braced to prevent displacement.

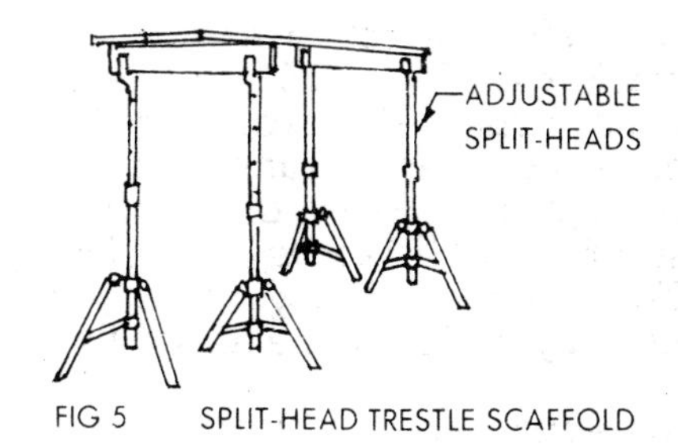

FIG 5 SPLIT-HEAD TRESTLE SCAFFOLD